Chronic Heart Failure

I. Quality of Life

H. Viefhues W. Schoene R. Rychlik (Eds.)

II. Nitrate Therapy

A. Kimchi B. S. Lewis M. Weiss (Eds.)

With 51 Figures
and 29 Tables

Springer-Verlag
Berlin Heidelberg New York
London Paris Tokyo
Hong Kong Barcelona
Budapest

Part I

Professor Dr. Herbert Viefhues
Furkenstrasse 35
D-4630 Bochum (W)
Federal Republic of Germany

Dr. phil. Wolfgang Schoene
Institut für Medizinische Soziologie
Westfälische Wilhelms-Universität
Domagkstrasse 3
D-4400 Münster (W)
Federal Republic of Germany

Professor Dr. Dr. med. R. Rychlik
Medizinische Forschung
Schwarz Pharma AG
Alfred-Nobel-Strasse 10
D-4019 Monheim (W)
Federal Republic of Germany

Part II

Professor Dr. med. Asher Kimchi
Cedars-Sinai Medical Center
University of California
8635 W. Third Street, Suite 1080 W.
Los Angeles, CA 90048, USA

Professor Dr. med. Basil S. Lewis
Cardiology Department
Lady Davis Carmel Hospital
7 Michal Street,
Haifa 34362, Israel

Dr. Marija Weiss
International Medical Affairs
Schwarz Pharma AG
Alfred-Nobel-Strasse 10
D-4019 Monheim (W)
Federal Republic of Germany

ISBN-13:978-3-642-76435-6 e-ISBN-13:978-3-642-76433-2
DOI: 10.1007/978-3-642-76433-2

Library of Congress Cataloging in Publication Data
Chronic heart failure. p. cm.
Includes bibliographical references and index.
Contents: 1. Quality of life/Herbert Viefhues, Wolfgang Schone and Reinhard Rychlik (eds.) – 2. Nitrate therapy/Asher Kimchi, Basil S. Lewis and Marija Weiss (eds.).
ISBN-13:978-3-642-76435-6
1. Congestive heart failure. 2. Heart failure. 3. Heart failure–Patients–Rehabilitation. 4. Nitrates–Therapeutic use.
I. Viefhues, Herbert.
[DNLM: 1. Heart Failure, Congestive–drug therapy. 2. Heart Failure, Congestive–psychology. WG 370 C557]
RC685.C53C47 1991 616.1'29–dc20 DNLM/DLC for Library of Congress 91-5006 CIP

Softcover reprint of the hardcover 1st edition 1991

27/3145-543210 Printed on acid-free paper

Contents

Contributors

Abrams, J.
Division of Cardiology
University of New Mexico
School of Medicine, Albuquerque
NM 87131, USA

Böhm, M.
Klinikum Großhadern
Medizinische Klinik I
Forschungslabor B
Universität München
Marchioninistrasse 15
8000 München 70 (W)
Federal Republic of Germany

Bullinger, M.
Institut
für Medizinische Psychologie
Universität München
Goethestrasse 31/I
8000 München 2 (W)
Federal Republic of Germany

Cintron, G. B.
Cardiology Section (111 A)
James A. Haley VA Hospital
13000 Bruce B. Downs Boulevard
Tampa, FL 33612, USA

Dews, I.
Department of Cardiology
Queen Mary Block
Oldchurch Hospital
Romford, RM7 OBE
United Kingdom

Erdman, R. A. M.
Department of Cardiology
Thoraxcentrum
Erasmus University
University Hospital Dijkzigt
P. O. Box 1738
NL-3000 DR Rotterdam
The Netherlands

Erdmann, E.
Klinikum Großhadern
Medizinische Klinik I
Forschungslabor B
Universität München
Marchioninistrasse 15
8000 München 70 (W)
Federal Republic of Germany

Feelisch, M.
Abteilung Pharmakologie
Schwarz Pharma AG
Alfred-Nobel-Strasse 10
4019 Monheim (W)
Federal Republic of Germany

Glasser, S. P.
Cardiology Section (111 A)
James A. Haley VA Hospital
13000 Bruce B. Downs Boulevard
Tampa, FL 33612, USA

Hauf, G. F.
Rehabilitationszentrum
für Herz- und Kreislaufkranke
Südring 15, 7812 Bad Krozingen (W)
Federal Republic of Germany

Hugenholtz, P. G.
SOCAR SA
Domaine de Leydefeur
CH-1261 Givrins
Switzerland

Jugdutt, B. I.
Division of Cardiology, 2C2, 43
Walter Mackenzie
Health Sciences Center
University of Alberta
Edmonton, Alberta T6G 2R7
Canada

Keck, E. W.
Kardiologische Abteilung
Universitäts-Kinderklinik
und Poliklinik
Martinistrasse 52
2000 Hamburg 20 (W)
Federal Republic of Germany

Langosch, W.
Rehabilitationszentrum
für Herz- und Kreislaufkranke
Südring 15, 7812 Bad Krozingen (W)
Federal Republic of Germany

Marks, C.
Department of Cardiology
Queen Mary Block
Oldchurch Hospital, Romford
RM7 OBE, United Kingdom

Marmor, A.
Cardiology Department
Rebecca Sieff Hospital
Sefat, Israel

Michorowski, B. L.
Division of Cardiology, 2C2, 43
Walter Mackenzie
Health Sciences Center
University of Alberta
Edmonton, Alberta T6G 2R7
Canada

Naber, D.
Psychiatrische Klinik
Universität München
Nussbaumstrasse 7
8000 München 2 (W)
Federal Republic of Germany

Noack, E.
Pharmakologisches Institut
der Universität Düsseldorf
Moorenstrasse 5
4000 Düsseldorf (W)
Federal Republic of Germany

Nötges, A.
Abteilung Kardiologie I
Herz-Kreislauf-Klinik Bevensen
Römstedter Strasse 25
3118 Bad Bevensen (W)
Federal Republic of Germany

Pavičić, L.
Faculty of Kinesiology
University of Zagreb
Yugoslavia

Pohl, W.
Institut
für Versicherungsbetriebslehre
Universität Hannover
Wunstorfer Strasse 14
3000 Hannover 91 (W)
Federal Republic of Germany

Popadić, M.
Department of Medicine
University Hospital
Faculty of Medicine Sarajevo
University of Sarajevo, Yugoslavia

Pöppel, E.
Institut
für Medizinische Psychologie
Universität München
Goethestrasse 31/I
8000 München 2 (W)
Federal Republic of Germany

Popović, G.
Department of Medicine
University Hospital
Faculty of Medicine
Zagreb, Yugoslavia

Potthoff, P.
Infratest Gesundheitsforschung
Landsberger Strasse 338
8000 München 21 (W)
Federal Republic of Germany

Rabinowitz, B.
The Chaim Sheba Medical Center
Heart Institute
Tel-Hashomer 52621, Israel

Reisin, L.
Cardiology Department
Barzilai Hospital
Ahskelon 78471, Israel

Rezaković, Dž. E.
Department of Medicine
Institute of Radiology
University Hospital
Faculty of Medicine
Sarajevo and Zagreb
Yugoslavia

Rychlik, R.
Medizinische Forschung
Schwarz Pharma AG
Alfred-Nobel-Strasse 10
4019 Monheim (W)
Federal Republic of Germany

Schneeweiss, A.
Sauerberg 12,
6208 Bad Schwalbach (W)
Federal Republic of Germany

Schulenburg, J.-M. von der
Institut
für Versicherungsbetriebslehre
Universität Hannover
Wunstorfer Strasse 14
3000 Hannover 91 (W)
Federal Republic of Germany

Sinn, R.
Abteilung Kardiologie I
Herz-Kreislauf-Klinik Bevensen
Römstedter Strasse 25
3118 Bad Bevensen (W)
Federal Republic of Germany

Štalec, J.
Faculty of Sciences
and Mathematics
University of Zagreb
Yugoslavia

Stephens, J.
Department of Cardiology
Queen Mary Block
Oldchurch Hospital
Romford, RM7 OBE
United Kingdom

Traber, U.
Abteilung Kardiologie I
Herz-Kreislauf-Klinik Bevensen
Römstedter Strasse 25
3118 Bad Bevensen (W)
Federal Republic of Germany

Troidl, H.
Chirurgische Klinik
Universität zu Köln
Ostmerheimer Strasse 200
5000 Köln-Merheim (W)
Federal Republic of Germany

Troschke, J. von
Abteilung
für Medizinische Soziologie
Stefan-Meier-Strasse 17
7800 Freiburg im Breisgau (W)
Federal Republic of Germany

Tymchak, W. T.
Division of Cardiology, 2C2, 43
Walter Mackenzie
Health Sciences Center
University of Alberta
Edmonton, Alberta T6G 2R7
Canada

Urbahn, D.
Infratest Gesundheitsforschung
Landsberger Strasse 338
8000 München 21 (W)
Federal Republic of Germany

VandenBurg, M.
Medical and Clinical Research
Consultants Ltd., Lewis House
1 Mildmay Road
Romford,
Essex RM7 7DA
United Kingdom

Viefhues, H.
Furkenstrasse 35
4630 Bochum (W)
Federal Republic of Germany

Wolf, R.
Abteilung Kardiologie I
Herz-Kreislauf-Klinik Bevensen
Römstedter Strasse 25
3118 Bad Bevensen (W)
Federal Republic of Germany

Woodings, D.
MCRC, Lewis House
1 Mildmay Road
Romford,
Essex RM7 7DA
United Kingdom

Introduction

A. Schneeweiss

Although the syndrome of congestive heart failure has been recognized many years ago, the approach for its evaluation and treatment has until recently, been partial and 'fragmentary'. Various aspects of the disease have been treated according to the evaluation tools and therapeutic measures available at each period. This approach resulted in some of the greatest achievements in the management of heart failure but also left many aspects neglected and also resulted in several paradoxes.

Examples of the achievements and limitations of the 'fragmentary' approach are the use of diuretics and hemodynamic measurements. The development of diuretics has provided us with an important tool for helping patients whose predominant problem was edema. The success of diuretics masked the fact that their use may often be hemodynamically unsound and that they may reduce cardiac output. Only many years after their introduction has the use of diuretics found its appropriate place.

Hemodynamic monitoring has gone via the same path. The great contribution of continuous bedside hemodynamic measurements to understanding heart failure resulted in over-usage by many clinicians, who found themselves treating hemodynamic charts rather than patients. It took almost a decade to realize that hemodynamic improvement, even in the chronic setting, does not necessarily mean symptomatic improvement or an increase in exercise capacity.

In recent years the complexity of the heart failure syndrome has been better realized and a more rational, multifactorial approach to its evaluation and management has been developed. End-points of therapy concentrate on the patient and not on parameters that are convenient to measure. Survival and exercise capacity are two important and easily evaluated end-points of treatment. An equally important goal, but more difficult to measure, is quality of life. The first part of the following presentation concerns mainly quality of life. The second part concentrates on therapy and particularly on the use of nitrates for heart failure patients. An important aspect in this topic is early prevention of the post-infarction changes in the left ventricle, leading to heart failure. Nitrates have been found effective in preventing these changes.

The material of this part is based mainly on presentations given at the first international meeting on heart failure. My colleagues and I appreciate the initiative of the Chairmen, Profs. Lewis and Kimchi, who provided that stage for our presentation.

I. Quality of Life

The Pathophysiological Basis, Clinical Presentation, and Therapy of Chronic Heart Failure

M. Böhm and E. Erdmann

Heart failure is characterized by the inability to deliver the volume of blood required by the organism due to a reduction in cardiac pumping performance. This limitation affects all the organs of the organism and results in symptoms typical of heart failure. These symptoms are caused either by reduced blood supply to the organs or by congestion upstream of the right or left ventricle, also referred to as "low output failure." In contrast, there are also forms of heart failure with a high cardiac output ("high output failure"). This less common form of heart failure is found in reduced afterload (e.g., in sepsis) or metabolic disorders such as hyperthyroidism, anemia, beriberi, Paget's disease or arteriovenous fistulas. Heart failure occurs in these cases as a consequence of inadequate elevation in cardiac output in response to increased peripheral demand. Furthermore, in both forms of heart failure compensatory reactions take place which in themselves lead to further pathological changes in the heart and other organs, and can thus exacerbate the basic disease and its symptoms. In the following sections, the fundamentals of cardiac muscle contraction, the regulation of contractile force by the sympathetic nervous system, and the specific cardiac changes which result from heart failure will be discussed. In addition, the causes of heart failure, its diagnosis, and the possibilities for drug therapy will be described. Instrumental and surgical therapeutic procedures will not be dealt with.

The Molecular Basis of Cardiac Muscle Contraction

The myocardial cell contracts upon electrical stimulation. The electrical stimulus is propagated along the cell membrane and the invaginations of the sarcolemma (the so-called transverse tubular or T system). Depolarization of the cell membrane opens voltage-dependent calcium channels, which leads to an increase in the slow inward calcium current [20]. The calcium ions which pass into the cell release further calcium from intracellular stores of the sarcoplasmic reticulum [13]. The calcium entering the cell from outside acts as a calcium trigger, whereas the calcium released from the intracellular stores mainly contributes to myocardial contraction.

Muscular contraction is described in terms of Huxley's theory of sliding filaments [17]. This theory rests on the interaction of thick (myosin, molecular weight approximately 500000) and thin (actin, molecular weight approximately 45000) myofilaments. The parallel axis arrangement of the myofilaments (sarcomeres), each of which forms an approximately 1-µm-thick myofibril, is illustrated in Fig. 1. Muscle contraction is based on shortening of a large number of serially arranged sarcomeres. This process is activated by the binding of free calcium to troponin C, a regulatory subunit of the contractile apparatus [17]. Shortening, i.e., contraction, is caused by the displacement of the myosin and actin filaments relative to one another. This is brought about by the so-called cross-bridges. The heads of the myosin molecule or cross-projections connect the actin molecule to the myosin filament (Fig. 2). On contraction, a narrowing of the angle between the myosin neck and the myosin

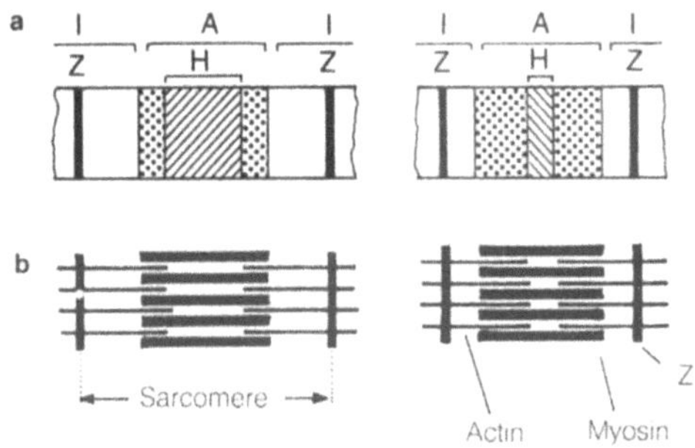

Fig. 1 a, b. Structure of the contractile apparatus of the heart, at rest *(left)*, and contracted *(right)*. In a myofibril there are about 1000 myosin molecules which are surrounded by around 2000 thinner actin molecules (**b**). The myosin molecules are attached to the Z line. The central portions of the sarcomere, which contain the thick myosin filaments with or without actin, appear under the light microscope as A bands (anisotropic picture). The lateral components which contain only actin are seen as I bands. The appearance of the central, actin-free H zone depends on the contractile condition of the myofibril. Microscopically, this molecular structure produces the typical picture of striated cardiac and skeletal muscle. (Modified from [28])

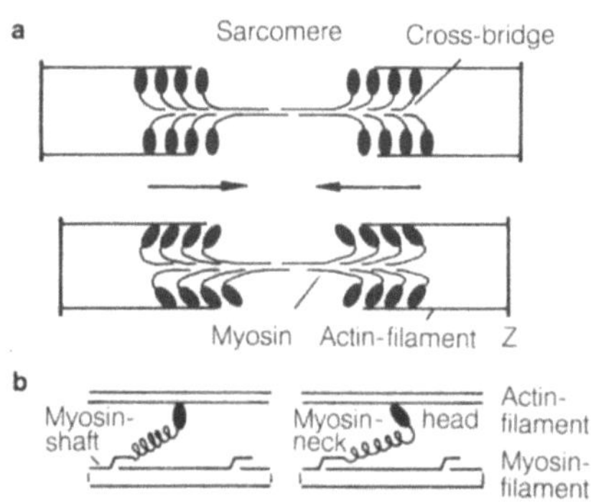

Fig. 2 a, b. Mechanism of the formation of cross-bridges in Mammalian hearts. **a** Myosin heads bind the myosin molecule to the actin filament. **b** Shortening of the muscle upon contraction occurs through a narrowing of the angle between the myosin head and myosin neck with a rapidly repeating binding and release of the cross-bridges from the actin molecule. (Modified from [28])

head occurs. Rapidly repeated binding and releasing of the myosin heads by the actin filament enable shortening of the muscle to take place [26].

The force of contraction is determined by the concentration of intracellular calcium available and the calcium sensitivity of the contractile proteins. Therefore, an increase in the force of contraction largely depends on increasing the inward flow of calcium and thus raising the intracellular calcium concentration [30]. This can be achieved either pharmacologically or through the physiological regulatory processes of the sympathetic nervous system. Upon exertion, the contractile force of the heart is augmented by activation of the sympathetic nervous system. Cardiac contractility is stimulated by sympathetic nerves and by the release of catecholamines from the adrenal medulla, kidneys, skeletal muscle, etc. The catecholamines released (norepinephrine and epinephrine) are able to activate membrane-based adenylate cyclase by binding to β-adrenoceptors in the myocardial cell membrane. Adenylate cyclase is an enzyme that, upon activation, forms increased quantities of cyclic adenosine monophosphate (cAMP) from ATP. The cAMP then activates cAMP-dependent protein kinase. This enzyme phosphorylates the sarcolemmal calcium channel, which leads to an elevation of the slow inward flow of calcium into the cell, and thereby to an increase in contractile force by facilitated calcium-induced calcium telease [27]. In diastole, about 88% of the calcium is transported back into the intracellular calcium store (sarcoplasmic reticulum) by a transport protein (phospholambam) in the membrane of the sarcoplasmic reticulum. Phospholambam can also be phosphorylated cAMP-dependently, a process which contributes to its enhanced activity during stimulation of β-adrenoceptors and leads to accelerated calcium uptake into the sarcoplasmic reticulum during diastole. The result is a more rapid relaxation of the cardiac muscle [34]. In addition to the reuptake of calcium into the sarcoplasmic reticulum mediated by phospholambam, there is also a sodium-calcium exchange mechanism which contributes to the elimination of calcium from the cytosol during diastole [19]. These mechanisms can be utilized in the treatment of heart failure with positive inotropic drugs (substances which augment contractile force; see below).

Etiology of Chronic Heart Failure

The most common cause of chronic heart failure is a decline in myocardial contractility as a consequence of long-term pressure or volume overload of one or both ventricles (arterial hypertension, recurrent pulmonary embolism, valvular defects) or dilated cardiomyopathy as a result of myocarditis. A further important cause is heart coronary disease, which causes a reduction in the oxygen supply to the myocardium. In addition to coronary heart disease, chronic anemia, hypoxemia due to pulmonary disease, hyperthyroidism, and direct impairment of myocardial contractility or a decline in the elasticity of the heart (e.g., myocardial inflammation, cardiomyopathy, amyloidosis, dys- or paraproteinemia, toxins, negative inotropically active drugs) can also be of significance. Bradycardial and tachycardial dysrhythmias can also result in impairment of cardiac pumping ability and, through a reduction in cardiac output, can lead to the typical symptoms of heart failure. Normally, however, several of the above-mentioned basic diseases are present simultaneously, and their cumulative effects on the failing heart can cause acute deterioration and exacerbate the course of the disease by creating a vicious circle [32]. In this respect, the basis for successful treatment of heart failure would be to eliminate the underlying disorder responsible for the reduction in pumping function. Unfortunately, this is only rarely possible.

In principle, the pathogenesis of myocardial failure can be divided into myocardial mechanical and cell physiological aspects. Mechanically, changes in preload, afterload, contractility or heart rate can lead to acute pump failure. Changes in preload and afterload are mostly a consequence of pressure overload. The possible causes can be summarized as follows:

1. *Changes in the preload:* Abnormal volume conditions (e.g., hypervolemia, aortic incompetence, mitral incompetence, reduced venous supply)
2. *Changes in the afterload:* Abnormal pressure conditions (e.g., hypertension, cor pulmonale, arteriolar vasodilatation)
3. *Changes in contractility:* Ischemic heart disease, dilated cardiomyopathy drugs or toxins with negative inotropic action
4. *Changes in heart rate:* As a result of bradycardial and tachycardial dysrhythmia (below and above the so-called critical heart rate)

The cell physiological aspect refers to the influence of noxious substances on subcellular structures, such as:

1. Effect on cell membrane-based receptors for hormones and drugs (e.g., thyroid hormones, somatotropic hormone, β-adrenoapter antagonists, cardiac glycosides)
2. Effect on passive cell membrane permeability for ions(e.g., antiarrhythmic drugs such as lidocaine, anticholinergics, calcium antagonists, uremic toxins, nickel, saponin, various snake venoms, bee or wasp venoms)
3. Effect on active ion transport (e.g. cardiac glycosides, lithium, potassium)
4. Changes in sarcoplasmic reticulum function (e.g., through reduction in extracellular calcium concentration, after blockade of oxidative phosphorylation, after release of membrane damaging lysosomal enzymes, membrane damage due to snake venoms)
5. Oxidative phosphorylation disturbances (O_2 deficiency, cobalt, lead, thallium, CN/CO/halothane)
6. Effects on regulatory and contractile proteins. Changes in the sarcomeres (e.g., increase in preload), abnormal fibril growth (e.g., hypertrophic cardiomyopathy, obstructive cardiomyopathy)
7. Reduction in calcium-dependent ATPase activity (e.g., acidosis, cobalt, etc.)
8. Damage to the lysosomes with release of lysosomal enzymes (lead, snake venoms, viruses)

9. Changes in protein biosynthesis (e.g., advanced age, antimetabolites, viruses, antiarrhythmic drugs, alcohol, diphtheria toxin, ionising radiation)
10. Disorders in mitochondrial ATP-ADP transport (due to autoantibodies, toxins)

Pathophysiology – Effects on The Heart

Impaired pumping performance due to one of mechanisms described above is accompamied by an increase in left ventricular end-diastolic pressure. This leads to increased stretching of the ventricle. This stretching of the left ventricle, however, has a direct influence on the development of the contractile potential of each individual myofibril [39]. The force developed depends directly on the degree of overlap of the actin and myosin filaments; at an optimal degree of overlap, a large number of cross-bridges can interact with the actin filaments, leading to maximal contractile force. This is known as Starling's law of the heart [33]. It implied that the development of tension created by the myofibril increases at an optimal overlap of myosin and action filaments. However, further stretching of the sarcomere leads to a decrease in developed tension [18]. Thus, when there is a loss of contractile force, the development of tension by individual fibrils is thought to be enhanced initially by an increase in sarcomere length. But a further increase in pretension due to impaired pump performance will lead to a decrease in developed tension. This mechanism is illustrated in Fig. 3.

A reduction in contractility with a reduction in cardiac output leads to activation of the sympathetic nervous system to maintain blood pressure and ensure the supply of blood to vital organs. Figure 4 shows that patients in this

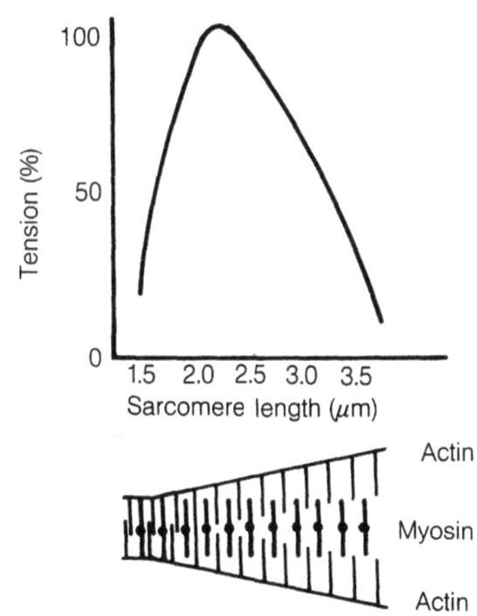

Fig. 3. Relation between the development of tension and sarcomere length and the dependency of tension and sarcomere length on the degree of overlap of the actin and myosin filaments

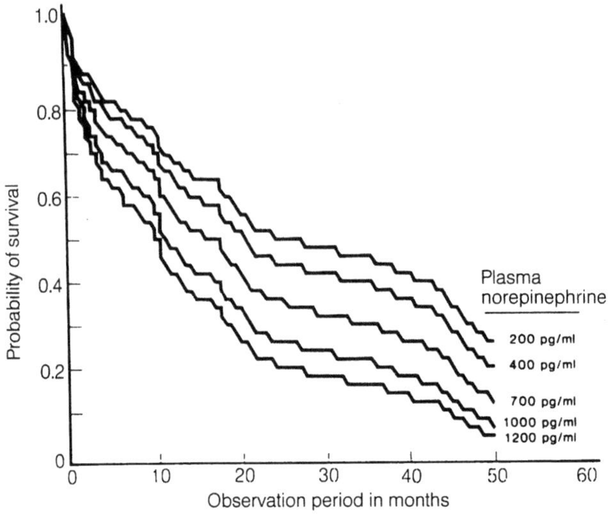

Fig. 4. Probability of survival in patients with heart failure in relation to plasma norepinephrine levels. Patients with high norepinephrine levels have a markedly lower probability of survival. Ordinate, relative probability of survival; abscissa, Observation period in months. (Modified from [8])

stage of heart failure a show elevated serum norepinephrine levels [7, 8, 36]. The higher the detected norepinephrine concentrations, the shorter is the average survival time of the patient [8].

The norepinephrine-related increased cardiac stimulation has serious consequences for the myocardial cell itself. Increased occupation of the β-adrenoceptor leads to receptor downregulation. Figure 5 shows the number of β-adrenoceptors in human myocardial tissue. Depending on the degree of heart failure, a marked decrease in the density of these β-adrenoceptors occurs [3, 5, 6]. The result is that the key mechanism by which circulating catecholamines exert their action on cardiac contractility is severely impaired. This is also of functional relevance for the heart, as shown in Fig. 6. Administration of the synthetic catecholamine dobutamine produces only a weak increase in the farce of contraction in the failing heart compared to nonfailing myocardium. The loss of efficacy is less pronounced in hearts of patients with moderate heart failure (New York Heart Association classification: NYHA II to III; see Table 2), and both groups are clearly less responsive to such β-agonists than healthy control hearts [3]. This means that the compensatory mechanism of increased activity of the sympathetic nervous system in the failing heart leads to a loss of catecholamine efficacy caused by a decrease in cardiac β-adrenoceptor sites. Figure 7 illustrates the consequences of a rise in serum norepinephrine levels in patients with heart failure. The reduction in β-adrenoceptors as a result of norepinephrine-induced desensitization is independent of the underlying heart disease and is probably only related to the severity of the heart failure and the related elevation in norepinephrine levels. In addition to the reduction in β-adrenocep-

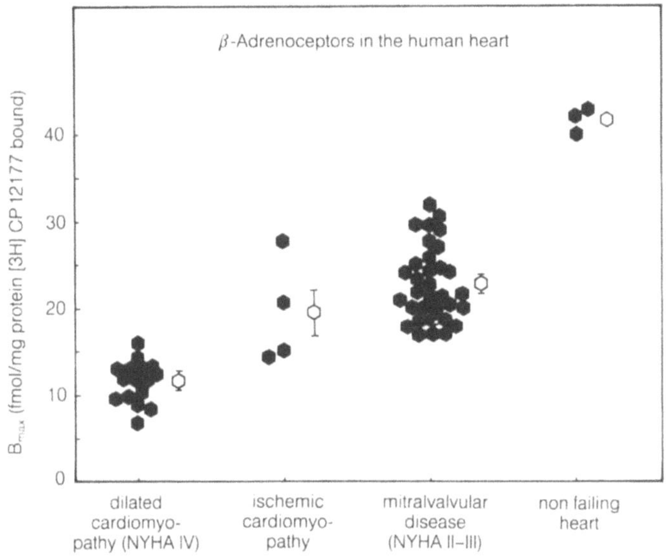

Fig. 5. β-Adrenoceptors in the myocardium of patients with severe (NYHA IV) and moderate (NYHA II–III) heart failure. The nonfailing myocardium originated from donor hearts which were not suitable for transplantation due to technical reasons. (Modified from [4])

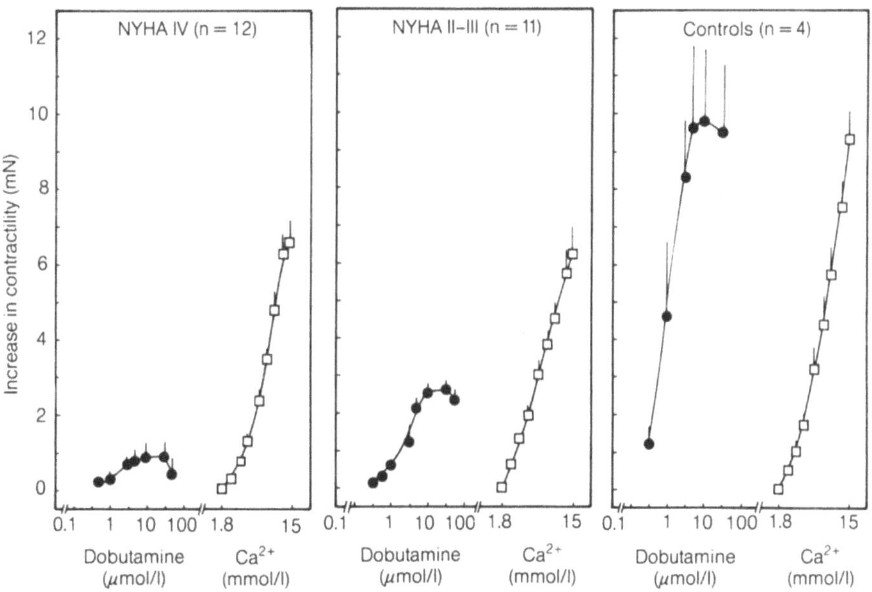

Fig. 6. Positive inotropic effect of dobutamine in the myocardium of patients with severe heart failure (NYHA IV), moderate heart failure (NYHA II–III), and in the nonfailing myocardium. One can clearly see that farce of contraction can still be enhanced by Ca^{2+} in heart failure, but not by dobutamine. This indicates that the contractile apparatus of the cardiac muscle is functional, but is not mobilized through the β-adrenoceptor-adenylate cyclase system. (Modified from [3])

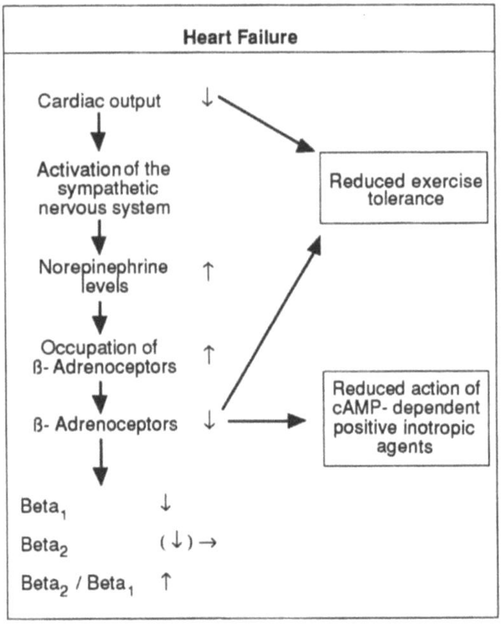

Fig. 7. Activation of the sympathetic nervous system and its consequences for the myocardium in heart failure

tors, a postreceptor defect of the cardiac adenylate cyclase has been reparted. Three independent groups of investigators [4a, 13a, 21a] have reported an increase of 40% in the inhibitory guanine-nucleotide leinding protein ($G_{i\alpha}$) in explanted hearts from patients with tominal heart failure. Most likely, this patholocheniscal alteration conntributes to the decline in adenylate cyclase and to the reduction in the inotropic effects of catecholaminess in heart failure.

A further possible compensatory mechanism of the myocardium is an induction of cardiac hypertrophy, i.e., an increase in contractile proteins. Myocardial hypertrophy, which manifests as increased thickness of the cardiac wall, is particularly pronounced in pressure overload of the left ventricle, e.g., in aortic stenosis or arterial hypertension. It should be noted, however, that all compensatory mechanisms play a role in the initial stages of heart failure and can occur concurrently. In dilated cardiomyopathies, the macroscopic findings are usually those of myocardial dilatation, whereas microscopically, hypertrophy of individual myofibrils is often seen.

Pathophysiology – Effects on Organ Perfusion

The renin-angiotensin-aldosterone system and the sympathetic nervous system are activated in order to maintain systolic blood pressure following a decline in cardiac contractility, leading to a rise in peripheral vascular resistance

Table 1. Distribution of resting cardiac ejection output in healthy and cardiac patients with and without clinically manifest heart failure. Modified after Wade and Bishop [38]

	Healthy	Cardiac Patients			
		Without heart failure		With heart failure	
	Blood flow (ml/min)	Blood flow (ml/min)	% of normal value	Blood flow (ml/min)	% of normal value
Intestines	1400	1200	80	800	53
Kidneys	1100	650	59	350	32
Brain	750	600	80	500	67
Coronaries	250	300	120	300	120
Skeletal muscle	1200	1200	100	1200	100
Skin	500	150	30	50	10
Other organs	600	300	75	200	50
Cardiac output	5800	4400	77	3400	60

[24]. Vasoconstriction takes place by activation of the sympathetic nervous system and through the effect of norepinephrine and the aemin-angiotensin system on vasculature. Because of the very nonuniform distribution of these vascular α-adrenoceptors this compensatory mechanism has important effects on the blood supply to individual organ systems. Table 1 compares the relative proportions of the cardiac output received by various organ systems at rest in patients with heart failure and in healthy controls. This investigation was conducted while the patients were physically at rest, since circulatory conditions in the individual organ regions vary greatly during exertion [38]. The data in this illustration demonstrate that, during heart failure, the supply of blood to the brain and heart remains virtually the same. The advantage of this compensatory mechanism rests in the redistribution of blood to vital organs, while circulation to the organs such as skin and kidneys, which are not acutely vital, is significantly reduced.

A drop in cardiac output leads to reduced blood flow to the kidneys, causing a fall in the glomerular filtration rate. This mechanism, in addition to actiavtion of the renin-angiotensin system, leads to retention of sodium and water, with an initial increase in the intravascular volume, and later to the development of edema by an expansion of extracellular fluid volume. Activation of the renin-angiotensin-aldosterone system primarily contributes to this dysregulation by causing peripheral vasoconstriction and by increasing sodium and water retention. The increased intravascular fluid volume and the increase in venous tone cause an increase in filtration of fluid into the extravascular spaces in the capillary bed and thereby lead to the development of peripheral edema [29]. In the lungs, left ventricular congestion and an elevation in extracellular fluid volume lead to the development of pulmonary edema (left ventricular failure) and/or pleural effusion (predominantly right ventricular failure).

History, Symptomatology, and Clinical Course

The symptoms, clinical course, and prognosis of chronic heart failure are primarily determined by the type and extent of the basic disorder, by the clinical severity, and by the possible occurrence of complications (e.g., thromboembolism, pulmonary edema, cerebrovascular complications, infections, dysrhythmias). The constellation of symptoms – reduced cardiac performance with congestion and possible generalized edema and enlarged heart – points to the common form of congestive heart failure (so-called wet decompensation: "backward failure"). But conditions of heart failure with hypotension, vertigo, intermittent abdominal complaints (postprandial), and muscular fatigue should also be considered as signs of heart failure ("forward failure").

The *history* of heart failure patients can itself provide important clues as to the cause, clinical severity, and prognosis of the disease. Previous complications and attempted therapies for heart disease are also recorded in the history. A history of known heart disease, including rheumatic heart disease in childhood or adolescence, provides evidence of the etiology of the underlying heart disease. In adults, arterial hypertension and coronary heart diesease should be considered the most frequent precursors of chronic heart failure. Typical risk factors and risk indicators (smoking, hypertension, hypercholesterolemia, diabetes mellitus, hyperuricemia, and obesity) can almost always be documented in coronary heart disease, and in the chronic heart failure which follows it. A past history of arterial embolism suggests either mitral stenosis, intermittent atrial fibrillation, dilated cadiomyopathy, atrial myxon endocarditis, or a previous myocardial infascition with farcnation of left ventricular aneurysm.

Paroxysmal and ultimately permanent atrial fibrillation or atrial flutter with absolute tachyarrhythmia and all forms of pathological bradycardia frequently accompany the clinical syndrome of biventricular failure. Septic infections, catarrhal infections – commonly in connection with physical overexertion – should also be considered as possible precursors of chronic endomyocarditis with subsequent heart failure. Bronchial asthma, chronic bronchopulmonary disease, pulmonary diseases of various etiologies and recurrent venous thromboembolism are frequent and familiar precursors of chronic right ventricular failure (chronic cor pulmonale).

The *symptoms* and clinical course of heart failure patients are uniformly characterized by decreased performance as a result of rapid fatigue, dyspnea, edema and, in certain cases, vertigo and sometimes cardiac pain. According to the New York Heart Association, four classes of clinical severity are distinguished on the basis of the information given by the patients. The categorization of a patient *with heart disease* is made in the light of his or her habitual extent of physical activity and symptoms. Therefore, this classification applies only in the case of proven heart disease. Even though there are many objections to using the patient's own assessment of his or her symptoms and their evaluation by the physician, this simple classification has proved valid for most clinical purposes (indications for surgery, assessment of therapeutic suc-

Table 2. Classification of clinical severity of heart disease according to New York Heart Association (1945)

Class I	Cardiac patients with no limitation of activities. On normal physical activity they suffer no dyspnea, anginal pain, or palpitations.
Class II	Patients with slight limitation of physical activity. These patients feel comfortable at rest and with mild exertion. Symptoms do not become apparent until a stronger degree of normal activity is involved.
Class III	Patients with marked limitation of physical activity. These patients only feel comfortable at rest, and symptoms arise even with a mild degree of normal activity.
Class IV	Patients to whom any physical activity brings discomfort. The symptoms of heart failure can even occur at rest and are intensified by physical activity.

cess, monitoring the course of the disease, expert reports or insurance inquiries, etc.). The criteria according to which clinical heart failure is classified by grade of severity (NYHA classes I–IV) are summarized in Table 2. All acute symptoms, such as cardiac asthma and pulmonary edema, syncope, and the symptoms of cardiogenic shock, must naturally be viewed separately from this classification of chronic heart failure. Patients should additionally be questioned about their inability to sleep lying flat, coldness in the extremities, tendency towards onychomycosis, and leg ulcers. Furthermore, pain in the upper right abdomen, flatulence, nycturia, and lack of appetite and nausea (congestive gastritis, congested liver) can be signs of heart failure.

Physical Examination

On physical examination, typical signs of heart failure can be identified: central and peripheral cyanosis, orthopnea, and rest and effort dyspnea, sometimes with prolonged expiration. Cold extremities, meteorism, generalized edema (mostly in dependent parts), subicterus, and reduced performance with dyspnea during examination are also common in these patients. The arterial pulses are weak, and as a rule the heart rate is slightly elevated. Arterial blood pressure is not characteristically changed. In advanced stages, however, reduced pulse pressure and hypotension are found. In the tachycardic form of absolute arrhythmia, there is a pulse deficit. In supine patients with right ventricular failure, the neck veins are grossly distended and the A wave of the jugular pulse is often excessively high. Upon compression of the liver, a hepatojugular reflux occurs with obvious tautness of the jugular vein. A systolic venous pulse together with palpable liver pulsation points to an additional, generally secondary tricuspid incompetence.

Percussion usually reveals cardiac enlargement and a third heart sound may be present in more severe cases. An apical heave and pathological murmurs and heart sounds frequently point to valvular defects. Pulmonary congestion can be identified during auscultation by mainly basal moist rales, sometimes sounding close to the ear, or by crackling expansion rales. A bronchospastic component with prolonged expiration and rhonchi is the rule in cardiac asthma. Congestive pleural effusions – both left and right – can freqently be found.

Technical examinations: Central venous pressure is elevated in right ventricular failure, as well as in biventricular failure. This increases further upon physical exertion and with intravenous volume replacement. Depending on the degree of pulmonary congestion, there may be some imbalance between alveolar ventilation and perfusion, sometimes with a drop in diffusion capacity. Arterial oxygen saturation is thus reduced, the oxygen saturation of mixed venous blood is greatly decreased and, as a consequence, the *arteriovenous oxygen differences* of the entire circulation are pathologically elevated. As a rule, the partial pressure of CO_2 is lowered (hypocapnia as a result of hyperventilation). On the other hand, the congestion can itself lead to an obstructive ventilatory disorder (cardiac asthma). In the decompensated state, the 24-h urine volume is reduced, the specific gravity of the urine is elevated, and low-grade proteinuria is regularly found. In rare cases, more severe proteinuria, or even a nephrotic syndrome, has been observed. As a consequence of secondary hyperaldosteronism, the serum sodium/potassium ratio drops to below 0.5.

The *electrocardiogram* reveals no specific changes in chronic heart failure. Disturbances of pacemaker function, conduction, and repolarization as well as pathological electrical axes (left and right axis deviation) and heart block all give clues to the primary disorder. Drug-related changes (cardiac glycosides, antiarrhythmics) are occasionally seen. Echocardiographic examination reveals an increase in the dimensions of the right and left ventricles. M-mode recordings are helpful in determining the dimensions in global hypokinesia of the left ventricle occurring in dilated cardiomyopathy. Two-dimensional echocardiography can be used to investigate disturbances of regional wall movements. In addition, Doppler echocardiography allows detection of valvular defects and pressure gradients.

Prognosis in Heart Failure

The practical treatment of heart failure is decisively dependent on its severity, and thus on the stage of the disease. The prognosis in heart failure is dependent on whether the therapy can achieve causal control of the underlying disease at an early stage. Early recognized and timely treated, congenital heart defects have a good prognosis. Improved life expectancy has become evident through consistent therapy of arterial hypertension. Apart from this, pacemaker implantation has been shown to extend life expectancy, and is the only possible way to achieve cardiac recompensation in left ventricular failure due to bradycardial dysrhythmias. The prognosis in patients with NYHA grades III and IV heart failure is much worse. This type of chronic heart failure is mostly a consequence of long-term coronary heart disease, of dilated cardiomyopathy secondary to myocarditis, of alcoholic cardiomyopathy, or of chronic cor pulmonale. The 1-year survival rate in seven independent studies on about 900 patients was 34%–58% [21]. In about 75% of the patients investigated, higher-grade arrhythmias were registered on the Holter ECG. Probably rhythm related, sudden cardiac deaths occurred in 33%–47% of the patients [31]. However, there are differences in prognosis depending on the type of drug therapy. Besides this, there are a number of newer drugs whose therapeutic benefit is presently being investigated in large studies.

Therapy of Chronic Heart Failure

Principles of Therapy

Overt heart failure should be treated because of the bad proposis and life quality. However, prophylactic therapy neither prevents nor delays the occurrence of symptoms. Whenever possible, the underlying disease should be treated first of all. Drug therapy should be preceded by general measures, such as reduction of physical activity and reduction of risk factors (e.g., nicotine) and weight. Symptomatic therapy in overt heart failure can be induced by increasing contractility with so-called positive inotropic drugs. Additionally, it is worthwhile to remove sodium and fluid by using diuretics. Reducing the work load of the heart by reducing the preload and afterloads is also very important.

The aims in the therapy of chronic heart failure can be summarized as follows:
1. Improved hemodynamics
   a) At rest
   b) During exercise
2. Improved exercise tolerance
   a) Submaximal
   b) Maximal
3. Improved quality of life
4. Reduction in arrhythmias
5. Reduced mortality

Improved hemodynamics at rest and during exertion mean that the patient is better able to tolerate exercise, which leads to improved quality of life. Furthermore, complications of overt heart failure, such as the occurrence of arrhythmias, must be treated with drugs. Reduced mortality parallels the decrease in complex arrhythmias. Depending on the clinical severity of myocardial insufficiency, stepwise plans have often been devised to schedule therapy. However, one must understand that, due to the great variability of the disease, such therapeutic schedules have only limited general application. In the individual case, the decision must naturally be made on an individual basis. In the initial stages (NYHA I–III) of heart failure, therapy should first concentrate on physical relaxation and reduction of risk factors. Prophylactic therapy with, for example, positive inotropic drugs cannot prevent progression of heart failure [10]. For this reason, these substances are not indicated in the initial stage (NYHA I). The next stage in class II heart failure is to take the dietary measure of reducing sodium intake. This should reduce sodium and water retention which is caused by heart failure. The next therapeutic measure in the schedule should be the introduction of diuretics (also vasodilators) which reduce cardiac preload and afterload and possibly counteract any existing sodium and water retention. Therapy with cardiac glycosides can already be initiated at this stage. Cardiac glycosides lead to an increase in contractility and, by slowing the heart rate, also help to reduce cardiac work, especially in atrial fibrillation [12]. In NYHA stage III, progression of heart failure can often be delayed by intensifying drug therapy and by further limiting physical exertion. Special types of therapy, such as positive inotropic substances which do not belong to the class of cardiac glycosides, and heart transplantation, are reserved for the most extreme forms of heart failure. Drug therapy, which is treatment for patients with chronic heart failure the most common, will be dealt with in greater detail below.

Drug Therapy

Diuretics

In chronic heart failure, the administration of diuretics leads to elimination of retained water and sodium. It is thus possible to reverse the elevation in extracellular fluid volume which accompanies heart failure. Basically, there are two groups of diuretics: those which promote potassium excretion and those which reduce it. Table 3 summarizes the various drug classes currently in use.

Table 3. Selection of diuretics currently in use

	Average total dose (mg)	Onset of effect	Duration of effect
1. Potassium-excretiony diuretics			
Benzothiadiazines			
Hydrochlorothiazide($t_{1/2}$ 2–5 h)	25– 50	2 h	12 h
Chlorthalidone ($t_{1/2}$ 36–89 h)	25–100	2 h	72 h
Loop diuretics			
Furosemide ($t_{1/2}$ 0.85 h)	40– 80	1 h	6–8 h
Ethacrynic acid	25–100	30 min	6–8 h
2. Potassium-sparing diuretics			
Triamterene	50 –100	2–4 h	7–9 h
Amiloride	2.5– 5	2 h	24 h
Spironolactone (aldosterone-dependent)	50 –100	1–2 days	2–3 days
3. Compound preparations			
Hydrochlorothiazide + amiloride	(e.g., 50 mg + 5 mg)		
Hydrochlorothiazide + triamterene	(e.g., 25 mg + 50 mg)		

Factors of clinical importance are the different half-lives and strengths of action of these substances. Of the potassium-loosing diuretics, the so-called loop diuretics, such as furosemide and ethacrynic acid, are characterized by short duration of action and quick attainment of their peak effect. They cause increased elimination of sodium, water, and protons. Their intensity of action and rapid effect dictate their indications and side effects. Along with nitrates, they are the drugs of choice in acute left ventricular failure, where quick elimination of sodium and water is desired. They may exert vasodilator properties in the peripheral vascular bed of the pulmonary and systemic circulation, leading to additional relief of the strain on the heart. Benzothiadiazines are medium-strength diuretics which also cause increased excretion of sodium, potassium, and water. The action of hydrochlorothiazide, for example, begins within 2 h of administration, with a peak effect at 2–6 h after intake. The duration of action is about 12 h. Mild diuretics such as the potassium-sparing diuretics triamterene, amiloride, and spironolactone cause excretion of sodium and water, but to a much lesser extent than the substances already mentioned. Natriuresis occurs without loss of potassium. Spironolactone is especially used in conditions which accompany elevated aldosterone levels, such as chronic liver disease.

Several side effects result from the profiles of action mentioned. Powerful diuretics can cause acute hypovolemia with a drop in blood pressure. The accompanying vasodilator action may contribute to this. Furthermore, dose-dependent damage to the inner ear can occur leading to temporary and sometimes even irreversible loss of hearing. Through the rapid elimination of potassium, hypokalemia can arise. Particularly in combined therapy with cardiac glycosides, such hypokalemia can lead to toxic side effects such as ventricular ectopy. Similar side effects can be found with the medium-strength diuretics. Additionally, benzothiadiazine diuretics lead to reduction in the supply of blood to the kidneys. Therefore, they should be used with caution in patients with impaired renal function. When mild diuretics or potassium-sparing diuretics are used, one should be aware of accumulation of potassium, especially in patients with impaired renal function. Here, hyperkalemia is the most important side effect to consider.

All of the above-mentioned diuretics lead to inhibition of renal elimination of uric acid. This should be taken into consideration when treating patients with gout. Diuretics also affect glucose tolerance. This can cause problems in adjusting blood sugar levels in diabetics.

The symptomatic benefit of therapy with diuretics is well-established. They appear to be particularly suited for reducing pulmonary congestion and initiating a reduction in peripheral edema. The use of diuretics alone in heart failure, however, is controversial. There have been investigations which have shown that diuretics result in a stimulation of neuroendocrine activity, a condition which is particularly undesirable in the treatment of heart failure. On the other hand, diuretics are well suited for combination therapy, where the dose of each individual substance can be reduced, thus reducing the risk of side effects.

Angiotensin Converting Enzyme Inhibitors

In heart failure, a compensatory activation of the renin-angiotensin-aldosterone system occurs [24]. In addition to sodium and water retention, this leads to peripheral vasoconstriction of the arteries and arterioles. The increased preload (due to venous constriction) and the increased afterload (due to arteriolar constriction) lead to further impairment of cardiac output. Thus, inhibition of angiotensin converting enzyme (ACE) appears to be a logical place to break this vicious circle. Enalapril and captopril were first introduced into therapy. These substances basically differ in their plasma half-life – the half-life of captopril is 1–2 h, which is much less than that of enalapril. Advantages and disadvantages in both substances justify their use in specific indications [25]. Captopril has advantages in acute cases due to its greater ease of monitoring. Enalapril plays an important role in the long-term therapy of heart failure, as a once to twice daily intake of the tablet suffices. Table 4 summarizes the pharmacodynamic data for captopril and enalapril. The most important side effect of these substances is a pronounced, early drop in blood pressure which is caused by vasodilatation. Therefore, this therapy should be started at a very low dose and the course should be very closely controlled. In ACE inhibitor therapy, it is also necessary to monitor renal function regularly since all of the substances available to date can lead to a reduction in glomerular filtration rate. This is particularly significant in bilateral renal artery stenosis. Leukopenia, rashes, headache, and hyperkalemia can also occur in rare cases. Coughing may also arise.

ACE inhibitors are the only substances used in the treatment of chronic heart failure which have been shown to both lead to symptomatic improvement and also increase the probability of survival [35]. These data do also possibly suggest that ACE inhibitors should be used early in the treatment of medium-grade heart failure. However, controlled studies on this issue are still lading at present. Although this drug class can cause side effects, only in rare cases is it necessary to discontinue medication. The question as to whether therapy with ACE inhibitors alone can be successful in the treatment of heart failure has not yet been answered. In the largest study available, all patients received additional treatment with positive inotropic substances (cardiac glycosides), diuretics, and sometimes other vasodilators [35]. Therefore, definitive statements cannot be made on this subject at present.

Table 4. Properties of the ACE inhibitors captopril and enalapril

	Captopril	Enalapril
Bioavailability (%)	60–70	60–70
Onset of action (min)	10–30	60
Time to plasma concentration (h)	1	4
Plasma half-life (h)	1– 2	11
Duration of action (h)	8–12	12–24
Daily oral dose (mg)	2–3 × 12.5–50	1–2 × 2.5–20
Content/tablet (mg)	25/50	5/10/20

Cardiac Glycosides

In 1785, the English physician William Withering (1741–1799) published his paper "An account of the foxglove, and some of its medical uses: with practical remarks on dropsy and other diseases." This book laid the foundations of digitalis therapy and its dosage (saturation and maintenance therapy). Today, several controlled studies are available on the efficacy of cardiac glycosides in the treatment of heart failure. A reduction in symptoms and an improvement in exercise tolerance have been well demonstrated.

Basically, there are two derivatives which are in therapeutic use today: digoxin and digitoxin. Strophanthin, used previously, no longer has any definite indications. The difference between digoxin and digitoxin is primarily related to their excretion. Digoxin is mainly excreted by the kidneys, whereas digitoxin can be excreted by the kidneys and the liver. This has the practical consequence that digitoxin finds greater application in patients with renal failure, whose dosage of digoxin would have to be reduced. On the other hand, the half-life and hence the rate of decay of digoxin is shorter than that of digitoxin. The pharmacokinetic data on these cardiac glycosides are summarized in Table 5.

Consideration of the pharmacokinetic and pharmacodynamic properties of cardiac glycosides is of particular importance because overdosage can very frequently cause side effects. The most important side effect is arrhythmia, which represents 70% of all side effects. Moreover, there are occasionally gastrointestinal disorders such as nausea and vomiting and disorders of the central nervous system which manifest themselves as fatigue, confusion and occasionally disordered colour perception. The most important adverse effects of cardiac glycosides are:

1. Arrhythmias; approx. 70% of side effects
   - Extrasystoles, tachycardias, atrial fibrillation
   - Bradycardias, AV block, paroxysmal atrial tachycardia with block
2. Gastrointestinal disorders; approx. 25%
   - Nausea, vomiting, pain, diarrhea

Table 5. Pharmacokinetic properties of cardiac glycosides

	Digoxin	Digitoxin
Bioavailability (%)	50–80	90–100
Usual route	i. v.; oral	i. v.; oral
Onset of action, i. v. (min)	20–30	30–120
Therapeutic plasma levels (ng/ml)	0.5–2.5	15–30
Elimination half-life	40–50 h	5–8 days
Decay rate/day (%)	20	7
Daily maintenance dose (mg) i. v.:	0.25–0.375	0.07–0.1
oral:	0.25–0.5	0.07–0.1
Dose reduction in renal failure	yes	no

3. Neurotoxic disorders; approx. 5%
   - Fatigue, confusion, disordered color vision

A cumulation of side effects occurs in electrolyte disorders (e.g., hypokalemia), in impaired myocardial perfusion or in old age. A decrease in glycoside sensitivity during fever or hyperthyroidism is known. Conditions in which sensitivity to cardiac glycosides is elevated or reduced are:

1. Elevated glycoside sensitivity
Hypokalemia – vomiting, diarrhea
Hypercalcemia
Hypomagnesemia
Hypoxia
Myocardial infarction
AV conduction disturbances
Amyloidosis, myocarditis
Hypothyroidism
Advanced age
Renal failure
Hypomotility in the gastrointestinal tract
Hypothermia

2. Reduced glycoside sensitivity
Fever
Hyperkalemia
Hyperthyroidism
Hypermotility in the gastrointestinal tract

Furthermore, it should be noted that many types of medication, given in combination with cardiac glycosides, can increase or decrease glycoside sensitivity. Possible drug interactions must be taken into consideration when deciding upon the dosage of cardiac glycosides. The types of medication which can cause enhanced sensitivity to cardiac glycosides include: antiarrhythmics, β-blockers, and Ca^{2+} antagonists (through a negative dromotropic action); benzothiazine diuretics, corticosteroids, laxatives, carbenoxolone, sympathomimetics, methylxanthines, Ca^{2+}; parasympatholytics (affecting absorption); quinidine and amiodarone (which elevate digoxin levels by influencing renal and nonrenal elimination); and verapamil, diltiazem, indomethacin, and triamterene (of little clinical importance). Drugs which reduce glycoside sensitivity include: K^+-sparing diuretics; and phentoin, charcoal, antacids, metoclopramide, neomycin, erythromycin and tetracyclines (which affect absorption by damage to intestinal flora).

Phosphodiesterase III Inhibitors

This drug class includes drugs such as amrinone, enoximone and milrinone which exert a positive inotropic action by inhibiting phosphodiesterase III, i.e., by reducing the breakdown of cAMP [30]. The pharmacological profile of these substances appeared theoretically favorable because they enhance contractility and in addition cause peripheral vasodilatation, which reduces afterload. However, experimental studies have shown that, as with catecholamines, the positive inotropic effect of phosphodiesterase inhibitors decreases with increasing severity of heart failure [4, 14]. The reason for this is the reduction of the cellular cAMP content in the myocardial cell present in severe heart failure. In chronic heart failure, a clinical study has shown that milrinone is able to improve the stress tolerance of patients [11]. However, it also increases the frequency of cardiac arrhythmias. On the whole, comparison between treatment groups showed no obvious advantage in the milrinone group compared to patients treated with digoxin alone. In evaluating this result, however, it should be borne in mind that the milrinone-treated group underwent a change in therapy from cardiac glycosides to milrinone, and also that the ejection fraction in these patients was slightly lower. Additional administration of milrinone combined with preexisting cardiac glycoside and diuretic therapy showed no advantages over treatment with cardiac glycosides and diuretics alone [11]. A recent doubblebehind placebo-controlled study with enoximone even showed a reduced survival rate if patients with chronic heart failure and no improvement of lencin tolerance [36a]. On the basis of the available investigations, no concrete conclusions can yet be drawn regarding the advantages of phosphodiesterase inhibitors over traditional therapy modalities in the treatment of heart failure have been demonstrated.

Prazosin

Another possibility for reducing afterload and thereby reducing stress on the heart is an antagonistic effect on α-receptors in the vascular musculature. Prazosin is one of the typical substances which have an antagonistic action on α1-receptors. However, studies have shown that the afterload can only be reduced initially and cardiac output thereby increased [9, 23]. The effect is only short-lived and is no longer detectable after a short period of therapy. Due to the development of tolerance, prazosin has not proved useful in the treatment of chronic heart failure.

Calcium Antagonists

Calcium antagonists also bring relief to the heart via a reduction in afterload [1, 22]. In clinical studies, however, no advantages were found, but rather disadvantages, compared to other vasodilators such as ACE inhibitors. These disadvantages may be due to the negative inotropic action of these drugs, which may be minimal but sometimes appear marked due to a reflex increase in heart rate. In any case, therapy with calcium antagonists cannot be considered to be of definite value.

β-Adrenoceptor Antagonists

β-Adrenoceptor antagonists have been used in patients with severe dilated cardiomyopathy for the treatment of heart failure. There are several reports describing hemodynamic and symptomatic improvement [2, 37]. The reason for this may lie in the inhibition of cardiac β-adrenoceptor reduction induced by elevated norepinephrine levels [16]. The reduction in heart failure-related tachycardia may also play a role. Moreover, the antiarrhythmic action of some β-adrenoceptor antagonists could prevent the occurrence of ventricular arrhythmias and thus contribute to the favorable effect of this drug class. It should be mentioned, however, that β-adrenoceptor antagonists can cause a critical decrease in cardiac output, if the myocardium is dependent on stimulation by endogenous catecholamines. For this reason, β-adrenoceptor antagonists are not indicated in all instances. They should only be given under controlled conditions in hospital, and in very low doses (e.g., 6.25 mg metoprolol orally).

Nitrates

Nitrates also cause a reduction in myocardial stress primarily through a decrease in preload [15]. However, it is well known that nitrates lead to tolerance build-up over long periods of use and are therefore not suited for the long-term-monotherapy of chronic heart failure. However, when nitrates (isosorbide dinitrate) and hydrolazime are given in addition to conzentional therapy with digoxin and diuretics they have been observed to improve left venticular function and survival. They do still have their place in the therapy of coronary heart disease and, if this plays a causal role in heart failure, then they are of course indicated.

Consequences

All of the therapies described above have disadvantages. This is particularly worth remembering in relation to the narrow therapeutic ratio of cardiac glycosides. Apart from this, side effects such as vasodilatation, electrolyte loss, or additional activation of the renin-angiotensin-aldosterone system frequently limit the use of an individual drug class. It has been shown here that the treatment of chronic heart failure as a rule consists of combination therapy in which the various modes of action supplement each other in a meaningful way. This allows a reduction in the dosage of each individual substance and thereby leads to a reduction in the frequency of potential side effects. Future research efforts will be directed towards the field of positive inotropic substances. There are numerous substances with new mechanisms of action, such as activation of sodium influx into the cell [30]. In addition, there are drugs which inhibit phosphodiesterase and also sensitize the contractile proteins for calcium [30]. The latter substances, however, are still in the experimental stage, and have not yet been investigated sufficiently for the therapy of chronic heart failure.

Major progress in recent years has been achieved by the combination of positive inotropic drugs (digitalis) and diuretics with ACE inhibiting vasodilators.

References

1. Agostoni PG, de Cesare N, Doria E, Polese A, Tamborini G, Guazzi MD (1986) Afterload reduction: a comparison of captopril and nifedipine in dilated cardiomyopathy. Br Heart J 47: 391–399
2. Aldermann J, Grossman W (1985) Are β-adrenergic-blocking drugs useful in the treatment of dilated cardiomyopathy? Circulation 71: 854–857
3. Böhm M, Beuckelmann D, Brown L, Feiler G, Lorenz B, Näbauer M, Kemkes B, Erdmann E (1988) Reduction of beta-adrenoceptor density and evaluation of positive inotropic responses in isolated, diseased human myocardium. Eur Heart J 9: 844–852
4. Böhm M, Diet F, Feiler G, Kemkes B, Kreuzer E, Weinhold C, Erdmann E (1988) Subsensitivity of the failing human heart to iso-prenaline and milrinone is related to β-adrenoceptor downregulation. J Cardiovasc Pharmacol 12: 726–732

4a. Böhm M, Gurschik P, Jakobs KH, Piske B, Schnabel P, Ungerer M, Erdmann E (1990) Increase of $G_{i\alpha}$ in human hearts with dilated lent not ischemic cardiomyopathy. Circulation 82: 1249–1265

5. Bristow MR, Ginsburg R, Minobe W, Cubiciotti RS, Sageman WS, Lurie K, Billingham MEI, Harrison DC, Stinson EB (1982) Decreased catecholamine sensitivity and β-adrenergic-receptor density in failing human hearts. N Engl J Med 307: 205–211
6. Bristow MR, Kantrowitz NE, Ginsburg R, Fowler MB (1985) Beta-adrenergic function in heart muscle disease and heart failure. J Mol Cell Cardiol 17 [Suppl 2]: 41–52
7. Chidsey CA, Harrison DC, Braunwald E (1962) Augmentation of the plasma nor-epinephrine response to exercise in patients with congestive heart failure. N Engl J Med 267: 650–654
8. Cohn JN, Levine TB, Olivary MT, Graberg V, Lura D, Francis GS, Simon AB, Rector T (1984) Plasma norepinephrine as a guide to prognosis in patients with chronic congestive heart failure. N Engl J Med 311: 819–823
9. Cohn JN, Archibald DG, Ziesche S, Franciosa JA, Harston WE, Tristani FE, Dunkman WB, Jacobs W, Francis GS, Flohr KH, Goldman S, Cobb FR, Shah PM, Saunders R, Fletcher RD, Loeb HS, Hughes VC, Baker B (1986) Effect of vasodilator therapy on mortality in chronic congestive heart failure. Results of a Veterans Administration Cooperative Study. N Engl J Med 314: 1547–1552
10. Colucci WS, Wright RF, Braunwald E (1986) New positive inotropic agents in the treatment of congestive heart failure. Mechanisms of actions and recent clinical developments. N Engl J Med 314: 290–299, 349–358
11. DiBianco R, Shabetai R, Kostuk W, Moran J, Schlant RC, Wright R et al. (1989) A comparison of oral milrinone, digoxin, and their combination in the treatment of patients with chronic heart failure. N Engl J Med 320: 677–683
12. Erdmann E (1988) Herzinsuffizienz – Digitalis – Diuretika – ACE-Hemmer. Therapiewoche 38: 1269–1277
13. Fabiato A (1983) Calcium induced release of calcium from the cardiac sarcoplasmic reticulum. Am J Physiol 245: C1–C14

13a. Feldman AM, Cates AE, Veazey WB, Hershberger RE, Bristow MR, Bunghman KL, Baumgartner WA Van Dop C (1988) Increase of the 40,000-mol wt pertussis toxin substrate (G-protein) in the failing human heart. J Clin Invest 82: 189–197

14. Feldman MD, Copelas L, Gwathmey JK, Phillips P, Warren SE, Schoen FJ, Grossman W, Morgan JP (1987) Deficient production of cyclic AMP: pharmacologic evidence of an important cause of contractile dysfunction in patients with end-stage heart failure. Circulation 75: 331–339
15. Franciosa JA, Nordstrom LA, Cohn JN (1978) Nitrate therapy for congestive heart failure. JAMA 240: 443–446
16. Heilbrunn SM, Shah P, Bristow MR, Valantine HA, Ginsburg R, Fowler MB (1989) Increased β-receptor density and improved hemodynamic response to catecholamine stimulation during long-term metoprolol therapy in heart failure from dilated cardiomyopathy. Circulation 79 (3): 483–490

17. Huxley H (1969) The mechanism of muscular contraction. Science 164: 1356–1365
18. Jewell BR (1977) A reexamination of the influence of muscle length on myocardial performance. Circ Res 40: 221–230
19. Langer GA (1982) Sodium-calcium exchange in the heart. Annu Rev Physiol 44: 435–449
20. McDonald TF (1984) Excitation-contraction coupling: relationship of the slow inward current contraction. In: Sperelakis N (ed) The physiology and pathophysiology of the heart. Nijhoff, Boston, pp 187–199
21. McKee PA, Castelli WP, McNamara PM, Kannel WB (1971) The natural history of congestive heart failure: the Framingham study. N Engl J Med 285: 1441–1446
21a. Neumann J, Scholz H, Döring V, Schmitz W, v. Meyerinck L, Kalmar P (1988) Increase in myocardial G-proteins in heart failure. Lacet II: 936–937
22. Packer M (1986) The role of vasodilator therapy in the treatment of severe chronic heart failure. Drugs 32: 13–26
23. Packer M (1987) Do vasodilators prolong life in heart failure? N Engl J Med 316 (23): 1471–1473
24. Packer M (1988) Neurohormonal interactions and adaptations in congestive heart failure. Circulation 77: 721–730
25. Packer M, Hung Lee W, Yushar M, Medina N (1986) Comparison of captopril and enalapril in patients with severe chronic heart failure. N Engl J Med 315 (14): 847–853
26. Potter JD, Gergely J (1974) Troponin, tropomyosin and actin interactions in the Ca^{++} regulation of muscle contraction. Biochemistry 13: 2697–2703
27. Reuter H (1983) Calcium channel modulation by neurotransmitters, enzymes, and drugs. Nature 301: 569–574
28. Rüegg JC (1977) Muskel. In: Schmidt RF, Thews G (eds) Physiologie des Menschen, 19th edn. Springer, Berlin, Heidelberg, New York
29. Scheppokat KD, Hammersen F, Walb D, Birks W (1977) Idiopathische Ödeme, Kapillaropathie und eiweißreiche Körperhöhlenergüsse. Klin Wochenschr 55: 1137–1147
30. Scholz H (1984) Inotropic drugs and their mechanisms of action. J Am Coll Cardiol 4: 389–397
31. Smith TW, Braunwald E, Kelly RD (1988) The management of heart failure. In: Braunwald E (ed) Heart disease. Saunders, Philadelphia, pp 485–543
32. Sodemann WA, Burch GE (1938) The precipitating cause of congestive heart failure. Am Heart J 15: 22–32
33. Starling EH (1908) The Linacre lecture on the law of the heart. Longmans, London
34. Tada M, Katz A (1982) Phosphorylation of the sarcoplasmic reticulum and sarcolemma. Ann Rev Physiol 44: 401–423
35. The Consensus Trial Study Group (1987) Effects of enalapril on mortality in severe congestive heart failure. N Engl J Med 316: 1429–1435
36. Thomas JA, Marks BH (1978) Plasma norepinephrine in congestive heart failure. Am J Cardiol 41: 233–243
36a. Uretsky BF, Jessup M, Konstam MA, Dec GW, Herrmann HC, Sandberg JA, Enoximone Multicenter Trial Group (1990) Multicenter trial of oral enoximone in patients with moderate to moderately severe congestive heart failure. Lack of benefit conpared with placebo. Circulation 82: 774–780
37. Waagstein F, Hjalmarson A, Varnauskas E, Wallentin I (1975) Effect of chronic beta-adrenergic receptor blockade in congestive cardiomyopathy. Br Heart J 37: 1022–1036
38. Wade OL, Bishop JM (1962) Cardiac output and regional blood flow. Blackwell, Oxford
39. Weber KT, Janicki JS (1980) The dynamics of ventricular contraction: force, lengthening and shortening. Fed Proc 39: 188–192
40. Wilson JR, Reichek N, Dunkman WB, Goldberg S (1981) Effect of diuresis on the performance of the failing left ventricle in man. Am J Med 70: 234–239

On the Epidemiology of Heart Failure

H. Viefhues

In this overview on the epidemiology of heart failure, we refer to the findings obtained between 1974 and 1986 which partly deal with the results of the extensive Framingham study on heart disease, a knowledge of which is assumed. We will illustrate the data from this study, compare it with our own more recent data or with additional factors from other regions, and discuss the epidemiological implications for future developments. In short, the investigations lead to the following epidemiologically relevant conclusions:

1. Heart failure is, as a rule, diagnosed on the basis of the symptoms evident in the patient's case history and using the routine cardiologic diagnostic instrumentation. In older patients, diagnosis is made more difficult by frequent multimorbidity.
2. The therapeutic possibilities are limited. Limitations on physical activity are supported by medication. These are mainly digitalis, diuretics and vasodilators.
3. The prognosis is poor once "heart failure" has been diagnosed; the disease usually takes a progressive course, and cure is out of the question. The lethality rate is approximately 50% at 5 years from first diagnosis.
4. Based on the various studies, heart failure proves to be a problem in older people above the age of around 45 but it is particularly extreme in persons above the age of 60. In men of all age groups, the risk is much greater than in women (2:1).
5. The major medical risks are hypertension, primary heart disease and, increasingly in Western industrial countries, diabetes mellitus.
6. The risks that lead to the development of the disease differ between the Western industrial countries and the developing countries. In the developing countries, heart failure is more often a consequence of rheumatic heart disease or of infectious diseases in general.
7. In the coming years, a further marked increase in cases of heart failure is expected as a consequence of the extreme changes in the age structure of society, a result of the increasing proportion of old people.
8. Exacerbated by age-related multimorbidity, heart failure is becoming, owing to the summation of individual risk factors, not only an important medical, but a predominantly geriatric problem.

Details of the epidemiological situation can be illustrated as follows. It should be noted that the majority of the investigations relate to the United States and the Western industrial nations. Owing to the fact that cardiovascular diseases are the major cause of death in these countries, socio-cultural interactions are also suspected. The results of all studies on heart failure show very congruent epidemiological trends. Only the focus of certain thematic relationships varies. The data of Margolis et al. [7], compiled in a short-term study over 1 year, thus confirm the prognoses and methods for the same region as the large Framingham study did over the compilation period from 1949–1966, i.e. the incidence of coronary heart disease and heart failure increases with increasing age (above the age of around 45) and is more frequent in men than in women. The risk increases considerably above the age of 60. A sex-related excess risk in men of 2:1 becomes apparent if the incidence of heart failure over the age of 74 is not included.

In line with the Framingham study, Smith and McFate [8] calculated the prevalence of heart failure in the United States in 1983 to be around 2.3 million cases. The predicted incidence rate in men was around 214000 and in women around 184000. A remarkable rise with increasing age and a higher percentage of men than women in all age groups was revealed. Underlying heart disease and age are the principal risk factors for heart failure. The major risk factor in 90% of heart failure patients is hypertension, ischaemic heart disease or a combination of the two. Diabetes mellitus increases the risk of heart failure in all age groups, but particularly in women. The prognosis of heart failure is poor and suggests damage to the cardiac muscle. This has consequences not only for the development of more effective medication for treating heart failure, but also for the prevention of causative progressive pathological processes. In relation to the major medical risk factor of hypertension, for instance, preventive therapy could take the form of antihypertensive medication. Even on the basis of the Framingham study, the prognosis for heart failure is generally poor. After 3 years, on average, only 50% of patients survived. The expected lethality rate 4 years after initial diagnosis is 52% in men and 34% in women.

Less recent figures which continue to be quoted regarding the distribution and frequency of heart failure assert that more than half of all patients with organic heart disease will develop heart failure [5]. Heart failure is the cause for 1%–2% of all admissions to selected hospitals in Australia, Canada, and the United States.

Two further surveys by Eisenberg et al. [1, 2] were conducted in King County, Washington, USA, and were related to the adult population with heart disease and children and adolescents up to 18 years of age. The first by Eisenberg et al. [1], which deals with the suburban area of King County, Washington, comprises 16 cities with 600000 inhabitants. Age and sex are comparatively balanced and representative. A monitoring system identified all deaths from cardiac arrest in people who received emergency treatment from the emergency medical service (EMS) outside the hospital. During an 18-month period, 649 cases of cardiac arrest were recorded (annual average 7.2 out of 10000). As in the Framingham study, the high number of cases with prior heart disease (81%) was also confirmed for this region; 57% of the cases had ventricular fibrillation. The highest success rate for resuscitation under optimal conditions was estimated at 2.0% out of 10000 cases.

The remarkable feature of this survey is that the incidence rates were recorded according to the month, the day of the week, as well as the time of day. No seasonal peaks for cardiac arrest were found, but there were connections between the cases of heart attack and deaths from primary heart disease. Peaks here were in winter, with a marked reduction in the summer. The mild winters in this region of the Pacific North-West may, however, also be the reason for these data. Peaks related to particular days of the week, and those (repeatedly postulated) related to the late night hours, were not seen in this survey.

Eisenberg et al. [2] investigated the causes of death of all cardiac outpatients aged under 18 for the same area over a period of 6.5 years. During the period of investigation, 19 cases of cardiac arrest were observed (annual frequency 12.7 per 100000 in people under 18). Sudden infant death syndrome (SIDS), at 32%, was the most frequent cause of death. Suffocation, 22%, was the second most frequent. In contrast to cardiac arrest in adults, the likelihood of successful resuscitation in children is very low. The authors attribute this to the different causes of death and the higher percentage of cases of asystole found in juvenile cases of cardiac arrest as compared with adults. Only 7% of cardiac arrests in children were successfully resuscitated. In comparison, the number of adults saved was 20%. This investigation is interesting with regard to the epidemiological distribution of heart failure combined with advanced age and heart disease only because there are virtually no epidemiologically relevant data available on children in this diagnostic area.

In his comparative study, Killip [4] refers to data on the causes of heart failure in China, England, Botswana and Sweden and compares them with data from Boston and Detroit. The pathogenesis of heart failure is shown not to be dependent on the socio-economic development of the respective countries, but on the accumulation of specific factors which lead to heart failure. It is certain that nutritionally related causes of heart failure appear to play an important role, particularly in developing countries, but hardly at all in Western countries. Rheumatic heart disease is also one of the greatest health problems in developing countries, but is less significant in the industrial countries: rheumatic syndromes after streptococcal infection have been brought more and more under control. Therefore, most new cases of heart failure in industrial countries are of non-rheumatic origin. This once again highlights the major risk factors for heart failure, as described in the Framingham study: hypertension and primary heart disease.

Since the types of therapy used in heart failure vary from case to case, as it can be caused by e.g. rheumatic disease, hypertension or viral infections, the importance of determining the pathogenesis becomes particularly clear.

With regard to the therapeutic measures adopted for heart failure, the National Disease and Therapeutic Index for 1983 shows the following picture for the United States: out of the large number of doctor-patient contacts;

10,935 million involved patients with heart failure. 49% of these patients were treated in hospital, 41% in office practices, 2% were house-calls, i.e. a total of 220,000 contacts; 42% of heart failure patients consult internists, 31% see family doctors and only 14% see cardiologists.

The desired therapeutic result was: improvement in myocardial function, 44%; improvement in diuresis, 33%; correction of calcium imbalance, 7%; controlling arrhythmia, 2%; vasodilators, 1%. The medication indicated for the treatment of heart failure was divided as follows: digitalis, 67%; diuretics, 59%; vasodilators, 10%.

The risk factor of diabetes mellitus is not to be underestimated. Patients with heart failure very frequently have diabetes mellitus and recurrent episodes of ischemic myocardial infarction or suffer from nontransmural myocardial infarction [4,9]. The effect of digitalis is as controversial. Carefully detailed studies have sometimes confirmed that high-dose treatment improves the prognosis. Diuretics are acknowledged to be the major effective therapy in heart failure. Since about 1975, diuretics and digitalis have been complemented by vasodilator therapy. The author feels it is important to establish the long-term prognosis of vasodilator therapy. At the moment, however, no epidemiological studies are available.

A paper by Klamerus [6] again refers to the Framingham study as well as to the Killip et al. study. The figures for 1986 show that 4 million Americans suffered from heart failure. The rate of hospital admission of patients with the initial diagnosis of heart failure had doubled in men and women above the age of 45; 52% of the men and 34% of the women died in the first 4 years after heart failure had been diagnosed.

Klamerus [6] rated the Framingham study as an important contribution towards understanding the epidemioloy of heart failure: although the results were published in 1971, they are known to be based on data compiled between 1949 and 1966. As a result, hypertension was listed as the most probable major risk factor. The author predicted that hypertension would probably decline as a major cause of severe heart failure because of improved diagnosis and the availability of effective treatment. However, more recent epidemiological data suggest that diabetes mellitus will increase as a major causative factor in heart failure. Although the Framingham study found diabetes mellitus in 16% of patients who later suffered from heart failure, Killip et al. [4] found that out of 100 cardiac patients, 31% had suffered from high blood pressure, 26% from diseases of the coronary vasculature, 25% from diabetes mellitus and 16% from miscellaneous myocardial disorders. In the pre-terminal stage 4 of heart failure, diseases of the coronary vasculature are the most common cause. The author predicted that the epidemiological characteristics in the Western industrial countries would change because the number of geriatric patients was increasing. Although there are more possibilities for improving ventricular function, and good medical and nursing care allow patients with a wide variety of cardiac diseases to live longer, the number of potential diseases related to heart failure is increasing. Klamerus [6] illustrated the pathophysiological causes of heart failure from the diagnostic and therapeutic view-points. Owing to the reduced contractility of the heart muscle, failure is caused by one or more of three factors:

1. Altered adrenalin metabolism and nerve supply
2. Reduced calcium delivery to the contractile elements of the heart
3. Reduced myocardial myosin-ATPase activity

The disease, as a rule, takes a progressively fatal course, and no therapeutical approaches have yet been found that effectively arrest this course. Heart failure is usually diagnosed on the basis of subjective signs and symptoms. The objective diagnosis is made by auscultation, ECG, X-rays, sonography, laboratory findings and similar routine cardiological procedures. The treatment of heart failure begins with limitation of physical activity. Pharmacological therapy begins either with digitalis or diuretics. Both can be administered concomitantly during the process of the disease. In patients who do not respond to this or similar forms of therapy, heart transplantation is the only alternative. In summary, Klamerus concludes that, in spite of progress in the treatment of heart failure, there has only been a very slight improvement in the general chances of survival, and, up to now, no therapy has slowed the progressive course of the disease.

The paper by Gerstenblith [3] focuses on heart failure as a geriatric problem, as it is expected that the percentage of the population over the age of 75 will increase by 75% in the next 15 years. Since many different studies show that heart failure is a disease of older people, Gerstenblith expects that the incidence in the United States will increase at the same rate as the rise in the average age of the population. The clinical diagnosis of heart failure in geriatric patients often proves to be difficult because the clinical symptoms of noncardiac-related diseases can influence the symptoms of heart failure. Treatment is independent of age. Ongoing studies suggest that medication with digitalis is unnecessary in the majority of older patients with heart failure. Careful dosage of diuretics and vasodilators is regarded as vital to survival. Gerstenblith pleads for future research into the care and treatment of heart failure to give greater consideration to the proportion of elderly patients, including those over 80. He feels that in this way the natural history and treatment of this disorder as a phenomenon of old age can be better understood.

As research data on the epidemiology of heart failure are not available in the Federal Republic of Germany, these studies can only offer a prognosis for the major risk factor of hypertension: on the basis of the results of the Munich blood pressure study, Wenzel and Laaser [10] calculated the population at risk, between 45 and 74 years of age, to be 4.6 million, divided into 2.2 million women and 2.4 million men. The figures cited above and current knowledge relating to risk factors indicate that we too will be faced with the problems predicted by Gerstenblith – both in relation to preventive medicine and to geriatrics.

References

1. Eisenberg MS, Bergner L, Hallström A (1975) Epidemiology of cardiac arrest and resuscitation in a suburban community. J Am Coll Epidemiology 8: 1
2. Eisenberg MS, Bergner L, Hallström A (1983) Epidemiology of cardiac arrest and resuscitation in children. Ann Emerg Med 12 (11): 672–674
3. Fleg JL (1986) CHF: reflections on current management of the older patient. Geriatrics 41 (9): 71–81
4. Killip T (1985) Epidemiology of congestive failure. Am J Cardiol 56 (2): 2 A–6 A
5. Klainer LM, Gibson TC, White KL (1965) The epidemiology of cardiac failure. J Chronic Dis 18 (8): 797–814
6. Klamerus KJ (1986) Current concepts in clinical therapeutics: congestive heart failure. Clin Pharm 5: 481–498
7. Margolis JA, Gillium RF, Feinleib M, Brasch RC, Fabsitz RR (1974) Community surveillance for coronary heart disease: the Framingham Cardiovascular Disease Survey. In: Methods and preliminary results. Am J Epidemiol 6 (100): 425–436
8. Smith W, Mc Fate (1985) Epidemiology of congestive heart failure. Am J Cardiol 55 (2): 3 A–8 A
9. Torp A (1978) Incidence of congestive cardiomyopathy. Postgrad Med J 54 (633): 435–439
10. Wenzel H, Laaser U (1988) Lebensjahre und Lebensqualität. Ein Konzept zur Kostenbewertung der Hypertonie. In: Nationale Blutdruckkonferenz (2): Laaser U, Wilhelmsen L (eds) Epidemiologie und Prävention der Hypertonie. pmi-Verlag Frankfurt, pp 105–117

Mortality Versus Quality of Life

R. Rychlik

Mortality as a Measure of Medical Treatment

In all countries which, by virtue of their economic development, have been able increasingly to utilize scientific findings to the advantage of their inhabitants, the general death rate (mortality), or crude death rate, has steadily declined over the course of this century. Constantly accelerating technical progress has drawn medicine's attention to this new technology itself and has, in a sense, made medicine dependent upon it as far as its power of judgment is concerned. This progress, which was made possible by a very mechanical and physiological way of looking at diseases, has also led to a distrust or neglect of information which is not produced by an analytical laboratory. This way of looking at things was supported by the success of formal experimental studies, without recognizing that artificially created study conditions no longer had anything to do with the real, uncontrolled process of patient care. Mortality, as a ratio of the number of deaths to the total population over a fixed period of time, thus soon became a standard measure for medical findings and types of treatment. As long ago as 1948, the World Health Organization (WHO) defined health as a state of physical, psychological, and social well-being and not only as an absence of illness. Over the past few years, there has been an increasingly intensive discussion on the evaluation criteria for medical measures, which has improved the WHO definition in many respects. The exclusive consideration of symptoms and rate of survival was criticized, and, in accordance with the WHO definition of health, a consideration, not only of patients' physical, but also their psychological and social, well-being, i.e., their quality of life in the broadest sense, was called for.

There is no generally valid definition of the term "quality of life." Current opinion is that quality of life refers to the emotional, functional, social, and physical aspects of human existence. Quality of life, however, cannot be observed directly but can only be deduced from various components, which are basically the following [8]:

- State of the patient's psychological health (e.g., anxiety, depression)
- Patient's functional capacity and performance in the various spheres of life (e.g., occupation, home, leisure time)
- Number and quality of relationships with other people (e.g., spouse, family, friends, colleagues)
- Patient's physical conditon (e.g., state of health, symptoms)

Interest in the concept of quality of life not only stems from an awareness of the psychosocial dimension of being ill, but also from the need to document the effects of physical methods of treatment and surgical operations as comprehensively as possible. In view of the increase in chronic diseases and progress in the field of surgery, the question arises as to the effect treatment or surgery has on the patient's everyday life [8]. According to Patrick and Erickson [23], quality of life and quantity of life can be viewed as distinct, but related concepts. They are used to evaluate the present and future condition of an individual or a group. Duration of life is expressed as survival time, i. e., the number of days a patient survives after a heart transplant, for example. Although the time of survival is easy to record objectively in retrospective studies, prospectively it can only be estimated.

The idea of introducing quality of life in comparison to mortality as a further parameter in the evaluation of medical treatment has, however, not yet penetrated general scientific thinking. To date, studies, for the most part, still deal with mortality as the main parameter of medical treatment. Yet, in most cases, no separate consideration is given to quality of life. The important thing still seems to be for the patient merely to live longer. The quality of life that this results in is thus of secondary importance. The title of a study published in 1986 in the United States is typical of this approach: "Years of life lost from cardiovascular disease." In 1985, Feinleib and Wilson [14] published health trends in the United States. They came to the following conclusion: "Six different measures of the health of Americans discussed include trends in life expectancy, which is related to the total mortality rate; trends in mortality by specific causes; trends in morbidity or the prevalence of various illnesses, including both acute and chronic conditions; trends in the estimates of various types of disabilities and impairments in the population; trends in expenditures for medical care and services; and, finally, some assessments of trends in several risk factors and physiological characteristics obtained by direct measurements of individuals." Quality of life does not get a single mention.

Foggin and Godon [15] investigated cardiovascular mortality in Quebec, Canada. They came to the following conclusion: "It would appear that outside the very large cities in areas of primary and industrial employment, men are at greater risk of cardiovascular mortality during their working lives than those who are over the age of 65." Similarly, there is no mention here of the fact that retired persons and the working population also have different characteristics of quality of life.

In 1986, Beaglehole and coworkers also published a study on cardiovascular mortality in New Zealand and Australia between 1968 and 1983. The essential objective was to explain diverging trends, an aim, however, which was not achieved [3]:

"From the limited cardiovascular morbidity and risk factor data available, it is not possible to identify the reasons for the differential rates of improvements in the two countries. Both nations require comprehensive and complementary studies which monitor changes in health practices such as diet, smoking, and exercise, as well as surveys that allow differences in the patterns of medical and surgical practices to be determined." Here, too, the influential variable – one could also say the disturbing variable – quality of life, is not mentioned. In 1988, the same group obtained results which delved deeper into the subject, but again the influential variable quality of life was ignored [4]:

"Much attention has been paid in recent years to the declining cardiovascular disease mortality rates in western industrialized countries; far less attention has been paid to other countries. A comparison of 1984 age-standardized mortality rates in five western Pacific countries indicates a wide variation for all cause mortality and all cardiovascular disease mortality, with Singapore having rates twice as great as Japan, which has the lowest rates. New Zealand has the highest rate of coronary heart disease, and Australia the lowest rate of stroke mortality. Mortality rates for each disease category studied were lower among women than men for each country. In the period 1968–1984 all countries experienced a decline in all cause mortality. All cardiovascular disease mortality rates have declined in all countries except Singapore; coronary heart disease mortality rates have increased in Singapore." The group also mentions the achievements of technological medicine: "The improvements in stroke mortality in Japan have been dramatic in comparison with the uric acid concentration and mortality was independent of age, body mass index, systolic blood pressure, adipose tissue distribution, smoking habits, serum cholesterol concentration, serum triglyceride concentration, serum creatinine concentration, serum calcium concentration, use of diuretics, and hematological disease. The increased mortality could not be explained by any increase in malignant neoplastic disease."

However, smaller countries also take part in the scientific collection and evaluation of mortality data. Brissonnette and Fareed [7], for example, studied the causes of death from cardiovascular disease on the island of Mauritius between 1972 and 1980. More unusual studies also crop up time and again, e.g., the study by Berkel and de Waard, who investigated mortality characteristics and life expectancy of Seventh-day Adventists (SDAs) in the Netherlands. They concluded [5]: "that evidence was found for the thesis that abstinence from cigarette smoking is the main factor explaining the low mortality from ischemic heart diseases among SDAs, while presumably an appropriate (prudent) diet confers additional benefit, for example, on colon cancer mortality."

Medical Treatment

Medical achievements based on scientific technology are generally seen as the major factor influencing the reduction in mortality. At least the author Wing [28], in his paper "The role of medicine in the decline of hypertension-related mortality," arrives at this conclusion: "While it is clear that factors other than medical intervention must have played the major role in hypertension-related mortality trends for most of the period of the decline, much scientific literature increasingly promotes the view that medical intervention is the crucial factor in mortality trends." The very success of treatment in coronary heart disease and hypertension supports this view. In this context, it is remarkable that heart failure, as a chronic and acute disease, is often seen merely as the terminal stage of the two primary diseases mentioned here. Basically, medical intervention in the area of cardiovascular medicine today can be seen as a means of preventing or halting the primary arteriosclerotic disease. Hence, the literature in recent years has also been concerned with medicoscientific discoveries which act at this point.

We will draw attention here to the paper by Amery and coworkers [1], which is an essential part of the European Working Party on Hypertension in the Elderly and is representative of the great number of publications on hypertension. While the influence of pharmacotherapy cannot be denied, this working party also obtained other remarkable results: "However, little benefit from treatment could be demonstrated in patients over the age of 80 years. The presence or absence of smoking had no influence on response to treatment in the elderly patients." Smoking can, to a certain extent, be viewed as a factor in quality of life, whereas, on the other hand, the life of very old patients (older than 80 years) would have required a separate analysis.

As early as 1983, Pickering rightly described marginal areas of pharmacotherapeutic intervention as being worthy of reconsideration [25]: "Treatment of mild hypertension and the reduction of cardiovascular mortality: the 'of or by' dilemma." Whether or not the treatment of mild hypertension really represents a dilemma depends on how the term "disease" is defined. Consideration of the life circumstances, however, is just as important here as consideration of the actual disease: if a patient is at risk of developing overt hypertension on account of his way of life, there is no objection to pharmacotherapeutic prevention. On the other hand, patients can certainly do without pharmacotherapy in certain cases if they readjust their way of life accordingly. This conclusion was also reached by Anderson and coworkers, who, in 1987, described the correlation between cholesterol and mortality [2].

With good reason, Heyden and Fodor raise the question in their report published in 1988 in New Zealand [16]: "Does regular exercise prolong life expectancy?" But the question of the extent to which regular exercise brings a gain for the patient in his quality of life must also be asked here. Regular exercise is certainly not justifiable as a type of therapy under the general conditions of life – it is common human behavior. Health is here only one of the many motivating factors.

In a study by Ordog and coworkers [19] in the United States, a further aspect of mortality becomes apparent: medical intervention can also have a negative effect on mortality. "This study shows a correlation between increasing digoxin levels and increasing mortality rates. We recommend the use of serum digoxin measurements to identify those asymptomatic patients with elevated levels. The physicians should seriously consider the indications for initiating or continuing digoxin treatment in any patient because of an increased mortality in patients with levels of more than 1.0 ng/ml." In this sense, as far as medical intervention is concerned, mortality would have to be viewed as the end point, as the negative end point, to be precise, in the quality of life.

As early as 1985, Verlangieri and coworkers investigated the influence of fruit and vegetable consumption on cardiovascular mortality [26]. "These data showed that the consumption of fruits and vegetables, particularly those rich in vitamin C, may have offered a protective effect against death from cardiovascular diseases."

In this context the results of a study by Wingard and Berkman [30] from the United States, who, in 1983, compared the risk of mortality associated with various sleeping patterns, also seem odd. The authors arrived at the surprising conclusion that a sleeping time between 7 and 8 h carried the lowest expectation of mortality, whereas a sleeping time of less than 6 or more than 9 h increased the mortality risk by a factor of 1.7.

Hence, there seems to be a great number of factors which determine a healthy life. First of all, one should get sufficient sleep. The influence of fruit and vegetables seems promising, as does physical exercise. Food rich in cholesterol should be avoided, and, when the disease occurs, it should be treated pharmacotherapeutically, but again, within certain limits. Overtreatment seems just as harmful as nontreatment.

Influence of Social Factors

In the context of the WHO definition mentioned at the beginning, apart from medical factors, the influence of socioeconomic factors, including different life-styles, has been mentioned increasingly in recent years as a factor influencing mortality. Thus, in 1984, in his publication "The sex differential in morbidity, mortality, and life-style" Wingard comments [29]: "In the United States women live longer than men, and they have lower death rates at virtually every age and for most causes of death. Similar relationships prevail in most developed nations. The sex differential in mortality has been increasing since the early 1900's, especially for those of age. The main difference in risk factor distribution was in diabetic men and women 20–49 years of age, of whom those with a family history of diabetes were older and had higher levels of fasting plasma cholesterol and triglyceride." Hence, there should have been little change in sex as a sociodemographic factor during the past 50 years.

The same seems to be true of age as a variable. An interesting study on this topic was conducted by Fedullo and Swinburne in 1983 [13]. They compared patients' ages with the costs which arise for the survival rate in an intensive care unit. They concluded: "Total hospital costs, total hospitalization duration, duration of MICU care, laboratory, radiology, and respiratory therapy costs did not vary with patient age. Because the outcome of MICU treatment is similar for all age groups, and because cost of this care for younger and older patients is the same, age should not be a factor in determining whether or not a patient received aggressive care in the MICU."

Other social factors, such as economic instabilities, particularly of unemployment rates, have also been studied to a greater degree in recent years [9, 11, 12, 21, 22]. In 1987, Brenner published a paper on the influence of unemployment on mortality in Scotland [6]. He demonstrated that: "... controlling for the significant effects of per capita cigarette, spirits, and fat consumption, and cold winter temperatures, there is in Scotland a significant long-term relation (at least a decade) between cumulative change in unemployment rates and mortality rates for all causes, for total heart disease, and in particular for ischemic heart disease."

Charlton's working group [10] produced contradictory results, at least for small areas, as in the case, for example, of Wales in the United Kingdom.

In any case, unemployment corresponds with the respective class affiliation. Rank or social class were included in the consideration of variables which influence mortality from an early stage. As early as 1984, Pearce's working group published a study on the correlation between social class and mortality in New Zealand [24]. The original resident population of New Zealand played a decisive role here: "Social class differences in male mortality in New Zealand were investigated separately for Maori, Pacific Island, and other New Zealand males aged 15–64 years. All three groups displayed strong social class mortality gradients but, for each class, the Maori mortality rates were approximately 50% higher than the rates for the 'other' category, while the Pacific Islander rates generally occupied an intermediate position."

In 1986, Lapidus and Bengtsson studied the influence of socioeconomic factors and physical activity on cardiovascular mortality [18]. There was a significant age specific correlation between low socioeconomic status according to the husband's occupation and myocardial infarction. No such association was seen between the socioeconomic status of the women themselves and myocardial infarction. Women with a low educational level had a significantly increased age specific incidence of angina pectoris. There was no significant correlation between marital status or number of children and incidence of ischaemic heart disease or overall mortality. Women who initially reported low physical activity at work during the last year had a significantly increased age specific 12 year incidence of stroke and death, as did those who reported low physical activity during leisure hours in whom the incidence of myocardial infarction and electrocardiographic changes indicating ischaemic heart disease were also increased.

Multivariate analyses showed that the association between low educational level and incidence of angina pectoris was independent of socioeconomic group, smoking habits, systolic blood pressure, indices of obesity, serum triglycerides, and serum cholesterol. Similarly, low physical activity during leisure hours seemed to be an independent risk factor for stroke, and low physical activity at work was an independent risk factor for overall mortality.

Other studies demonstrate the great value of social relationships. A Finnish study [17], dating from 1988, arrived at the following preliminary result: "For men, there was a graded association between extent of social connections and mortality." Similar studies were conducted in 1987 in Sweden [20] and England [27]. In the latter study, the following hypothesis was established: "Social activities may have a modifying effect on life stresses and risk factors associated with mortality."

Summary

Although factors influencing mortality are, to begin with, limited mainly to medical factors, in recent years an increasing influence of social factors has been demonstrated. These preliminary results certainly lead to a view of mortality which no longer considers only mortality or the rate of survival. In the meantime, medicosociological research has also demonstrated that quality of life and mortality are closely related and by no means mutually exclusive. Unfortunately, too little attention has been paid to the second factor to date, and this is certainly also due to the difficulty of the methodical approaches and the weaknesses in documentation. Yet this is the starting point for research in the next few years, for quality of life has always been a goal of medicine which not only the patient, but also the public health system recognizes as such. Similarly, the relevant health authorities should not only ask for survival data, which demonstrate that a drug is only effective if it prolongs the rate of survival. Drugs which improve the quality of life without directly influencing mortality deserve more notice than they have received in the past. Neither from a socioeconomic nor from an ethical point of view does a prolongation of life without an accompanying stabilization of quality of life, or even its improvement, seem justifiable.

References

1. Amery A, Birkenhaeger W, Brixko P, Bulpitt C, Clement D, de Leeuw P, de Plaen JF, Deruyttere M, De Schaepdryver A, Dollery C et al. (1986) Influence of antihypertensive drug treatment on morbidity and mortality in patients over the age of 60 years. European Working Party on high blood pressure in the elderly (EWPHE) results: sub-group analysis on entry stratification. J Hypertens [Suppl] 4 (6): S642–647
2. Anderson KM, Castelli WP, Levy D (1987) Cholesterol and mortality. 30 years of follow-up from the Framingham study. JAMA 257 (16): 2176–2180
3. Beaglehole R, Bonita R, Jackson R, Stewart A (1986) Cardiovascular mortality in New Zealand and Australia 1968–1983: how can the diverging trends be explained? NZ Med J 99 (794): 1–3
4. Beaglehole R, Ronita R, Stewart A (1988) Cardiovascular disease mortality trends in the western Pacific, 1968–1984. NZ Med J 101 (849): 441–443
5. Berkel J, de Waard F (1983) Mortality pattern and life expectancy of seventh-day adventists in the Netherlands. Int J Epidemiol 12 (4): 455–459
6. Brenner MH (1987) Economic instability, unemployment rates, behavioral risks, and mortality rates in Scotland, 1952–1983. Int J Health Serv 17 (3): 475–487
7. Brissonnette LG, Fareed DS (1985) Cardiovascular diseases as a cause of death in the island of Mauritius, 1972–1980. World Health Stat Q 38 (2): 163–175
8. Bullinger M, Pöppel E (1988) Lebensqualität in der Medizin: Schlagwort oder Forschungsansatz? Dtsch Ärztebl 11: S 679–680
9. Carr-Hill RA (1987) Mortality and social class (letter). Lancet 2 (8554): 338
10. Charlton JR, Bauer R, Thakhore A, Silver R, Aristidou M (1987) Unemployment and mortality: a small area analysis. J Epidemiol Community Health 41 (2): 107–113
11. Duh RW, Asal NR (1984) Mortality among laundry and dry cleaning workers in Oklahoma. Am J Public Health 74 (11): 1278–1280
12. Fanning D (1988) A mortality study of lead workers, 1926–1985. Arch Environ Health 43 (3): 247–251

13. Fedullo AF, Swinburne AJ (1983) Relationship of patient age to cost and survival in a medical ICU. Crit Care Med 11 (3): 155–159
14. Feinleib M, Wilson RW (1985) Trends in health in the United States. Environ Health Perspect 62: 267–276
15. Foggin P, Godon D (1986) Cardiovascular mortality as it relates to the geographic distribution of employment in non-metropolitan Quebec. Soc Sci Med 22 (5): 559–569
16. Heyden S, Fodor GJ (1988) Does regular exercise prolong life expectancy? Sports Med 6 (2): 63–71
17. Kaplan GA, Salonen JT, Cohen RD, Brand RJ, Syme SL, Puska P (1988) Social connections and mortality from all causes and from cardiovascular disease: prospective evidence from eastern Finland. Am J Epidemiol 128 (2): 370–380
18. Lapidus L, Bengtsson C (1986) Socioeconomic factors and physical activity in relation to cardiovascular disease and death. A 12-year follow-up of participants in a population study of women in Gothenburg, Sweden. Br Heart J 55 (3): 295–301
19. Ordog GJ, Benaron S, Bhasin V, Wasserberger J, Balasubramanium S (1987) Serum digoxin levels and mortality in 5,100 patients. Ann Emerg Med 16 (1): 32–39
20. Orth-Gomer K, Johnson JV (1987) Social network interaction and mortality. A six-year follow-up study of a random sample of the Swedish population. J Chronic Dis 40 (10): 949–957
21. Paffenbarger RS Jr, Hyde RT, Hsieh CC, Wing AL (1986) Physical activity, other lifestyle patterns, cardiovascular disease and longevity. Acta Med Scand [Suppl] 711: 85–91
22. Paffenbarger RS Jr, Hyde RT, Wing AL, Hsieh CC (1986) Physical activity, all-cause mortality, and longevity of college alumni. N Engl J Med 314 (10): 605–613
23. Patrick DL, Erickson P (1988) Assessing health-related quality of life for clinical decision making. In: Walker SR, Rosser RM (eds) Quality of life: assessment and application. Kluver Academic
24. Pearce NE, Davis PB, Smith AH, Foster FH (1984) Mortality and social class in New Zealand. III: male mortality by ethnic group. NZ Med J 97 (748): 31–35
25. Pickering TG (1983) Treatment of mild hypertension and the reduction of cardiovascular mortality: the 'of or by' dilemma. JAMA 249 (3): 399–400
26. Verlangieri AJ, Kapeghian JC, el-Dean S, Bush M (1985) Fruit and vegetable consumption and cardiovascular mortality. Med Hypotheses 16 (1): 7–15
27. Welin L, Tibblin G, Svaerdsudd K, Tibblin B, Ander-Peciva S, Larsson B, Wilhelmsen L (1985) Prospective study of social influences on mortality. The study of men born in 1913 and 1923. Lancet 1 (8434): 915–918
28. Wing S (1984) The role of medicine in the decline of hypertension-related mortality. Int J Health Serv 14 (4): 649–666
29. Wingard DL (1984) The sex differential in morbidity, mortality, and lifestyle. Annu Rev Public Health 5: 433–458
30. Wingard DL, Berkman LF (1983) Mortality risk associated with sleeping patterns among adults. Sleep 6 (2): 102–107

Quality of Life Assessment in Congestive Heart Failure: Points of View from a Cardiologist and a Psychologist

P. G. Hugenholtz and R. A. M. Erdman

Introduction

The term "quality of life" (QoL) is in vogue. It is fashionable to mention it in conversation between colleagues, to "measure" it as part of scientific investigations and to use its concept as part of the ever-increasing number of cost-benefit analyses. This is because, as physicians, we would like to ascertain the overall condition of our patients, rather than to restrict ourselves to an assessment of symptoms elicited upon questioning, and because patients in this enlightened era are supposed to be able to give a more accurate account of their real feelings when not "led on". And so, since the 1970s, the questionnaire-based boom in QoL assessment has taken off [1–4].

The purpose of this article is to review critically the concept that a patient-completed, relatively undirected expression of his own state of health has a validity comparable with the traditional history elicited by the experienced doctor, and to see whether both approaches might be combined to give a multi-dimensional appraisal of the patient's condition. A scientific and intellectual analysis will be applied to the concept of QoL and to the techniques currently available for its estimation; it will soon be apparent that QoL assessment means different things to different people.

In cardiovascular diseases, the goals of treatment are directed both towards restoring normal life expectancy and to the alleviation of symptoms, or better still maintenance of normal function. However, the ways people cope with their cardiac disease and their actual circumstances in life play an important role in improving quality of their lives and in that way in their recovery. In our opinion, it has become obvious that, in assessing the effect of treatment on restoring cardiac function, a wider perspective of functioning is needed which includes many dimensions of patients' lives. The essential questions are therefore: (a) what is meant by QoL anyway; (b) are QoL measurements in congestive heart failure relevant, and, if so, what aspects of QoL of should be measured; (c) what are the necessary (psychometric) characteristics of good measurements; (d) which QoL methods or tests are available for practical use; and, most importantly, (e) what is the difference between the assessment of QoL and the concept of quality-adjusted life year (QALY) [5]? Finally, we shall conclude with a number of recommendations for those cardiologists who, after all these considerations, still want to measure quality of life in their clinical practice.

What is QoL?

Essentially one could say that QoL assessment probes the patient's own perception of his state of health. It is opposed to the assessment made by other observers, e.g. doctors, psychologists, social workers, etc. It investigates with a range of answers the general question: "How are you feeling?" Thus an answer might be "lousy" or "on top of the world". The single, or graded answer to the global question represents the integration by the patient of a large number of specific aspects of his life – energy, sex-drive, appetite, etc. It is logical to strive for a more precise estimate of QoL by teasing out all the individual filaments and addressing them one by one. However, any assessment, no matter how detailed, remains a cross-section in time of the patient's illness – a kind of static snapshot which tells us nothing of future developments. In this it may differ from the doctor's interrogation of the patient, which may yield information of prognostic importance, such as symptoms of heart failure in a patient following myocardial infarction. QoL scores, on the other hand, may even be positively misleading in terms of prognosis, as with the hypertensive patient who feels worse on treatment than off it, while presumably improving his life expectancy and future quality of life by lowering his blood pressure. Thus questions used to elicit QoL scores must be tailored in some degree to the individual patient. A monk may enjoy an excellent QoL despite scoring rather low on the sexual activity scale.

An attempt has been made by Kind et al. [5] to link the necessarily static QoL score to a prognostic index assessed by the doctor, and to produce from this a kind of matrix from which a composite index score, supposedly embodying both well-being and prognosis, can be calculated: the so-called QALY. Unfortunately the numerical values assigned to the grades do not stand in arithmetical relationship to one another (thus a rating of IV is not *twice* as bad as a rating of II). This invalidates any attempt to manipulate the matrix mathematically, and the single index value obtained, which is supposed to express somehow both the patient's well-being and his prognosis, has as a consequence little objective validity. Unfortunately, this genuine attempt at simplification has been seized upon by the cost accountants, who build the "concrete" quantifications thus derived into economic models, with predictably spurious (and often pernicious) results. Furthermore, the QALY expresses the concept of "value for money", whereas the QoL reflects "value to the patient".

Nevertheless, the rapidly growing literature on QoL in relation to cardiovascular diseases with its QoLs and QALYs appears to confirm the importance of this concept in the field of health [1–22]. To complicate matters, and contrary to general expectations, no significant association between the severity of cardiac disease and the state of psychological well-being of the patient has been detected. This may, in fact, have been one of the reasons why, in the last decade, the attention given to QoL measurements after medical treatment has been growing steadily. For example, for a rather long time it was thought that coronary bypass surgery would lead to rapid work resumption, and that the improvement in physical capacity would contribute to a productive life after the operation. However, both advanced surgical treatment as well as traditional pharmacological therapies have yielded disappointing results in this regard, even though they clearly improved patient symptoms [5]. In contrast to cardiac parameters, psychosocial variables, such as pre-operative work status, "non-work" income, blue or white collar occupation, difficulty in re-entering the job market after a prolonged absence, age, perception of health and education, all appear to be more important for estimating the likelihood of employment status after surgery than the success or failure of the procedure itself [6, 7].

In fact, it is our opinion, given the fact that this report is restricted to cardiology, that today there are no suitable techniques or tests available which fulfil all the aims formulated in the paragraphs above. Rather, a new test instrument indicating the "value added" (for the patient) to life by the cardiological intervention (whether drug therapy or "interventional") should be developed, for which a proposal is attached. Before this is discussed, the main aim of this study, the assessment of congestive heart failure and its current concepts must be reviewed.

What Is Congestive Heart Failure?

If, from the earlier paragraphs, the reader has become uncertain whether QoL is measurable at all, the matter becomes even more complicated when reviewing the current concepts and definitions of congestive heart failure. We quote from several leading authorities the exact wording of their definitions:

Braunwald and Smith

> Heart failure may be defined as the pathophysiological state in which the abnormality of cardiac function is responsible for the failure of the heart to pump blood at the rate commensurate with the requirements of the metabolizing tissues or can do so only from an abnormally elevated filling pressure.

Hurst [24] states:

> Clinically, the term congestive heart failure is used to describe a complicated and variable symptom-sign complex or syndrome that usually, but not necessarily, includes dyspnea and increased fatiguability, tachypnea and tachycardia, pulmonary râles, cardiomegaly, ventricular gallop sound(s), and peripheral edema.

A European committee consisting of Denolin, Kuhn, Krayenbühl, Loogen and Reale attempted yet another definition of heart failure, as they felt that Braunwald and Smith [23] had emphasized the cardiac components and Hurst [24] the peripheral consequences too much. They proposed the following definition:

> Heart failure is the state of any heart disease in which, despite adequate ventricular filling, the heart's output is decreased or in which the heart is unable to pump blood at a rate adequate for satisfying the requirements of the tissues with function parameters remaining within normal limits.

Apropos these efforts, Julian in his Editor's Comment [8] observed:

> "*Omnis descriptio periculosa est*" (Every definition is dangerous) – Erasmus. Whether an agreed definition of heart failure will be achieved remains to be seen. The Editor hopes that his vanity will be excused if he quotes himself: "Failure in anything implies expectations unfulfilled, and one's definition of heart failure depends upon what one expects of the heart. No single definition suffices because the clinical and physiological criteria necessarily differ. The clinician regards his patients as having heart failure when there are symptoms or physical signs attributable to inadequate cardiac performance. The physiologist regards the heart as failing when the contractility of the ventricles or the cardiac output fall outside the statistically defined normal range . . .".

It is evident that the concept is described differently by each of these leading authorities, and this is not at all surprising, for it depends on which part of the deregulated cardiovascular system one wishes to lay the emphasis. If the main physiological derangement is lack of forward output, *fatigue* will be the leading symptom. If, on the other hand, "backward failure" with high filling pressures and congestion of the lungs are the predominant derangement, *shortness of breath* will be the symptom the physician looks for. Yet, to the patient, fatigue, lack of energy and shortness of breath are often very close, if not similar, and usually assessed as discomfort, although they affect QoL in different ways. The quality of his daily life and indeed his lifestyle may either be markedly improved subjectively when a diuretic lessens the "chest tightness" despite a worsening of his leg fatigue, or he may report the same experience as a worsening QoL because he walks less easily. How does one assess the truth when one of these changes may lead the questionnaire to indicate worsening while an objective test of cardiac performance may indicate improvement?

We now approach the core of the problem: how much validity does a QoL score possess when it seems to be at odds with the more important objective findings? In other words, how much weight can be given to a non-specific QoL assessment compared with a carefully documented list of symptoms obtained by an experienced interviewer who can combine this with inspection and examination. What we have seen in the last 2 decades of research into QoL assessment tools is an effort to replace the carefully conducted interview by the expert (of which there are indeed fewer and fewer) by an "objective" instrument that can be obtained by a lesser trained individual (or provided by the patient himself). We submit that this substitution is a fallacy and at the root of the lack of a usable cardiologically validated questionnaire. At issue is not whether a carefully taken history can actually be replaced by a self-administered questionnaire, but whether even the most experienced physician can, on the basis of symptoms alone, describe exactly the cardiovascular status or prognosis. Yet this appears to be the aim of those advocating a more objective measurement of the QoL. Also, they seek not only to assess the overall degree of well-being of the patient, but, in addition, his response to medical intervention and therefore change in prognosis. It is this intermingling of aims and the implicit, but unsubstantiated, promise of a "better" tool that has led to the confusing stampede into the concept of the QoL questionnaire.

This confusion is compounded when the available QoL assessment instruments are taken as an established scientific tool and connected to the issue of cost-benefit analysis [22–32]. On the face of it, clearly an intervention providing a better prognosis, disappearance of symptoms and improved lifestyle, such as successful bypass surgery for a 90% mainstem obstruction in the coronary system, has much more to offer in terms of cost-benefit ratio than a similar operation on a patient with three-vessel disease in whom left ventricular dysfunction has objectively been demonstrated. However, since for both these patients the QoL has been improved by the alleviation of their anginal symptoms, this "measurement" has not added one iota to the overall cost-benefit evaluation for that individual. Both patients will report an improved QoL, but the QALYs are quite [26, 29, 30] different. For them, it is the relief of symptoms which initially marks the interpretation of the QoL form and *not* the additional years of life gained. Those utilizing the QALY concept should not be surprised that misleading information creeps into their forms!

It is true that in congestive heart failure, still the main subject under discussion, the issue is indeed more complex since symptoms – as argued before – are less specific and have fewer prognostic implications. But even here, the substitution of a complicated measurement tool, such as any of the questionnaire instruments to be described for a relatively simple interviewer-directed assessment form, appears to render no increased benefit, confuses the issue and actually provides those who are looking for a meaningful cost-benefit relationship with erroneous information. In fact, it is our observation that the patient will, in order to shorten the visit to the medical "system", often want to please that "system" by "faking" positive answers to the questionnaire just as he did previously to the treating physician so as to please him. In contrast, the carefully instructed interviewer, whether nurse, physician or psychologist, is better able to elicit the required truth, since a trained person can combine the oral response with direct clinical inspection and examination to verify the veracity of the response. One should of course bear in mind that even this method is not devoid of traps: an introvert may be suffering more but give the impression that he is suffering less than the extrovert who expresses symptoms readily. The most sensible approach therefore is to combine the observations of an expert with a means of allowing the patient's own feelings to be given due weight, and this forms the basis of the new assessment and information form (NAIF) proposed in a subsequent section.

Relevance and Aspects of QoL Measurements

As pointed out in the last World Congress of Cardiology in 1984, during the Symposium on Quality of Life in Cardiovascular Disease [33], part of the recent success story of dissolving the coronary artery clot in myocardial infarction has not been told. That is, what happens in the months and years following intervention, once a strategy of early reperfusion in myocardial infarction has been implemented? Has QoL indeed improved? There is little doubt about the initial benefits of thrombolytic therapy. In selected patients the infarct-related vessel can be opened, the death rate can be halved, infarct size can be limited by a considerable amount, while ventricular function and normal heart size can be maintained, provided that the strategy to reperfuse is instituted early.

But the real issue we face is, is the QoL improved? In order to answer this question, the data from the randomized trial by the Netherlands Interuniversity Cardiology Institute [34] were analyzed by a model very similar to that applied to the metoprolol study in Stockholm [35]. The outcome of the Netherlands trial, which randomly selected between conventional treatment in the coronary care unit and immediate reperfusion by intracoronary streptokinase, shows an early reperfusion rate of 85% in the streptokinase group [34, 36]. The results are summarized in Table 1. In patients with both inferior and anterior infarction, we observed a reduction in infarct size from myocardial alpha-hydroxy-butyrate dehydrogenose release and improvement in global and regional left ventricular function after thrombolytic therapy. At 1-year follow-up, recurrent infarction appeared to be frequent in patients with inferior in-

Table 1. Data from the Netherlands Interuniversity Cardiology Institute trial [34]. Summary of the clinical course in hospital in the two groups of patients

Clinical course in hospital	Controls (D)	Thrombolysis (D)	p
Patients	264	269	
Hospital mortality (14 days)	26	14	0.05
Recurrent infarction (14 days)	9	12	
Angina pectoris	55	57	
Heart failure coronary care unit			
Mild	55	54	
Severe	12	10	
Shock	24	13	
Dopamine/dobutamine treatment	42	26	0.03
Respiratory support	11	6	
Intra-aortic balloon pump	10	16	
Heart failure reconvalescence	53	37	0.05
Ventricular fibrillation	61	38	0.01
Pericarditis	46	19	0.0004
Bleeding	7	53	0.0001
PTCA	9	59[a]	
Bypass surgery	16	29	

[a] Percutaneous transluminal coronary angioplasty (PTCA) was performed more frequently in the thrombolysis group when the 46 PTCA immediately after thrombolysis are included. Only p values < 0.05 are reported.

Table 2. Cost effect analysis of thrombolytic therapy

	Anterior MI[a]		Inferior MI[a]	
	Control (D)	Lysis (D)	Control (D)	Lysis (D)
Patients	116	130	148	139
Previous MI	29	28	31	28
Median cumulative HBDH release (U/L)	1280	840	970	670
Median angiographic LVEF, day 10–40 (%)	43	50	50	58
Total 1-year mortality	26	14	17	12
Acute catheterization		117		124
Catheterization after 10–40 days	85	100	120	99
Recurrent MI	5	10	9	26
Late PTCA	6	14	7	9
Bypass surgery	11	22	21	22
Days in hospital	2430	2680	3054	2877
Average costs/patient in Dfl	19000	28000	20000	27000
Average costs/patient in US dollars	8300	12200	8700	11700

MI, myocardial infarction; HBDH, alpha-hydroxybutyrate dehydrogenase; LVEF, left ventricular ejection fraction; PTCA, percutaneous transluminal coronary angioplasty; Dfl, Netherlands guilder.

[a] Outcome of thrombolytic therapy (lysis) and conventional treatment in 533 patients. Number of patients and interventions are presented as well as duration of hospital admission.

farction after thrombolytic therapy, but less so in patients with anterior infarction. A significant reduction in 1-year mortality was observed only in patients with anterior infarction upon admission. Note also the significant reduction in late(r) occurrence of congestive heart failure during convalescence. There were no differences in duration of hospital admission in either group, although bypass surgery was performed less frequently in conventionally treated patients with anterior infarction.

In Table 2, all the subsequent steps and interventions within the 1st year after recovery from their infarction are presented in both groups – those given conventional therapy and those assigned to thrombolytic therapy. We tabulated every admission and calculated the costs of each of these interventions [36, 37]. The average costs per patient are presented in both Dutch guilders and US dollars (at the rate of 1 US$ = 2.2 HFl). There are obvious differences between the absolute figures in terms of the variations in average health care costs in the United States and the Netherlands. Expressed in US dollars, $ 8,300 were required for the average patient treated conventionally and followed up with conventional therapy (which in some cases included percutaneous transluminal coronary angioplasty and coronary artery bypass surgery). Were that same patient assigned to the reperfusion group, $ 12,200 would be expended. Thus, the additional costs of thrombolytic therapy by intracoronary streptokinase and subsequent additional procedures are approximately 50% of the costs of conventional treatment. Figures for inferior infarction were similar.

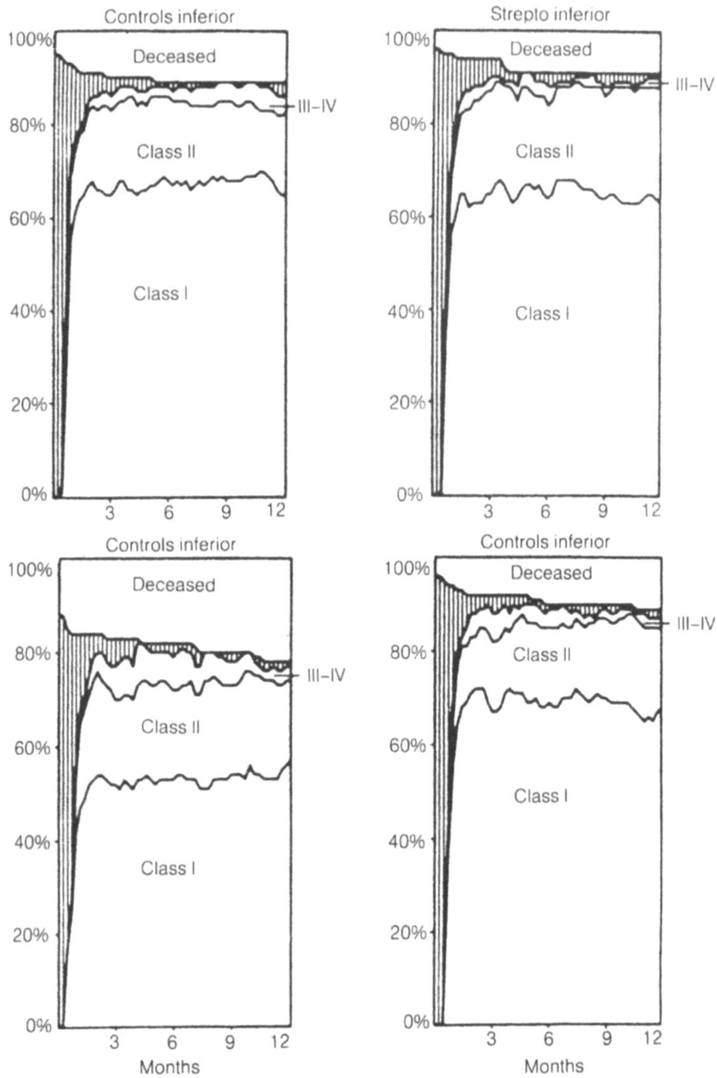

Fig. 1. Clinical sequelae over the first 12 months after onset of therapy with either standard therapy (in 1981–1983) or thrombolysis with streptokinase. *Class I,* total asymptomatic, no further visits to physician required; *Class II,* occasional visit to physician; *Class III–IV,* repeated readmissions to the hospital for various procedures. *Hatched area,* time spent in hospital. It is evident that in the group with anterior wall infarctions treated with streptokinase, the area under the curves of Classes I and II is much larger than that under standard therapy, in addition to the markedly decreased number of deaths over the first 12 months. No such clear-cut differences can be seen in patients with the, usually smaller, inferior located myocardial infarctions

In order to assess QoL after conventional treatment or thrombolysis, functional impairment by angina or heart failure was registered at visits to the outpatient clinic during the 1-year follow-up (New York Heart Association, NYHA, classes I–IV). Furthermore, hospital admissions and interventions were registered as shown in Table 2. In patients with inferior infarction, no differences were observed between conventional treatment and thrombolysis in duration of hospital admission or the average period free of symptoms after hospital discharge (class I). In patients with anterior infarction, however, there were both significantly fewer deaths and more people in classes I and II than in the control group, and survival in class I was significantly higher after thrombolytic therapy (Fig. 1); thus the QoL improved.

There are two factors that should be taken into account when thrombolytic therapy is considered in a patient with acute myocardial infarction: the extent of myocardial ischemia and the interval from onset of symptoms to therapy. Clearly, if one has a patient who comes in late and has a relatively small area of infarction, very little is gained from reperfusion and other possible interventions (which also raise costs considerably). His QoL after recovery will be unimpaired, but also unimproved while costs may be high. Patients who will benefit most are those with larger infarcts who are treated early. The long-term benefits of thrombolytic therapy are evident in patients with anterior infarction, while in patients with inferior infarction the initial salvage of myocardium is negated by a high incidence of reinfarction.

In other words, for an increment of approximately $ 4,000 reperfusion therapy can be implemented, with the benefits mentioned earlier which are particularly applicable for the larger, usually anterior wall infarctions. In inferior wall infarction, however, the cost-benefit ratio is unfavourable because we achieved little improvement in the QoL in spite of the additional expenditures.

These examples show in patients prior to the onset of congestive heart failure how cost-benefit assessment, a QoL questionnaire and the QALY measurement are inextricably interrelated. Again, this relationship points to the need of a better QoL measurement tool in cardiology prior to the QALY determination.

QoL Tests Now in Use

Despite the lack of a scientifically based methodology, several measures to assess quality of life have been proposed. From the many QoL tests which have been proposed in cardiovascular disorders, five will be discussed in order to exemplify differences and variations in methodological design and practical use: the Karnofsky Performance Status (KPS) [13]; the Heart Patient Psychological Questionnaire (HPPQ) [14]; the Disability and Distress Index [15]; the Sickness Impact Profile (SIP) [16]; and the Nottingham Health Profile (NHP)

Table 3. Data on reliability concerning five quality of life tests

Test instrument	Reliability	Reference
1. Karnofsky Performance Status (KPS)	Test-retest $r = 0.69$ Interagreement 72%	Yates et al. [19] Hutchison et al. [18]
2. Heart Patients Psychological Questionnaire (HPPQ)	Test-retest Well-being 0.73 feelings of being disabled 0.85 Displeasure 0.80 Social inhibition 0.77	Erdman et al. [15]
3. Disability and Distress Index (DDI)	Test-retest agreement 97% Interagreement 88%	Kind et al. [20]
4. Sickness Impact Profile (SIP)	Test-retest range 0.88–0.97 Internal consistency range: 0.81–0.97	Bergner et al. [16]
5. Nottingham Health Profile (NHP)	"Reliability scores are high"	Fletcher et al. [1] p 562

[17]. Table 3 represents a survey of reliability data concerning the five QoL tests mentioned here, and Table 4 shows their specific dimensions.

The KPS is measured by the patient's ability to carry on his normal activities versus the degree of dependency on help and nursing care. This is expressed in percentages (0%–100%), and the performance status is estimated (usually by the physician) before and after treatment. The estimations vary from dead (0%) to normal (100%) – equivalent to no complaints or no evidence of disease. The KPS is often used in clinical situations especially for practical reasons: it is easy to fill in. The KPS has at least three disadvantages: (a) the test measures only the physical status of the patient; (b) it pays no attention to subjective evaluations of the patient himself; and (c) it has weak psychometric aspects. The agreement between KPS scores determined by physicians and patients' self-assessment is very low – 38%; and the agreement score between physicians appeared to be poor also [18]. Further, Yates et al. [19] found a relatively low test-retest correlation with the KPS: 0.69. So, the reliability of the KPS is not impressive, and as a consequence there must be doubt about its validity.

The HPPQ is an example of a psychological well-being scale – a mood questionnaire, if you will. One of the advantages of this test is that it was standardized on 1,649 cardiac patients undergoing cardiac rehabilitation after cardiac surgery or myocardial infarction in the Netherlands. The HPPQ measures four aspects of psychological well-being: (a) anxiety; (b) feelings of being disabled; (c) discontentment; and (d) social inhibition. The test contains 52 items of the yes-no type, and it takes approximately 10 min to complete. Test-retest indices vary from 0.73 to 0.85 – all four scales appear to be stable. The HPPQ can be used by cardiologists as a diagnostic aid for each individual

Table 4. Five quality of life tests and their dimensions

Specification of areas covered	KPS (11 items)	HPPQ (52 items)	DDI (30 items)	SIP (136 items)	NHP (45 items)
Cardiac symptoms					Energy/pain
Satisfaction with results of treatment	Ambulation Mobility		Disability	Sleep/rest Eating Ambulation Mobility	Sleep/physical mobility
Side effects of drugs					
Psychological well-being		Well-being/ anxiety Feelings of being disabled Displeasure	Distress	Alertness Emotional behaviour Communication	Emotional reactions
Cognitive and/or sexual functioning					Sex life
Performance at work or home			Performance at work slightly or severely limited	Work Home management	Employment Household work
Leisure time social relations		Social inhibition	Social disability	Recreation/ pastimes Social interaction	Social isolation Hobbies Holidays Personal life
Self-care	Self-care			Body-care	

patient in comparing the test scores with the reference group of 1,649 cardiac patients. Furthermore, the test can be used in effect measurement before and after treatment, for example, before and after participation in a cardiac rehabilitation programme [14].

The DDI was designed by Kind, Rosser and Williams [20] to establish profiles of QoL, specifically for physical mobility and freedom from pain. Their eight levels of disability vary from state 1 – no disability, by way of state 4 – choice of work or performance at work severely limited, housewives and old people able to do only light housework but able to go out shopping, to state 8 – unconscious. They use four levels of distress – none, mild, moderate and severe. The authors attempt to incorporate these scores in a global measurement of QoL using a valuation matrix (1 = healthy, 0 = dead). Concerning the psychometric aspects of the DDI, the test-retest agreement is 97%, and the interrogator agreement appears to be 88%. Validity was investigated by establishing the consistency ratings in ten subjects concerning the valuations of the health status, scored previously in the valuation matrix; 90% of the assessments appeared to be consistent. Further research is needed to establish other validity aspects of the DDI, which appears to be rather restricted to pain and suffers furthermore from the questionable tendency to manipulate mathematically data which do not represent arithmetical quantities.

The SIP contains 136 items, divided into 12 categories (sleep and rest, eating, work, home management, recreation/pastimes, ambulation, mobility, body care and movement, social interaction, alertness behaviour, emotional behaviour and communication). It was developed to provide a measure of perceived health status that is sensitive enough to detect changes or differences in health status that occur over time or between groups. It provides one single score for the overall sickness impact. The test-retest agreement is high (0.97), and the degree of internal consistency also appears to be high (0.97). The test can be administered in 20–30 min by an interviewer or by means of self-assessment. Validity was determined by correlating SIP scores with overall ratings of dysfunction, sickness and days with restricted activities. The results supported its validity [16]. The SIP is a measure whose psychometric properties have been extensively studied and which provides the possibility of investigating efficiently a broad category of relevant consequences of diseases. Limited further evidence of the validity of the SIP is offered by Read et al. who found correlations between the SIP and two other important measures of overall health [22]. The Pearson correlation between the SIP and the General Health Rating Index is 0.52, and between the SIP and the Quality of Well-being scale 0.55. However, it remains somewhat limited, by addressing itself to the overall sickness rather than extent of well-being.

The NHP consists of two parts. Part I sets out to measure subjective health status by means of yes/no responses to 38 selected simple statements, relating to six dimensions: energy, pain, emotional reactions, sleep, social isolation and physical mobility. All items concern limitations on activity or aspects of distress. Part II of the NHP relates to seven areas of task performance most affected by health: occupation, jobs around the home, social life, home life, sex life, hobbies and holidays. The NHP has been validated and shown to discriminate between patients with a variety of conditions [17]. Reliability scores are high. The NHP has been used to measure QoL in a major cost-benefit study of cardiac transplants in England and also in coronary artery bypass surgery [12]. Although the NHP is able to show changes of considerable magnitude due to transplant and coronary bypass surgery, more data on its reliability, validity and sensitivity to less dramatic forms of therapy, such as pharmacological treatment of congestive heart failure, are needed.

In summary, we conclude that (a) the KPS has, from the psychometric point of view, weak features; (b) the HPPQ provides information about psychological mood states and is reliable, valid and well standardized for a post-myocardial infarction population; (c) the DDI needs further psychometric investigation and fails in many areas of cardiovascular assessment, although it is best at evaluating effects of therapy on prognosis in various disorders; and (d) the SIP does not include many relevant aspects of overall well-being; while (e) the NHP best emphasizes the breadth of the term QoL. However, while the

SIP is psychometrically well defined and easy to use, little is known about the psychometric qualities of the NHP, a significant weakness. Thus, all fail in major aspects, and a real cardiovascular-specific instrument may be in order, provided it incorporates most of the considerations outlined earlier.

The New Instrument (NAIF)

The new test instrument has been detailed in Appendix A. Obviously no practical experience is available at present but the system is now being tested in two large-scale clinical trials.

Summary

No linear relationship between a given medical or surgical intervention and the state of psychological well-being of the patient has thus far been demonstrated. Even today it remains unclear how convalescence could be influenced by, for example, personality factors such as ego strength and coping mechanisms, or situational factors such as emotional support by family and friends, or prior work and social conditions. The QoL of an executive with a "golden handshake" and three-vessel disease and failed percutaneous transluminal coronary angioplasty may be better than that of a carpenter, who after a small inferior infarct thinks he is disabled for life. Even more difficult is the assessment of the patient with congestive heart failure.

As the main aim of treatment of patients with cardiovascular disease is to reduce cardiac complaints, such as chest pain and shortness of breath, and to improve physical activities, the benefits of treatment are relatively easy to detect. However, they might also be assessed by improvement of other areas, for example, performance at home or at work, or by psychological well-being. It remains unclear which parameter(s) have the greatest prognostic information.

However, cardiologists who, next to their time-honoured methods, will want to measure QoL in their clinical practice today will have to make a choice on the basis of existing tests. We want to emphasize, from a somewhat sceptical point of view, that, before doing so, one has to first formulate for oneself what is meant by QoL, and secondly, which aspects of QoL must be improved in this particular patient. Having established this requirement, five recommendations are made:

1. QoL assessment methods have their limitations; do not expect too much, this field is just starting to develop.
2. Pay attention to validity, reliability and sensitivity of QoL methods before selecting the one you can use best.
3. Employ the method which can be practised by both the patient and the doctor, and choose a "combination instrument", such as provided in NAIF, provided the latter is properly validated, a process presently being undertaken.

4. Always compare your own clinical evaluation with the patient's QoL score.
5. When the patient's score on a QoL index is lower than you would expect from an objective assessment, this may be an indication to obtain a psychiatric opinion.

QALY as a concept and instrument would appear to require further work before it can be used in the assessment of efficaciousness of therapy in congestive failure, although in other disorders the QALY has proven its value.
A new instrument (NAIF) avoiding most of the objections to previous tests and including the best parts of previous procedures has been provided for your evaluation (Appendix A).

Acknowledgement. The authors are grateful to A. Williams, J. Lennane, P. Poole-Wilson, D. Julian, and M. v. d. Brand, for critically reviewing and adding their thoughts to this manuscript.

Appendix A

The New Assessment and Information Form (NAIF) to Measure Quality of Life in Cardiac Patients

Instructions

This is a combination of the patient "self-rating" questionnaire which, at the same time, is supervised by the interviewer. In other words, the patient is supplied with a (patient) form to assess, without due influence from the interviewer, on an analogue scale how he would rate the response to a given question. He transmits this assessment to the interviewer who then marks this response on his own (interviewer) form, with or without his own comments as he sees the interview progress. The aim of this dual approach is to couple the primary response by the patient with the overall evaluation, such as visual expression, interest, veracity, and other characteristics which an experienced interviewer ought to be able to assess. Another major item in the design of the assessment is the reversed order of grading of questions marked by an asterisk (*). In fact, the instructions to the patient on the patient form will be:

You will find in this questionnaire, which is aimed at assessing the state of your health at present, 32 questions grouped in 4 × 8 blocks relating to your physical mobility, your emotional status, your social function, and your economic status. You can choose in your response in a general manner by choosing between seven options. Very much means that you very much agree with that particular question (1), and not at all means that you do not agree with it at all (7). In other words, there are no wrong answers, every answer reflects your current feeling about this particular question. Make sure it is your opinion as you feel it now, but transmit this to the interviewer. The actual figures that will be filled in in the various boxes may not appear in logical order – they are! Please do not miss a single question, as the total score is the important outcome.
In short, please indicate the box that you think reflects best your current feeling.

Physical Mobility

	Very much – not at all						
1. I am short of breath on physical exertion.	1	2	3	4	5	6	7
2. I have been low in energy throughout the day.	1	2	3	4	5	6	7
*3. I spend a lot of time on my hobbies.	7	6	5	4	3	2	1
4. In carrying out my daily duties and chores I am hindered by my state of health.	1	2	3	4	5	6	7
5. My state of health impairs my ambition to obtain profitable employment.	1	2	3	4	5	6	7
*6. I readily take initiatives.	7	6	5	4	3	2	1

Emotional Status/Sexual Function

7. I am tense.	1	2	3	4	5	6	7
8. I feel that no one cares.	1	2	3	4	5	6	7
*9. My health allows me to indulge in as much sexual activity as I like.	7	6	5	4	3	2	1
10. This situation has deteriorated recently.	1	2	3	4	5	6	7
11. I feel in low spirits and depressive.	1	2	3	4	5	6	7
12. I am readily irritated.	1	2	3	4	5	6	7

Social Function

*13. I actively participate in sports.	7	6	5	4	3	2	1
14. My state of health prevents me from going on holidays.	1	2	3	4	5	6	7
*15. I have regular contact with friends and family members.	7	6	5	4	3	2	1
16. I am often home alone.	1	2	3	4	5	6	7
*17. I like to engage in new social contacts.	7	6	5	4	3	2	1
*18. I often go to film, theatre and sports events.	7	6	5	4	3	2	1

Cognitive Function/Economic Status

*19. I think my memory functions well.	7	6	5	4	3	2	1
*20. I can make decisions quickly.	7	6	5	4	3	2	1
21. My state of health leads to financial problems.	1	2	3	4	5	6	7
*22. I have a very quick grasp of what is said to me.	7	6	5	4	3	2	1
23. I have noted recent deterioration in my intellectual powers.	1	2	3	4	5	6	7
*24. I can usually concentrate well.	7	6	5	4	3	2	1

Instructions to the Interviewer

The NAIF questionnaire is based on the principle that the patient can express himself on a visual analogue scale, but that you, as a trained observer, fill in the rating by crossing off the appropriate box

References

1. Fletcher AE, Hunt BM, Bulpitt CJ (1987) Evaluation of quality of life in clinical trials of cardiovascular disease. J Chronic Dis 40: 557–566
2. O'Brien BJ, Buxton MJ, Ferguson BA (1987) Measuring the effectiveness of heart transplant programmes: quality of life data and their relationship to survival analysis. J Chronic Dis 40: 137 S–153 S
3. Van Dam FSAM, Somers R, Van Beek-Couzijn AL (1981) Quality of life: some theoretical issues. J Clin Pharmacol 21: 166 S–168 S
4. Kaplan RM (1988) Health-related quality of life in cardiovascular disease. J Consult Clin Psychol 56: 382–392
5. Kind P, Rosser R, Williams A (1982) Valuation of quality of life: some psychometric evidence. In: Jones-Lee MW (ed) The value of life and safety. Elsevier/North Holland, Amsterdam
6. CASS Principal investigators and their associates (1983) Coronary artery surgery study (CASS): a randomized trial of coronary artery bypass survival data. Circulation 65: 939–950
7. Oberman A, Wayne JB, Kouchoukos NT et al. (1982) Employment status after coronary artery bypass surgery. Circulation 65: 115–119
8. Julian DG (1983) Cardiology, 4th edn. Bailliere Tindall, London, p 9
9. Calman KC (1987) Definitions and dimensions of quality of life. In: Aaronson NK, Beckmann J (eds) The quality of life in cancer patients. Raven, New York, pp 1–9
10. Aristotle (1976) Ethics. Penguin, Harmondsworth
11. Spitzer WO, Dobson AJ, Hall J et al. (1981) Measuring the quality of life of cancer patients: a concise QL-index for use by physicians. J Chronic Dis 34: 585–597
12. Buxton M, Acheson RM, Caine N, Gibson S, O'Brien B (1985) Costs and benefits of the heart transplant programmes at Harefield and Papworth Hospitals, London. Her Majesty's Stationery Office, London
13. Bulpitt CJ (1983) Randomized controlled clinical trials. Nijhof, The Hague, pp 85–86
14. Karnofsky DA, Burchenal JA (1949) The clinical evaluation of chemotherapeutic agents in cancer. In: Macleod CM (ed) Evaluation of chemotherapeutic agents. Columbia University Press, New York, pp 191–205
15. Erdman RAM, Duivenvoorden HJ, Verhage F, Kazemier M, Hugenholtz PG (1986) Predictability of beneficial effects in cardiac rehabilitation: a randomized clinical trial of psychosocial variables. J Cardiopulmonary Rehabil 6: 206–213
16. Bergner M, Bobbitt RA, Carter WB, Gilson BS (1981) The sickness impact profile: development and final revision of a health status measure. Med Care 19: 787–805
17. Hunt SM, McKenna SP, McEwen J (1989) A quantitative approach to perceived health. J Epidemiol Community Health 34: 282–285
18. Hutchinson TA, Boyd NF, Feinstein AR (1979) Scientific problems in clinical scales as demonstrated in the Karnofsky Index of Performance Status. J Chronic Dis 32: 661–667
19. Yates JW, Chalmer BR, McKegney FP (1980) Evaluation of patients with advanced cancer using the Karnofsky Performance Status. Cancer 45: 2220–2224
20. Kind P, Rosser R, Williams A (1982) Valuation of quality of life: some psychometric evidence. In: Jones-Lee MW (ed) The value of life and safety. Elsevier/North Holland, Amsterdam
21. Bulpitt CJ, Fletcher AE (1987) Measurement of the quality of life in angina. J Hypertens 5: S 41–S 45
22. Read JL, Quinn RJ, Hoefer MA (1987) Measuring overall health: an evaluation of three important approaches. J Chronic Dis 40: 7 S–21 S
23. Braunwald E, Smith TW (1982) The management of heart failure. Heart disease. Saunders, New York
24. Hurst JW (1986) Pathophysiology of heart failure. In: Schlant, Sonnenblick (eds) The heart, 6th ed. p 319
25. Rosser R, Kind P (1978) A scale of valuations of states of illness: is there a social consensus? Int J Epidemiol 7: 347–358

26. Gudex C (1986) QALYs and their use by the Health Service. University of York, Centre for Health Economics (discussion paper)
27. Rosser R, Kind P (1978) A scale of valuations of states of illness: is there a social consensus? Int J Epidemiol 7/4: 347–358
28. Williams A (1984) Medical ethics. In: Culyer AJ (ed) Nuffield/York Portfolios. Nuffield Provincial Hospitals Trust
29. Williams A (1985) The economics of coronary artery bypass grafting. Br Med J 291: 325–329
30. Williams A (1987a) The importance of 'quality of life' in policy decisions. In: Walker SR (ed) Quality of life: assessment and application. MTP Press, Lancester
31. Williams A (1987b) The cost-effectiveness approach to the treatment of angina. In: Patterson D (ed) The management of angina pectoris. Castle House
32. Symposium on the quality of life (1986) World Congress of Cardiology Washington, DC. Am Heart [Suppl]
33. Gorlin R, Wilhelmsen L, Hugenholtz P (1987) Quality of life in cardiovascular disease: interaction of pharmacotherapy, disease, and life-style. Am Heart J 114 (1, 2):
34. Simoons ML, Serruys PW, Brand M vd et al. (1986) Early thrombolysis in acute myocardial infarction: limitation of infarct size and improved survival. J Am Coll Cardiol 7: 717–728
35. Olsson G, Lubsen J, Es GA van, Rehnqvist N (1986) Quality of life after myocardial infarction: effect of long term metoprolol on mortality and morbidity. Br Med J 292: 1491–1493
36. Vermeer F, Simoons ML, Baer FW et al. (1986) Which patients benefit most from early thrombolytic therapy with intracoronary streptokinase? Circulation 74: 1379–1389
37. Vermeer F, Simoons ML, Zwaan C de et al. (1988) Cost benefit analysis of early thrombolytic therapy with intracoronary streptokinase. Br Heart J 59: 527–534

From Heaven Through the World to the Earth

H. Viefhues

The History of Mentality Towards "Quality of Life"

"Perhaps you do not know, Your Highness", says Marcel Proust's archivist to the Duke of Guermantes, "that there is a new word for 'way of thinking'. It's now being called "mentality". "Oh, I see!, mentality – I must make a note of that so I can use it again", says the Duke. Since our takeover from France of research on these lines into the history of mentality[1], we have also "wanted to use the word again". In this sense, we would like, in the following, to give a rough outline of the shifts in modes of thought through the ages, as they affect the concept of "quality of life", i. e. the related ways of thinking and mental attitudes. We shall make use of Weber's ideal typology, which is not meant to be a reflection of historical reality, but a classifying pattern in which the "metaphor", as it were, of an epochal interpretation of man and the world is used to illustrate such periodic changes in modes of perception.

Heaven – Godliness

We begin with a reminder of the mode of thought in the Middle Ages, in which we will apply the religious concept of "godliness" in relation to our problem of quality of life. This is not the place to give an account of the *conditio humana* in the Middle Ages. In those days, man was from his birth until his uncertain, but ever-imminent departure, so mercilessly exposed to the capriciousness of life in the form of disease and the omnipresence of death, to chaos and neglect, that life on earth was experienced more like that of a "slave of death, a guest in the room, a wanderer on the road"[2]. The short time spent

[1] Mentalities are even considered by French orthodox-marxist historians themselves to be complex epochal phenomena, which cannot be simply seen as epiphenomena, as a "superstructure" of material foundations, but as independent historical manifestations which cannot only be traced back to the individual "reasons", but "arise" from the whole chain of historical events, take shape and become capable of elucidation. This history of mentality has its precursors in the thoughts of the Vico's "Senso commune", which finds a "judgement without all reflection, generally felt by an entire group, an entire nation or all of mankind" [2] and how he was appealed to in Montesquieu in the *manière des pensées* and in Germany of the eighteenth century with the concept of *Denkungsart.*

[2] "Attitude towards life", "world outlook" and "ethos" belong in this conceptual field. It is a certain way "of seeing reality as valid", a certain "spiritual *habitus* ", which corresponds to a certain type of ethics.

in this "terrestrial vale of tears" produced a worldly pessimism, punctuated by intervals of garish merry-making at feast-day celebrations, which certainly provided some variation, but was no miraculous cure-zell. The permanent dangers of life and the helplessness of mankind were countered by well-drilled behavioural patterns; a "form of living", of which the formulated goals – however much they fluctuated in radicalism over the centuries – lay in the world after death. Despite all the grandiose attempts to make *caritas* on earth a reality, the only solace was provided by those who promised the citizens of this world the highest fulfilment of all their hopes and desires in the next world. For the late Middle Ages also, as Johan Huizinga shows, the glaring contrast between the unbearable adversity of real existence and the beautiful pretence-world of courtly culture was not to be bridged by any mediating world ethics. This is what the very different lifestyles of the various classes had in common: for the clergyman, for the nobleman and for the serf, only in a form of transcendental existence was happiness possible, and this was their common aim. The success of all action and all gains of mankind on earth were out of human hands. All past successes and future planning had always a provisional character. Only death could bring about the final transformation from the qualitative negation of this life to an eternal transcendental existence of maximal quality, subsumed under the word "heaven", the *visio beatifica Dei,* the "vision of God". It is the eternal communion of man with God: "but we know that when He appears, we shall be like Him, for we shall see Him as He is" (I John 3:2) "face to face" (I Cor. 13:12). This means that the aim of our existence, and thereby the ultimate fulfilment of meaning, are revealed to us after death in the unfathomable mystery of sharing in the total fullness of God. The mystics experience this unification with God in a witness-like anticipation, "in eternal love", as still formulated recently in Question 240 of the old school catechism. Man's most general question in the still fundamentally unquestioned Christian culture of the Middle Ages is: "How do I get to heaven?" The road to heaven, symbolised by climbing a ladder, consists in the belief in and imitation of Christ in accordance with the commandments, as in the teachings of the Church. In writings on the rungs to "eternal life", such as Bonaventura's *Itinerarium mentis in Deum,* or Bernard de Clairvaux's *Liber de modo bene vivendi,* are shown some of the many ways to attain the goal of qualitatively consummate existence, "heaven" as the goal of a *scala coeli.*

"If you only will be earnest, then yours is heaven.
How infinitely rich can the poorest be". (Angelus Silesius)

The attainment of godliness is completely and equally open to every man, a concept of happiness that has been spiritualised since antiquity, and which is now situated in transcendence. Godliness as the "eternal quality of life". "He who has God is blissful", cries Augustine.

World Happiness

Humanism and the Renaissance gave birth to a completely new mentality: mediaeval godliness became secularised as happiness. The "new man" recreated himself by creating his own world. In this self-created word lies a self-determined goal, i. e. his "salvation", happiness in life here on earth, for which man himself sets his own criteria. The striving for supreme happiness, which earlier aspired to a "heavenly good", now pursues earthly goods. A consummate situation in life, the ultimate quality of life is achieved when one feels good. Happy is he who is safe and sound, unharmed, who can protect himself from suffering, in short, he who "feels good". This situation arose as the result of the revival of classical antiquity. This was not just imitative, however, but, by confronting antiquity and its mentality, and particularly its values, the individual became aware of his own special and contemporaneous new qualities. As the Romans adapted the Hellenistic mentality in order to create forms of "Roman mentality", the "new man" of the Renaissance became newly aware of himself when contemplating the key values of classical antiquity. The road to happiness now runs via the ancient basic virtues of, for example, the three great schools of philosophy, but the ancient teachers of happiness were newly paraphrased in the light of *uomo unico:* of the autonomous subject. Alongside the cardinal virtues, the ideal of *serenitas* appears within the scope of defining quality of life; the brightness of fine weather, the pleasure of a long meal, the lucidity of social intercourse, the warmth and security of being among trusted friends – one immediately recognises that serenity is the happiness ideal of the cultured and educated elite. The prince is the guarantor for the transmission of earthly happiness to the major classes of society. Moreover, he is bound by a further catalogue of virtues which lie outside personal, individual virtues and, "detached" from these ethics, he is responsible for the welfare of the entire *res publica,* as Machiavelli observed. But he also remains the guardian of the welfare of his subjects in that he initiates "charity" and the "right morality" of traditional virtues. He is hereby acting – the last transcendent connection with the past – in the name of God, "by God's mercy".

Absolutism totally rationalised this mediation of happiness in which it is – in modern language – "bureaucratised". In Germany, the class of "cultured and educated public servants", individuals who were academically trained and educated irrespective of birth, became the social vehicle of this agency, because the nobility which had up to then dominated could not, on their own, reinstate welfare and morals and the basis for happiness – especially after the catastrophe of the Thirty Years' War. Alongside and in place of the "nobility by birth", the "nobility of the mind" emerged as a class in the service of princes and subjects, developing at the same time an increasingly aristocratic way of looking at themselves. They became bearers of enlightened absolutism and also, above all, of the Enlightenment itself.

To speak two years after the bicentenary of the French revolution about the shift in mentality of 1789 would be like taking cockades to Paris. We will come back to one of the aspects of happiness promised by the revolution at a later stage. The crisis of the Enlightenment in Germany at the end of the eighteenth century differentiates the concept of happiness according to the class structure. The courtly, aristocratic concept of happiness stems from the courtier of the Renaissance, i.e. the cultured and educated humanist, and can be traced back to the ancient concept of the "right life", "eubiosis" versus that of "eudaemonia". After the social unification of nobility by birth with the nobility of mind into "the cultured and educated", it became their catalogue of virtues: it was a humanistic ideal of life with a Christian-protestant or natural-religious-deistic approach. Culture and education, as Wilhelm Meister learnt, is *savoir-vivre* in a set of ancient-heroic virtues, acquired through the canon of ancient times as the Truth, which also comprises the Good and the Beautiful, as presented by Shaftesbury. This acquisition of a mentality is a form of dietetics – which at the turn of the 18th century had not yet become a branch of medicine – and guarantees the attainment of eubiotics as the optimal life situation.

Alongside welfare and "standard of living", happiness had always possessed as an inner, spiritual component "morality" as a humanistic or Christian conviction, and thus had an "ethos" that could not basically be considered moralising. Particularly in the petit bourgeois-pietistic circles, it is just this that is stressed: the aspired quality of life "Happiness" is bourgeois affluence, which nevertheless also implicitly demands devoutness: "happiness favours only the virtuous". A remainder of the aspiration to piety is still present in the sense of spiritual happiness. There can "also be happiness in misery" even as far as "dying happy". The right conviction, the motive, and finally, in Kant, the "duty" – the fulfilment of which is happiness – all have a certain primacy in continental European thinking.

Earth – the Pursuit of Happiness

The turning point occurs in the eighteenth century with English Utilitarianism and the "declaration of human rights". In 1670, Richard Cumberland in his book *From the Laws of Nature* defined correct action as that which achieves public welfare via the shortest route. The only things that count here are the end, and the means to achieve that end. The motives for action are disregarded, as in the "Fable of the Bee" where evil intentions can have a good effect [4]. In 1726, Francis Hutcheson defined the ethical goal of action as the greatest happiness for the greatest number of people. The beginning of the American Declaration of Independence of July 4, 1776, refers to "the certain unalienable rights of all men", namely "life, liberty and the pursuit of happiness". In Thomas Jefferson's formulation, "the pursuit of happiness" is declared to be one of the three most important fundamental rights of man, alongside life and liberty. The "inner component", the mental or spiritual content of "happiness" is placed above the individual and is no longer a part of a general norm which belongs to "happiness". Everyone may and should search for happiness and the government should provide help in finding individual happiness. In eighteenth century England, the then richest country in the world with its equally wealthy American states, a world power with the highest standard of living and with a degree of freedom for gentry and commerce which was unimaginable for continental Europe, it was not a question of the standard of living, but of the quality of life. In the continental, hierarchically structured, absolutistic world, a revolutionary transformation was needed in order to make the so-called "unconditional happiness" a governmental goal. On the whole, the Germans deviated from the Anglo-Saxon utilitarian assessment of rationality in what Niklas Luhmann called the "German's special way of transcendentalism". Happiness was not the optimal satisfaction of needs, with the criterion of calculable "success", but was the fulfilment of responsibility; an "austerity vision" against the eudaemonism of wealth. However poor a man might be, as a cultured and educated public servant he is proud that economic aspects are irrelevant. Happiness lies in participation in the neo-humanistic "Bildung" [the German concept of being cultured and educated], in the universal norms of reasonableness and the awareness of fulfilled duty.

It also remained the duty of the constitutional monarch of the nineteenth century to promote the happiness and the welfare of each individual citizen by governmental measures. However, there was no clear doctrine of inner happiness within the meaning of optimal satisfaction of needs in any egalitarian and democratic sense. Eudaemonism, as defined by the Brockhaus Encyclopedia of 1833, is "that standpoint which sees man's happiness as the ultimate or highest inspiration for all responsibilities and virtuous conduct and consequently raises it to the highest principle of morality". Naturally, pursued happiness – Brockhaus continues – can also be "material or spiritual well-being, or both, or include only one's own happiness or also that of others". The latter formulation would, to a certain degree, correspond to the concept of happiness in the Declaration of Human Rights. But, this concept plainly has something objectionable about it for Brockhaus lexicographers, which only then disappears if one pursues happiness "through a condition of the soul expected from a sensible life", as the ancient Stoics and Epicureans experienced life. Essentially, all four ideal-typical situations of life continued to be considered worthwhile in Germany up to the first half of the twentieth century: the petit bourgois-pietistic combination of wealth and piety, the aristocratically cultured and educated formula of correct living from the neo-humanistic tradition, the state-mediated welfare for attaining happiness and, finally, the democratically guaranteed pursuit of individual, earthly happiness.

The way of life in an industrial society is based on the following triad:

1. Continuous mechanisation
2. Continuous economisation and thereby quantification
3. Continuous politicising of all relationships

This finding basically shows a fundamental scientifically orientated stance in the sense of ends-means analysis, establishing cause-effect relationships. It is this "godless world" which is "to be conceived from within itself", "as a world of factuality, which exists without the participation of transcendental reality".

Thus, the established concept of the state is that of a functional organisation similar to the model of a partnership contract. The "law of God" is no longer binding in the way we deal with one another, but rather the rules as applied in this contract. Adherence to these rules is no longer the virtue of obedience, and the failure to obey is no longer a sin, but reveals itself as a social freedom from problems or as a social violation of the rules laid down for achieving earthly happiness in freedom. This new mentality also prevails in the second half of the twentieth century in the Federal Republic of Germany. It is no longer the case that unalterable divine moral laws demand obedience; obedience is derived, on purely goal-orientated grounds, from contractual organisations with regulations which can be changed if they can no longer fulfil their purpose, and/or if there is a majority in favour of these changes.

Furthermore, the attitude derived from an agrarian culture of accepting "strokes of fate" was replaced in the industrial culture by the desire to create external reality. In the same way, happiness can also be defined as a practicable, sociotechnical phenomenon. After the secularisation of the good life by the Enlightenment, and the secularisation of both the concept and pursuit of happiness – now defined as a human right – a modern interpretation remains from the earlier history of mentality: the individual's pursuit of happiness on his own responsibility or through collective regulation. The need to create a new world is considered pressing if the old world is absent, as in the United States, or if it has collapsed, as in Europe after 30 years of war in the twentieth century. Worlds are not complete entities, but are associations which make up a whole. Among these, according to our current mental horizons, two goals can be identified, the funding of which will be discussed:

1. The "standard of living" in the highly industrialised West has reached a level between satisfaction and oversupply. This can only be slightly increased, since all basic material needs are satisfactorily met.
2. "Quality of life": the goal of qualitative improvement after attaining a certain "basic wealth". This was already defined by the second president of the United States, John Adams (1735–1826), as a further stage in the quality of life. He added "ease, comfort and security" to the recognised basic rights of life, liberty and the pursuit of happiness. This roughly comprises the quality of life that we are aiming at in today's discussion.

A further step would consider the quality of life a problem of aesthetics. Problems of image arise here for the individual, as there is no longer any objective aesthetic. The question of the quality of a good style of life has now become one of self-realisation by self-expression, and with this the development of a self-image. In modern societies, the quality of life therefore includes two dimensions:

1. Active participation in the "social world" through satisfying fulfilment of the roles offered – the functional dimension.
2. Active creation of "one's own world" through satisfying "you-relationships" and self-realisation – the empathic dimension.

A certain degree of participation and self-determination, which must not be lacking (e.g., as a consequence of disease and handicap) if a negative, deficient quality of life is not to be experienced, brings forward the need to quantify the quality of life[3].

Quality of Life in Medicine

The common goal of medicine is "health". This is defined as a long and painless life. For the happiness which is to follow, a further complementary common goal is added, that of the economy: the common goal of the economy is wealth, which has also been quantified under the concept of standard of living. When the common goals of medicine and the economy achieved a certain basic level for our whole population, then the question logically arose as to the quality over and above this standard. The objective, which should transpose the achieved high standard of living into such an optimal quality of life, should also include the health-care political component. Health-care political goals are given which

1. Do not allow disease to adversely affect the ability to achieve optimal satisfaction of needs (prevention movement).
2. Minimise or even eliminate disease-related loss of optimal satisfaction of needs.

This is where the concept of the quality of life is incorporated into therapy and rehabilitation. The therapeutic goal generally defined by Freud is delineated according to these two features: the following should namely be reinstated:

1. Performance capacity.
2. Capacity for enjoyment.

According to the sociotechnical approach for producing happiness, there is either

1. Rehabilitation of the capacity for happiness, that is "pursuit action" (individual approach), or
2. Rehabilitation of happiness itself (collective approach).

Because of their sociotechnical disposition, both forms provided for the reinstatement of the loss of optimal satisfaction of needs contain an operational quantification, i.e. it must be possible to quantify the quality of "happiness" in order to correlate it scientifically with other aspects of medicine. This is also necessary if we want to have a universalistic medicine, in the sense of the

[3] Numerical self-assurance of earthly happiness compared to heavenly eternities began particularly in diary literature of the eighteenth and nineteenth centuries, a statistical eudaemology which briefly lists the growth of wealth, the productivity of the pen, the number of acts of love – in short, everything that one can count (e.g. G. C. Lichtenberg, Jules Michelet, and many others).

desired "world civilisation", which, disregarding cultural differences, can, for the most part, be generalised. There are indeed such fundamental requirements of "keeping oneself in existence" which one can ascribe to all people in a secular world civilisation. The requirement here, however, is the purely secular conception of "suffering", namely that of a purely negative phenomenon, as the sensless, though potentially compensatory defect in human existence. If we now bring the question of the meaning of suffering into play, thus allowing suffering also to take on a positive meaning, then not only does the community of world civilisation disintegrate into cultural identities, the participation of which would permit various interpretations of the loss of need satisfaction, but the generally quantifiable concept of quality of life falls apart as well.

Quality of Life:

Semantic History, Political and Scientific Terminology, Measurement and Methodology, Medical Research

H. Viefhues

A Semantic History of the Term 'Quality of Life'

There is disagreement with regard to the first appearance of the term "quality of life". Although the Oxford English Dictionary (1971) is generally referred to, with its evidence that Robert Beye coined the phrase in 1784, Kann [94] interprets it rather as an "accidental formulation without ideological or political background". It is uncontested that the political and scientific concept of quality of life dates from the second half of the twentieth century. The lack of continuity in the semantic history from the eighteenth to the twentieth century reinforces Kann's supposition. Semantic history is primarily the scientific or political history of a term. "In today's terminology, the term has double usage in American English, presumably because of the prominent television commentator Eric Sevareid, who in 1956 analysed the presidential candidacy betwen Adlai Stevenson (later Ambassador to the UNO) and Dwight D. Eisenhower in a radio commentary" [94]. Sevareid: "I put in the phrase 'the quality of life' to try to describe his approach to the national condition as it differed from the stock New Deal approach which, it seemed to me, was pretty directly concerned with more quantity of every group" [94].

The usage of the phrase "quality of life" continued in the 1960s to describe new political goals and, with the advent of Lyndon B. Johnson's presidency, took on a central position in the political programme. Johnson spoke in 1964 of new goals of politics: "These goals cannot be measured by the size of our bank balances. They can only be measured in the quality of lives that our people lead" quoted in [115].

In German, "quality of life" is translated as *"Qualität des Lebens"* or *„Lebensqualität"*, whereas already in 1971 "the bettering of the quality of life" was translated as *"Qualitätsverbesserung des Lebens"*. That same year, Willy Brandt turned "Quality of life" into *"Lebensqualität"* (quoted in [94]):

1. Foreign word: „Standard of living and quality of life are simply not the same";
2. Loan translation 1 (semantic, morphemic and morphological): "A society that asks about the quality of life";
3. Loan translation 2 (only semantic and morphemic): "the collective component of the standard of living – or the quality of life.

When Willy Brandt, in his governmental declaration in 1973, most likely inspired by Erhard Eppler, used the term "quality of life", it became an integral part of the German vocabulary.

'Quality of Life' as a Political and Scientific Concept

The parallel between the introduction of the term "quality of life" or "life quality" as a part of a new political programme and the scientific development of the concept is striking. This applies to the United States in the 1960s and to the Federal Republic of Germany in the 1970s. K.W.Thomas sees the German reception of the term as the importation of a "battle cry ... of American sociologists, ecologists and their young followers against the consumer society, with their goal of existence in the quantity of life" (quoted in [94]). At the beginning of the 1970s, the "disastrous political career of the slogan quality of life" [128] came to an end in the United States. Quality of life, however, remains a subject of macrosociological discussion.

The scientific development of the concept is based upon behavioural changes in sociology and political science (cf. Dahl), as well as on the processing of system- and theory-driven macrotheories (German, Galbraith, Parsons, N.Luhmann). In the microscopic area, behaviourism inspires a stronger incorporation of psychological theories and provides concepts for the methodology of behavioural measurement. The macroscopic approach is based on the controllability of societal change, whereby controlling resources become more and more generalised. On the whole, the political system is seen as a controlling or planning centre for social development. The controlling resource which develops in this process is the information itself [128, 190].

The political programme of quality of life has its origin in the Federal Republic of Germany in connection with planned social politics and a policy of internal reforms. In a political connection, reorientation is striven for:

- Freedom of the individual cannot only de defined in terms of a market economy
- Social self-initiative should be strengthened
- The state itself must take on the task of infrastructural improvement (for example, education) in order to reform society from within.

At the beginning of the 1970s, the traditional political parties were also part of the "social democratic consensus" (Dahrendorf) associated with these politics. In this consensus, not only etatistic but also liberal movements combined to form a consensus. After deterioration of this consensus, partly as a side effect of reforms and partly through objective developments, the political programme of quality of life faded into the background of the political debate in the mid-1970s.

Measuring the Quality of Life

An Introduction to the Discussion

Measuring the quality of life should
- Explain the reasons for and the direction of social change [6]
- Place the results (output) of social activities in comparison to expenditures (input) in the foreground [190]
- Serve "very direct political and administrative intentions, namely the identification of welfare deficits in underprivileged groups, regions and areas of life [192]
- Contribute to the evaluation of political accomplishments [190].

The measurement of quality of life is separate from the economic system as a whole, a system which reduces social statistics to market-orientated, i.e. monetary, dimensions (cf. [36]).

Concepts and Methodology

Quality of Life Seen Objectively: Basic Needs

McCall [115] states in his definition "that QoL consists in the obtaining of the necessary conditions for happiness in a given society or region". The necessary conditions, on the other hand, are defined as "general happiness requirements" [118]. These have an objective (requirement) character, cannot be defined in terms of satisfactory/unsatisfactory and are different from wants and desires. McCall considers it important that the basic requirements, the fulfillment of which makes up the quality of life, are not of psychological dimensions, but rather of general anthropological ones. He subscribes to Maslow's requirement hierarchy. McCall sees the measurement of objective quality of life or its deficits as being possible largely by using objective indicators, but also by using subjective indicators: "a closer examination of how subjective indicators behave, and how they relate to 'objective' data, shows that it is possible to combine, within a single conceptual or methodological framework, the notion of a subjective QoL indicator with that which is constitutive of QoL, the latter being wholly non-subjective".

Quality of Life Seen Subjectively: Well-being

Campbell [36] criticises an objectivistic concept of the quality of life: "if we believe that the quality of life lies in the objective circumstances of life, . . . (objective social indicators) will tell us all we need to know; but if we believe, as I assume most psychologists do, that the quality of life lies in the experience of life, then these are surrogate indicators". Campbell distinguishes between three dimensions of a concept of quality of life:

- Cognitive comparison (judgement) on the part of the individual of the actual situation and own goals (Cantril)

- Affective balance of positive and negative affects (Bradburn)
- Processing and observation of psychological and emotional stress

Bradburn's concept (affective balance scale) is widely applied (cf. [192]). Diener [49] elaborates on the framework: Bradburn covers the basic aspects of the concept, the fact that negative and positive affects in certain domains or situations are independent of each other, a rate of positive and negative affects. The difference results in the affective balance – if positive effects predominate (satisfaction), well-being exists. Diener explains that there is still uncertainty about the validity of the existing measuring instruments.

Compromise One: Perceived Quality of Life

Andrews and Withey [6] "describes ... a two-dimensional conceptual model which proposes that one's overall sense of quality of life is understandable in terms of a combination of affective responses (i.e. evaluations) of life 'domains'. Life domains are of two types: role-situations and values". Thus, the objectivity vs. subjectivity alternative gradually disintegrates. Although Andrews and Withey's concept has greater affinity to the concept of well-being, the emphasis that perception of quality of life is measured in the affects points to the objective or collective dimension of quality of life in the selection of life domains.

Compromise Two: Welfare

Zapf [191] also understands the perceived quality of life as "the assessments those concerned make about their situation in life, and their satisfaction". These are equalised by objective, semi-scientifically measurable living conditions: "observable circumstances of the individual's life, for example income, living conditions, health, education, social contacts". Perceived quality of life and satisfaction are then again components of a concept of welfare. "An 'objectivistic position' states that welfare is based on the satisfaction of clearly defined needs and that subjective reflexes can be neglected. The 'subjectivistic position' insists that welfare is essentially based on the perception of those concerned and that each individual or household can better judge their welfare than can any progressive or conservative politicians." With this concept, Zapf tries to mediate: "To our purposes, we understand the term quality of life first of all to mean good living conditions, which are tied together with a positive, subjective sense of well-being. In a more general definition, the quality of life of individuals and groups is determined by the constellation (level, distribution, correlation) of each of the living situations and the components of subjective well-being" [192].

Special Case: Qualitative Growth

Economic criticism of the VGR and of a subjectivistic welfare economcy (cf. [190]) considers problems of the quality of life in vertical disparities of care. Whereas the market provides the individual with sufficient and high quality goods, there is a lack of public goods and of the quality of such goods.

Research Question and Results

Andrews [6] examines which domains (measured according to "affective responses" or satisfied domains of life) most influence the perceived quality of life of the individual, and, in the language of the test, which of these make it most predictable. He supports the model of a totality of quality of life which is measured on an analog scale in response to the question "how do you feel about your life as a whole?". In a number of stages, 800 concerns (derived from a pretest) are allocated to a list of 123 domains which presumably influence the quality of life. The results of the survey are used to answer the question, "which domains are most important in predicting life quality?" [6]. Twelve domains are determined, which in an additive model, have the highest power of prediction. "The 12 selected domains were chosen with several criteria in mind: (1) demonstrated predictive power . . .; (2) dispersion in the multidimensional space . . .; and (3) potential policy relevance" [6]. The 12 domains are: house/apartment, spare time activities, things to do with family, health, amount of fun, time to do things, job index, national government index, efficacy index, family index, consumer index and money index.

Campbell [36] has developed three measurements, analogous to his three-part concept of quality of life:

- Cognitive measure of satisfaction with life as a whole (a question of "overall satisfaction" and a combination of ten domains): "index of satisfaction with life domains".
- Affective measure of satisfaction with life (pleasantness-unpleasantness); paired adjectives: "index of general affect".
- Measure of perceived stresses: "index of perceived stress".

He uses the data on satisfaction with life (state-wide survey with a structured interview) obtained with these instruments and compares them with objective data about the person questioned (income, age, race, etc.). He comes to the conclusion that:

- The objective data say little about subjective satisfaction with life, which has mainly psychological determinants.
- "If we assume that these three measures are all tapping into that attribute of human experience called sense of well-being, or more commonly, happiness, then we must conclude that this experience has more than one dimension" [36].

Berger and Mohr [15] report on the welfare surveys from 1978 to 1984, which were carried out with conceptional points developed by Zapf [190, 191]: quality of life or welfare are combined with the components of objective conditions of life and subjective well-being. Welfare types are formed which systematise the correlation between objective conditions of life and subjective well-being:

- Well-being: well-being under objectively good conditions.
- Deprivation: non-/well-being under objectively bad conditions.

- Adaptation: well-being under objectively bad conditions (Zapf [191] calls this welfare type "resignation").
- Dissonance: non-/well-being under objectively good conditions.

The investigation is directed towards

- Explaining the distribution of the welfare types.
- Measuring and explaining the occurrence of the inconsistent welfare types, adaptation and dissonance.
- Measuring the changes chronologically.

The structural data is surveyed sociodemographically. Cognitive components are formed to measure well-being (cf. [193]:

- Satisfaction with individual domains.
- Satisfaction in comparison with the average citizen.
- Satisfaction in comparison with friends and acquaintances.
- Satisfaction in comparison with the life situation 5 years ago.
- Satisfaction with the anticipated future [15].

Proceeding on the assumption that "dissatisfaction in subjective domains of life can be explained through mental processes" [193], affective processes are thus eliminated. The description of the distribution of welfare types results (Welfare Survey 1979, [191]) in:

- "A clear correlation of status situations, risk groups and welfare types."
- "Satisfaction with public services is clearly less than with private circumstances in life. The disparities in living conditions are greater than in perceived quality of life. And: disparities are not the exception, rather the norm. More than seventy percent of the population show one or more clear 'jumps' in the ten investigated dimensions of objective and subjective welfare."
- There is clear deprivation in the Federal Republic of Germany.

The question as to how to explain the frequency of inconsistent welfare types remains open: Zapf [193] adheres to a principal correlation between objective domains and subjective well-being. The results of the time series analysis are not very clear-cut. Berger and Mohr [15] refer to the general reform atmosphere of the 1970s and the problem atmosphere of the 1980s, in which subjective welfare nevertheless remains stable.

Medicine and Measuring the Quality of Life of the Patient

Motives and Lines of Discussion

Structural correlations between politics and medicine give rise to assumptions about the background of the discussion of quality of life in medicine:

- Some uncertainty prevails about the formulation of the health goals.
- Existing indications for medical intervention in the life of the patient need to be reflected upon.

- The biomedical and work- or disability-oriented criteria for success are no longer sufficient.

The background to the discussion about the quality of life derives from the attempt to rationalise medicine in the direction of qualitative growth of medical intervention. The connection with the macrosociological discussion about quality of life establishes itself methodically as well as conceptually. The multilevel concept, objective aspects and subjective (cognitive as well as affective) perception of quality of life, the interdisciplinary structure (psychology, sociology, statistics) will be taken up by medical research. Displacements, however, do occur:

- The main interest is not placed on the consistent "types of welfare" (cf. [191], but on a medical adaptation: coping with the disease
- In general, it is assumed that illness changes the way of life and thereby impairs the quality of life [9]
- The question is not directed at the holistic satisfaction with life, but at the changes in quality of life caused by disease: "The components that constitute the quality of life in the healthy individual differ markedly from those that apply to patients afflicted by symptoms of disease" [169].
- The predominantly three-part sociological model of quality of life (affective, cognitive and stress) is extended by objective dimensions (1) of the physical condition; and (2) deviations in social behaviour.

The methodological connection is obtained from the behaviouristic construction of the measuring instrument. Displacements result

- Through a strong preference for objective indicators (critically cf. [125]).
- Through the small population of patients with specific clinical pictures.
- From the short intervals in time sequences.
- From ethical problems.
- Further problems in socioepidemiological research: underlying diseases, prospective v. retrospective, control groups, etc.

Najman [125], in his overview of the literature on measuring the quality of life in medicine, mentions further topics for discussion:

- Doubts about the efficacy of medical progress and medical intervention (Fuchs, Carlson, Illich).
- The search for additional indicators of medical success: "Foremost amongst these is the impact of modern medical technology on the quality of life".

The increase in the incidence of chronic diseases is mentioned as a further reason to expand the sociological aspect of medical evaluation. This aspect also presents new challenges to therapy and rehabilitation [183]. While emphasis in measuring the quality of life is placed on the evaluation of the medical care of certain clinical pictures, further impulses arise from the discussion surrounding the measurement of the health status of certain populations (cf. [96]) and the evaluation of medical care (for example, medical audit for in- and outpatient care: cf. [42, 73]). From practical experience in measuring the quality of life in general and clinical medicine, there are further special interests to be formulated:

- Nursing personnel and doctors should be able to use the measuring instruments and apply the results [75, 149, 182].
- Therapeutic alternatives should also be comprehensible to the patient in order to increase compliance [171].
- The patient's conceptions of life which result from the quality of life measurements should lead to interventions which change behaviour [182].

Concepts and Research Questions

"Functionality" Model

The main question as to how the quality of life can be measured is: to what extent is the capability of carrying out daily activities restricted? Quality of life is reduced according to the degree of functional restriction. This functional restriction can be assessed objectively and subjectively. Taylor [169] proposes a three-part model for measuring the quality of life (Table 1).

The level of the symptoms and the functional restrictions can be objectively assessed (cf. construction of the measuring instruments). Objective data are given a patient-oriented, subjective weighting by assessing the patient's well-being (regarding this concept, cf. especially [182]).

Potthoff [138] points out that the functionality model is based on behavioural disease terminology: he distinguishes "three manifestations of morbidity: clinical, subjective and behavioural evidence of morbidity. Under these three aspects of health and disease, the latter in particular has frequently been used for constructing measures of medical success. 'Health' is defined as absence of disease-related restrictions in daily activities of life and social role expectations. These include the ability to exercise a social role appropriate to age and sex, along with the ability to function physically and the ability to be self-sufficient".

Input, Process, or Outcome Studies

In the methodology of quality assessment of medical achievement, one distinguishes between input (structure), process, and output studies (cf. [42]). In general, quality of life measurements count as output studies: successes in

Table 1. Three-part model for measuring quality of life (from [174])

Symptoms	Patient perception	Functional capacity
Frequency	Degree of well-being	Normal daily activities
Severity	General health status	Social life
Emotional tone	Satisfaction with life	Emotional tone
Duration	Anxiety regarding health	Intellectual performance

therapy or rehabilitation (impact on patient) should be measured on patients. Furthermore, outcome studies are concerned with a prognosis on the proportion of patients with certain functional restrictions which can then be retrospectively compared with the actual data. (Such a prognosis study is not available in our selection of literature, but cf. [152]).

However, if the evaluation of certain therapeutic alternatives, especially in the case of drug or surgical procedures, is in the foreground, one must also discuss input studies: the measured or estimated quality of life serves as an indicator for assessing the medical technical input. The omission of prognosis transforms many studies into process studies: patient management, treatment alternatives and courses of therapy are evaluated in the chronological order of quality of life measurement.

Research Questions

In the development of measuring instruments, descriptive aspects move to the forefront: the state of health of certain patient populations at certain times. The measuring instruments are highly developed, but problems occur in translating these into sociomedical research (see below). Typical questions are:

- Which areas for possible therapeutic success can be gained from a description of the quality of life of the patients [61]?
- How can the general practitioner judge the effects of his or her efforts in the treatment of a broad spectrum of disease [73]?
- How can new health practices be evaluated [96]?
- How can long-term consequences of certain diseases be measured and evaluated [105]?
- How can a rational choice between therapeutic possibilities be substantiated [149]?

Measuring Instruments

There is an extensive range of literature validating the measurement instruments listed (Table 2). The instruments mentioned are generally rated as reliable. When practical application is considered, the following questions arise:

- Is it possible to measure short-term changes in patients (improvement/deterioration of health) (cf. [68])?
- How can comprehensive standard instruments be tailored to meet the demands of specific patient populations and diseases?
- How can the expectations of patients and experts in therapy and rehabilitation practices under investigation be reconciled?

The discussion on these points is not yet concluded. A summarised presentation of the standard instruments that are applied in measuring quality of life is given by Wenger [183]:

The *Sickness Impact Profile* deals with patient reports of dysfunctional behavior attributed to sickness.
The *McMaster Health Index Questionnaire* evaluates patient reports of performance, rather than capacity; it addresses the patient's feelings and thought, but does not ask the respondent to relate this to sickness.
The *Nottingham Health Profile* evaluates symptomatic evidence of illness and associated changes in daily life activities.
The *Psychological General Well-Being Index* quantifies six categories of psychological general well-being, rather than assessing deviations from normal health due to disease and its treatment.
The *General Health Rating Index* is completely subjective and does not refer to health components (physical, mental, or social) or health behavior. Self-perception of health and physical impairment gauged by this index has predicted the use of medical services.
The *Quality of Well-Being Scale* assesses objectively reportable facts in the preceding 6 days for elements of self-care, mobility, institutionalization, social activities, and symptoms and problems.

The standard instruments mentioned here mostly arose from attempts to measure health status (for example, the Sickness Impact Profile within the scope of the Health Maintenance Organizations, and the General Health Rating Index (also the Rand Health Insurance Battery) within the scope of the evaluation of insurance benefits for particular populations). If one follows Spitzer's [158] suggestion to differentiate the measurement of health status from the measurement of quality of life in clinical tests in such a way that the concept "measurement of the quality of life" characterises the latter, problems in applying standard instruments become evident. The difficulties in transposing an instrument, for example, the Sickness Impact Profile, into a disease-specific measurement series in clinical tests lead not least to the inflation of further measuring instruments, which are each supposed to cover one clinical picture under the specific conditions of the clinical test. A synopsis of these instruments (investigated dimensions, validity, reliability, application, etc.) is unfortunately not available. However, it is urgently required. The following instruments for specifically measuring quality of life are particularly used and discussed in the literature.

Activities of Daily Living Scale (ADL). Six sociobiological functions are included in order to determine the degree of limitation in the specific functional areas and in general [99].
Anamnestic Comparative Self Assessment (ACSA). In a self assessment of general well-being, the patients are encouraged to construct an individual scale of quality of life [19].
Katz Adjustment Scale. Patients' self assessment and assessment by relevant third parties provide a means for socially categorizing the patient.
Linear Analogue Self Assessment (LASA). [140]
Spitzer Quality of Life Index (QLI). Activity, every-day life, health, environment, the future: according to these aspects, doctors estimate the

Table 2

Instrument cont.	Reference	Type	Dimensions Investigated		Technical Information
Sickness Impact Profile (SIP)	Bergner [16] Bergner et al. [17, 18]	General Profile Assessing	Physical Psychological Social	Functionality Patient perception Functionality Patient perception Functionality Patient perception	136 questions with the groups: psychosocial (e.g. communication), physical (e.g. mobility), others (e.g. recreational pastimes) Self-report or support from the interviewer Approx. 30 min to fill in
McMaster Health Index Questionnaire	Chambers [38] Chambers et al. [39]	General Index Behavioural	Physical Psychological Social	Functionality Functionality Functionality	59 questions on the following domains: physical (e.g. global physical functioning), social (e.g. social role performance), emotional (e.g. critical life events) Self-report Approx. 20 min to fill in
Nottingham Health Profile (NHP)	Hunt [84, 85] Hunt and McEwan [86] Hunt et al. [87, 88]	General Profile Assessing	cf. SIP		45 questions on the domains "experience" (e.g. social isolation) and "daily life" (e.g. employment) Approx. 10 min to fill in
Psychological General Well-Being Index (PGWB)	Fazio [58]	General Index Behavioural	Psychological	Functionality Patient perception	22 questions (e.g. life satisfaction) Self-report or support from the interviewer Approx. 12 min to fill in
General Health Rating Index (Rand Health Insurance Battery, Excerpt)	Brook and Avery [25] Brook et al. [26]	General Index Assessing	Physical Psychological	Patient perception Patient perception	29 questions (e.g. past, present, and future perception of health) Self-report or support from the interviewer Approx. 7 min to fill in
Quality of Well-Being Scale (Index of Well-Being)	Kaplan and Bush [95] Kaplan et al. [96]	General Index Behavioural	Physical Psychological Social	Functionality Functionality Functionality	50 questions (e.g. social activities) Trained interviewer Approx. 12 min to fill in
Karnofsky Index	Karnofsky	Specific Index Behavioural	Physical Psychological Social	Functionality Functionality Functionality	11 categories (e.g. impairment of function in physical activities) Rating

Table 2 (continued)

Instrument cont.	Reference	Type	Dimensions Investigated		Technical Information
Spitzer Quality of Life Index (QLI)	Spitzer [158] Spitzer et al. [159, 160]	Specific Index Behavioural	Physical Psychological Social	Functionality Patient perception Functionality Patient perception Functionality Patient perception	Five components (e.g. relationship to environment) Rating (Bullinger [27] refers to self-assessment version)
Linear Analog Self-Assessment (LASA)	Priestman and Baum [140]	Specific	According to the construction of the scales		Self-report on visual analog scales
Symptom list (e.g. New York Heart Association Functional Classes)	Olsson et al. [130]	Specific Behavioural	Physical Psychological Social	Functionality Symptoms Functionality Symptoms Functionality Symptoms	Rating
Anamnestic Comparative Self Assessment (ASCA)	Bernheim et al. [19]	Specific Index Behavioural	Psychological	Perception	11 stage scale with which the patients rate their current quality of life with regard to the "best" vs. the "worst" times of their lives
Activities of Daily Living Scale (ADL)	Katz [98] Katz and Akpom [99]	Specific Index Behavioural	Physical Social	Functionality Functionality	Six dimensions of every-day activities (e.g. dressing) are described as independent and dependent Rating
Katz Adjustment Scale	Katz et al. [100] Katz and Lyerly [97]	Specific Profile Behavioural Assessing	Psychological Social	Functionality Symptoms Patient perception Perception by others Functionality Symptoms Patient perception Perception by others	Five patient and five "significant other" scales which describe and assess the symptoms and expectations, performance in socially expected activities and recreational activities Trained interviewer

quality of life of the patients in order to obtain an index ([160]; in German: [143]).

Spitzer [158] classified the measuring instruments from the point of view of practical application: health status measures (populations of sick and healthy persons) vs. quality of life measures (populations of sick persons which allow clear findings). He also differentiates the measurement of quality of life from the broad area of the clinical measurement of function, using hypotheses (hypothesis-determined measure) in which the measuring instruments tend to be "custom designed" [13]. Bullinger suggests a further classification [27]:

- General vs. disease-related instruments.
- Indices vs. profiles.
- Self vs. outside assessment.
- Behaviour-related vs. assessment-related instruments.

A classification of measuring instruments according to investigated dimensions of quality of life proves to be extremely difficult to construct; while a consensus on the fundamental conception remains to be achieved, such an attempt is dependent on listing the dimensions of the individual instruments (cf. [183]), without being able to guarantee direct comparability. A pragmatic method is derived by using the simplified three-part model by Taylor [169], where quality of life is categorised under (1) functionality, (2) patient perception; and (3) symptoms. Combination of these dimensions, with the intrinsic aim of measuring quality of life to determine the effects of "disease" on physical, psychological and social "functioning" (cf. [61]), results in an understandable, if somewhat unspecific, picture. Physical, psychological and social aspects can be covered by the dimensions of functionality, and by patient perception of infringements on life and symptoms. Bullinger [27] provides information on psychometric and sociometric testing of the individual instruments.

A Few Examples and Results of Quality of Life Measurement

In his overview of the literature, Najman [125] arrives at a very critical conclusion: i.e. that the 23 articles on measuring the quality of life found for the period of 1977–1981

- Were methodologically sloppy.
- Were based almost entirely on objective indicators.
- Did not make the assessment of "normal functioning" very clear.

He comes to the conclusion that great uncertainty about the quality of life prevails. He approaches research on quality of life from the aspect of legitimation of medical-technical progress: "it is consequently difficult to escape the impression that any of these studies were undertaken to support or advocate an already preferred type of medical intervention" [125]. In summary, Najman postulates that the new investigative trend provides no new answers: medical-technical development continues to be looked on as positive – now also under the aspect of enhanced quality of life. Patient perspectives (environment, psychological, cognitive as well as affective situations and expectations of the rehabilitation process) are faded out, the concept of dys/functionality remains unclear and matter of judgment, and possibilities for control are lacking. In the methodological area and in the abundance of literature at least, the picture has improved since 1981. A short illustration of two examples should provide insight into the current discussion. In 1986 a retrospective report, Klonoff [105] investigated the quality of life of patients with severe injuries to the head, 2–4 years after the injury. The population comprised 63 patients. Structural data were compiled:

- Pre-injury (sex, age, marital status, education).
- Injury-related (reason for injury, Glasgow coma scale, multiple trauma, further injuries).
- Post-injury (post-case period, marital status, age).

In patient polling (self-administered questionnaire), the following were recorded:

- Physical and psychological complaints (22 items).
- Household chores and work history post-injury (25 items).
- Indices according to Sickness Impact Profile.

Relatives ("significant others") were questioned, using the relative scales of the Katz adjustment scale. Following the Sickness Impact Profile standard, values from a normal and a psychiatric control group were included. In the discussion of results, the relevance of the long-term consequences of head injuries for the perceived quality of life becomes apparent. Social disorientation ("failure to cope with the disease") in the experimental group is comparable with the psychiatric control group. The adjustment to the significant others poll confirmed the self-assessment on the part of the patients. The "findings ... indicate that behavioural changes reported during the acute stage post-injury continue and may even be exacerbated during later, post-injury phases" [105].

In a report by Selby [149], 246 patients with breast cancer requiring hospitalisation were questioned as to the extent to which the disease had caused behavioural changes and impaired their capacity to function. The following were compiled:

- A general assessment of the perceived quality of life as a whole (uniscale).
- Self-assessed influences of the disease using various analogue Sickness Impact Profile scales.
- Twelve items which derive from clinical experience (e.g. pain, alopecia), polled by the doctors.

Five patient groups were formed and evaluated with measurement instruments (feasibility, reliability, validity). The main concern of the authors was the development of a workable instrument for clinical personnel. "The measurement instrument described here seems suitable for general descriptive purposes and its evaluation suggests that quantitative assessment of aspects of the quality of life of cancer patients is possible with relatively simple methods. The information obtained about some complex areas such as emotional disorders is limited but further developments and refining of such methods may provide a valuable additional endpoint in the investigation of therapy for cancer, particularly when such therapy may be toxic and the outcome is often palliation" [149].

Heart Diseases and Quality of Life

In heart failure, measuring quality of life with regard to assessing the success of therapy and rehabilitation has obviously become an established procedure in the United States, and a journal entitled Quality of Life and Cardiovascular Care has been published since 1985. In 1987, a conference was held in Washington on the subject "Quality of Life in Cardiovascular Disease" (cf. publications in the American Heart Journal, 118).

Publications in Anglo-American journals, which report on research in the Federal Republic of Germany into heart and cardiovascular diseases, are trying to find support for the discussion on quality of life [9, 152]. A standardisation of investigations comparable to the position taken in American discussions is not apparent. The Freiburg Personality Inventory and the Questionnaire on Satisfaction with Life [55, 124] point in this direction. Siegrist [152] also provides a thematic exception:

- Within the framework of a combined prospective and retrospective study he considers "impaired quality of life" as a risk factor (not as an outcome) in heart disease.
- He has developed an "inventory of negative life events" [153] which measures objective and affective as well as cognitive aspects.

The German contributions to the available literature, when compared to the American literature on quality of life, are more epidemiologically oriented (cf. [185]):

- Impaired quality of life as a risk factor [152].
- Social support as a factor in genesis of the disease and coping [90].
- Satisfaction with life as a factor in successful rehabilitation and for the patient's career [124].

In addition to the work of Badura [9], who (at least in abstract) uses conceptual terminology from the American discussion for illustration of the Oldenburg study and presents scales for measuring social support and recognisable care in order to process collective (not, in a strict sence, disease-specific) aspects of quality of life, there is, in connection with the Oldenburg study, that of Kaufhold [101], who uses the concept of the type-A behavioural pattern and investigates disease-coping strategies within the predetermined personality structures of the patients.

In the light of the limited amount of literature examined, a short illustration of the contributions to the conference on Quality of Life in Cardiovascular Disease can only provide a few examples and give a summarised report on the status of the discussion. Taylor [169] is involved with measurement problematics in general. He mentions the known levels: symptoms, patient perception, functional capacity. He demands measuring instruments which are designed specifically with heart disease in mind and concludes: "As yet, however, no study with a formal instrument to measure quality of life has been reported with any drug in an acceptably sized group of patients with angina pectoris" [169].

Julian [93] sees tension between the primary concern on the part of the doctor to maintain the life of the patient after an infarction and possible aspects of the quality of life. The main concern is to reduce pain and eliminate fear; to treat this, more analgesics should be used. Hjalmarson [80] is investigating the use of beta-blockers in postinfarction prophylaxis with regard to the aspect of quality of life. He mentions the following dimensions for measurement:

- Calculation of life expectancy (without stating the instrument).
- Determination of retention time in the New York Heart Association (NYHA) functional classes.
- Return to working life.

He refers to clinical tests with beta-blockers and considers their worth as documented. The recommended measurement dimensions are related to the Stockholm metoprolol trial: grouping the NYHA class retention time is done objectively and leads to determination of professional reintegration in a 3-year follow-up. "One can assume that being in a better functional class and having fewer side effects or complications increase quality of life" [80].

Scheidt [145] views a generally poor state of data for assessing the effect of individual methods of treatment on quality of life; he does, however, assume that being forced to change one's lifestyle limits this. He suggests that probable objective correlates to quality of life be investigated: "Data that might be expected to correlate with quality of life include numbers of anginal episodes or sublingual nitroglycerin tablets consumed; exercise tolerance; extent of limitation of activity, both work related and recreational; and symptoms or side effects related to the therapy rather than the disease being treated" [145]. Storstein [165] provides an overview of the measuring instruments available and the methodological problems involved (cf. [184]).

On Research in the Federal Republic of Germany

The literature evaluated deals with

- A Munich study group on quality of life in medicine.
- A study group on quality of life of the European Organization for Research and Treatment of Cancer (EORTC) (cf. [2]).

The Munich Study Group

The Munich study group on quality of life is made up of scientists from interdisciplinary fields: doctors, sociologists, psychologists and statisticians from the Biometric Centre for Therapeutic Studies and the Institute for Medical Psychology of the University of Munich [28]. The coworker teams put forward topics of emphasis:

- The measurement of quality of life focuses primarily on therapeutic studies, particularly in the clinical trial phase.
- One area of emphasis is the biometric and psychometric testing of measuring instruments.
- Special attention is devoted to the psychologic dimension of quality of life.

Analogous to the Anglo-American discussion, the determination of patients' quality of life is viewed in the light of the potential for optimising medical care: therapeutic alternatives can be evaluated, decisions on medical interventions can be made and treatment can be directed [29].

Bullinger [28] reports on the objectives and the first results of the work. The target areas are as follows:

- Documentation of approaches towards measuring quality of life in a literature data bank.
- Research into the fundamentals of quality-of-life assessment.
- Development of an operational conceptual model.
- Translation and adaptation of existing (Anglo-American) measuring instruments for German-speaking populations.
- Accompanying therapeutic studies.

In the area of basic research, a pre-test is reported on in which definitions of quality of life were sought in an open format. After classification of the answers, allocation by experts and a test of adequacy, 15 categories were filtered out. The results are consistent with the consensus of quality-of-life research. A conceptual model was developed (Table 3).

Table 3. Quality of Life Matrix Model (From [30])

Components	Personality	Disease	Living Conditions		
Evaluative dimensions	Physical state	Psychological well-being	Social relations	Every-day-life functions	
Emotional	Fatigue	Depression	Loneliness	Dependence	
Behavioural	Functional status	Problem solving	Social activities	Household/ work	Global quality of life rating
Cognitive	Relevance satisfaction	Relevance satisfaction	Relevance satisfaction	Relevance satisfaction	
Resources / strains	Symptoms and side effects	Coping mechanism	Social support	Daily hassles	

The translation of published scales is not without its problems. Other socio-legalistic structural requirements and societal hierarchical value structures make it necessary to test the quality of the translated scales extensively. Notwithstanding this great effort, Bullinger recognises the merit of translation for the purpose of international comparability of research results. The EORTC questionnaire, for example, has been translated into 11 languages and will be applied in an international research programme on lung cancer. The Munich study group translated the Psychological General Well-Being Index (cf. [58]) – an instrument for measuring psychological condition (cf. [60]) – into German. In a questionnaire on quality of life from the Biometric Centre for Therapeutic Clinical Studies, published scales were used: EORTC, Spitzer Quality of Life Index ([160], in German [143]); Psychological General Well-Being Index, Nottingham Health Profile, Profile of Mood States (cf. [118], in German [20]). Questionnaire modules were developed for the following areas: complaints, condition, every-day life, assessments, social support, coping, sense of control, and situation in life.

The accompanying supervision of therapeutic clinical studies under the aspect of quality of life comprises various types of studies (cohort polls, cohort observations, clinical tests) for investigating the various methods of treatment of cancer and heart diseases. A table in the appendix of the text [28] provides an overview.

The close connection seen by the Munich study group between quality of life measurement and clinical tests of therapies has not assumed the versatility of the American discussion. Although the subject of quality of life was initially established in large clinical trials, important conceptual and methodological initiatives have already arisen from research into health indicators (Sickness Impact Profile and Quality of Life Index) and from the efforts to allow quality of life aspects to be incorporated into daily medical routine (self-evaluation of therapy and rehabilitation). Similarly, the preference for psychological indicators is just as controversial (cf. [86]).

Additional National Efforts

In describing the research into quality of life in the Federal Republic of Germany, Bullinger's focus [28] is not only directed to the experiences of the Munich group; she has also indicated other research efforts:

- In rehabilitation research (Fahrenberg, Myrtek, Muthny).
- In the development of health indicators (Potthof – translation of the Sickness Impact Profile and the Quality of Well-being Index).
- In sociopsychological studies of well-being (Glatzer/Zapf).

The important works by Siegrist and co-workers [152–155] also deserve mention. However, the reference lists in German publications point unmistakably to Anglo-American predominance in research into quality of life.

References

1. Aaronson NK, Beckmann J (eds) (1987) Quality of life of cancer patients. Raven, New York
2. Aaronson NK, Bullinger M et al. (1988) A modular approach to quality-of-life assessment in cancer clinical trials. In: Scheurlen H, Kay R, Baum M (eds) Cancer clinical trial: a critical appraisal. Springer, Berlin Heidelberg New York, pp 231–249
3. Alexander JL, Willems EP (1981) Quality of life: some measurement requirements. Arch Phys Med Rehabil 62 (6) 261–265
4. Anderson F (1976) The effect of screening on the quality of life after seventy. J R Coll Physicians Lond 10: 161–169
5. Andrews FM, Inglehart RF (1979) The structure of subjective well-being in nine western societies. Soc Indicators Res 6: 73–90
6. Andrews FM, Withey St (1974) Developing measures of perceived life quality. Results from Several National Surveys. Soc Indicators Res 1: 1–26
7. Applefeld MM (1986) Chronic congestive heart failure. Where have we been? Where are we heading? Am J Med 80: 73–77
8. Astrand PO (1987) Exercise physiology and its role in disease prevention and in rehabilitation. Arch Phys Med Rehabil 68 (5) 305–309
9. Badura B, Waltz M (1984) Social support and the quality of life following myocardial infarction. Soc Indicators Res 14: 295–311
10. Bailit H, Lewis J et al. (1975) Assessing the quality of care. Nurs Outlook 23 (3) 153–159
11. Balaban DJ, Sagi PC et al. (1986) Weights for scoring the quality of well-being instruments among rheumatoid arthritics: a comparison to general population weights. Med Care 24 (11) 973–980
12. Barofsky I, Cohen SJ et al. (1984) Selecting a quality of life assessment instrument: standardized tests, clinical assessment, or custom-designed instruments. In: Wenger NK, Mattson ME (eds) Assessment of quality of life in clinical trials of cardiovascular therapies. Le Jacq, New York, pp 239–249
13. Berdit M, Williamson JW (1973) Function limitation scale for measuring health outcomes. In: Berg RL (ed) Health status indexes. Chicago, pp 59
14. Berg RL, Hallauer DS et al. (1976) Neglected aspects of the quality of life. Health Serv Res 11: 391–395
15. Berger R, Mohr H-H (1986) Lebensqualität in der Bundesrepublik 1978 und 1984. Soziale Welt 37 (1): 25–47
16. Bergner M (1984) The sickness impact profile. In: Wenger NK, Mattson ME (eds) Assessment of quality of life in clinical trials of cardiovascular therapies. Le Jacq, New York, pp 152–159
17. Bergner M, Bobbitt RA et al. (1976) The sickness impact profile: conceptual formulation and methodology for the development of a health status measure. Int J Health Serv 6 (3) 393–415
18. Bergner M, Bobbitt RA et al. (1981) The sickness impact profile: development and final revision of a health status measure. Med Care 19 (8): 787–805
19. Bernheim J, Avonts G et al. (1985) Anamnestic comparative self assessment (ACSA) to correlate medical and psychosocial variables with subjective quality of life of cancer patients. Proc Am Soc Clin Oncol 4: 248
20. Biehl B, Dangel S et al. (1986) Profile of mood states. CIPS (Collegium Internationale Psychiatriae Scalarum): internationale Skalen für Psychiatrie. Beltz, Weinheim
21. Binswanger U (1979) Indikationen zur Haemodialysebehandlung und Nierentransplantation. Schweiz Med Wochenschr 109 (10): 339–342
22. Blades BC et al. (1979) Quality of life after major burns. J Trauma 19 (8): 556–558
23. Böckenförde E-W (1973) Qualität des Lebens – Aufgabe und Verantwortung des Staates?. Neue Gesellschaft 20: 261
24. Breslow L (1972) A quantitative approach to the World Health Organization definition of health: physical, mental and social well-being. Int J Epidemiol 1: 347–354

25. Brook RH, Avery AD (1977) Quality of medical care assessment using outcome measures. National Center for Health Services Research: Research Summary Series. Hyattsville, Md: Department of Health, Education and Welfare
26. Brook RH, Ware JE et al. (1979) Overview of adult health status measures fielded in Rand's health insurance study. Med Care 17 [Suppl] (7), 1–131
27. Bullinger M (1988a) Forschungsinstrumente zur Erfassung der Lebensqualität bei Krebs – ein Überblick. TS, Munich
28. Bullinger M (1988b) Quality of life research in West-Germany – the Munich experience. TS, Munich
29. Bullinger M, Pöppel E (1988c) Lebensqualität in der Medizin: Schlagwort oder Forschungsansatz? Dtsch Ärztebl 85 (11) 679–680
30. Bulpitt CJ (1982) Quality of life in hypertensive patients: In: Amery A (ed) Hypertensive cardiovascular disease: pathophysiology and treatment. Nijhoff, Den Haag, pp 929–948
31. Bush JW (1984) General health policy model/quality of well-being (QWB) scale. In: Wenger NK, Mattson ME (eds) Assessment of quality of life in clinical trials of cardiovascular therapies. Le Jacq, New York, pp 189–199
32. Bush JW, Chen MM et al. (1973) Health status index in cost effectiveness: analysis of PKU program. In: Berg RL (ed) Health status indexes. Chicago, pp 172
33. Bush JW, Chen MM et al. (1974) Statistical models of social preferences of constructing a health status index. National Technical Information Service: Publ No 236155/8ST. Springfield
34. Bush JW, Blischke WR et al. (1975) Health indices, outcomes, and the quality of medical care. In: Yaffe R, Zalkind D (eds) Evaluation in health services delivery. Engineering Foundation, New York, pp 313–339
35. Calman KC (1987) Definitions and dimensions of quality of life. In: Aaronson NK, Beckman JH (eds) The quality of life of cancer patients. Raven, New York, pp 1–10
36. Campbell A (1976) Subjective measures of well-being. Pschol 31: 117–124
37. Campbell A, Converse P et al. (1976) The quality of American life. Perceptions, evaluations, and satisfactions. Russel Sage Foundation, New York
38. Chambers LW (1984) The McMaster health index questionnaire. In: Wenger NK, Mattson ME (eds) Assessment of quality of life in clinical trials of cardiovascular therapies. Le Jacq, New York, pp 160–164
39. Chambers LW, MacDonald LA et al. (1982) The McMaster health index questionnaire as a measure of the quality of life for patients with rheumatoid disease. J Rheumatol 9 (5) 780–784
40. Chen MM, Bush JW (1976) Social indicators for health planning and policy analysis. Policy Sci 6: 71
41. Chen MM, Bush JW (1979) Health status measures, policy and biomedical research. In: Mushkin SJ, Dunlop DW (eds) Health what is it worth? Measures of Health Benefits. Pergamon-Verlag, New York
42. Christoffel T, Loewenthal M (1977) Evaluating the quality of ambulatory health care. A review of emerging methods. Med Care 15 (11) 877–897
43. Croog SH, Levine S et al. (1986) The effects of antihypertensive therapy on the quality of life. N Engl J Med 314 (26) 1657–1664
44. Culyer AJ (ed) (1983) Health indicators. Robertson, Oxford
45. Dean H (1985) Choosing multiple instruments to measure the quality of life. Oncol Nurs Forum 12 (5) 96–100
46. Department of Health Services (ed) (1977) Sickness Impact Profile. University of Washington
47. Deyo RA (1984) Measuring functional outcomes in therapeutic trials for chronic disease. Controlled Clin Trials 5: 223–240
48. Deyo RA, Inui TS et al. (1983) Measuring functional outcomes in chronic disease: a comparison of traditional scales and self-administered health status questionnaire in patients with rheumatoid arthritis. Med Care 31: 180–192
49. Diener E (1984) Subjective well-being. Psychol Bull 95: 542–575

50. Donabedian A (1966) Evaluating the quality of medical care. Milbank Mem Fund 44: 166–206
51. Donabedian A (1976) Measuring and evaluating hospital and medical care. Bull NY Acad Med 52: 51–59
52. Drenick EJ (1981) Risk of obesity and surgical indications. Int J Obes 5 (4) 387–398
53. Dupuy HJ (1984) The psychological general well-being (PGWB) index. In: Wenger NK, Mattson ME (eds) Assessment of quality of life in clinical trials of cardiovascular therapies. Le Jacq, New York, pp 170–183
54. Egger J (1983) Altersabhängige Krankheitsverarbeitung bei Patienten mit kardiovaskulaeren Erkrankungen. Aktuel Gerontol 13 (3) 96–100
55. Fahrenberg J, Myrtek M et al. (1986) Multimodale Erfassung der Lebenszufriedenheit. Eine Untersuchung an Herz-Kreislauf-Patienten. Psychother Med Psychol 36: 347–354
56. Fanshel S (1972) A meaningful measure of health for epidemiology. Int J Epidemiol 1 (4): 319–337
57. Fanshel S, Bush JW (1970) A health status index and its application to health services outcomes. Oper Res 18: 1021
58. Fazio AF (1977) A concurrent validational study of the NCHS general well being schedule. Vital and health statistics. National Center for Health Statistics: Publication No (HRA) 78–1347. Hyattsville, MD
59. Flanagan JC (1978) A research approach to improving our quality of life. Am Psychol 138–147
60. Fletcher AE, Bulpitt CJ (1986) Assessment of quality of life in cardiovascular therapy. Br J Clin Pharmacol 21 [Suppl 2]: 173S–181S
61. Fletcher AE, Hunt BM (1987) Evaluation of quality of life in clinical trials of cardiovascular disease. J Chronic Dis 40 (6): 557–569
62. Foxman B, Lohr KN et al. (1982) Conceptualization and measurement of physiological health in adults. V. Congestive heart failure. Rand Corporation, Santa Monica
63. Friedman M, Thorensen CE et al. (1984) Alteration of type A behavior and reduction in cardiac recurrence in postmyocardial infarction patients. Am Heart J 108 (2): 237–248
64. Furberg CD (1985) Assessment of quality of life. In: Friedman L, Furberg CD (eds) Fundamentals of clinical trials. PSG, Littletown, pp 161–170
65. Glatzer W, Zapf W (eds) (1984) Lebensqualität in der Bundesrepublik. Objektive Lebensbedingungen und subjektives Wohlbefinden. Campus, Frankfurt
66. Gofin R (1982) The health status of elderly men: a community study. Public Health 96 (6): 345–354
67. Gonnella JS, Zeleznek C (1974) Factors involved in comprehensive patient care evaluations. Med Care 12: 928–934
68. Gore SM (1988) Integrated reporting of quality and length of life – a statistican's perspective. Eur Heart J 9 (2): 228–234
69. Granger CV, Greer DS (1976) Functional status measurement and medical rehabilitation outcomes. Arch Phys Med Rehabil 57: 103–109
70. Greene R (1976) Assuring quality in medical care: the state of the art. Ballinger, Cambridge
71. Greenwald HP (1987) The specificity of quality-of-life measures among the seriously ill. Med Care 25 (7): 642–651
72. Guyatt GH (1987) Measuring quality of life: a review of means of measurement in clinical trials of new medicines. Pharm Med 2: 49–60
73. Hall J, Hall N et al. (1987) Measurement of outcome of general practice: comparison of three health status measures. Fam Pract 4 (2): 117–122
74. Hansson L (1985) Assessment of the patient's response. J Hypertens 3 (2): S65–S69
75. Harvey RM, Doyle EF et al. (1974) Editorial: Major changes made by the Criteria Committee of the New York Heart Association. Circulation 49: 390
76. Hays RD, Ware JE Jr (1986) My medical care is better than yours: social desirability and patient satisfaction ratings. Med Car 24 (6): 519–525

77. Herndon DN, LeMaster J et al. (1986) The quality of life after major thermal injury in children: an analysis of 12 survivors with greater than or equal to 80% total body, 70% third-degree burns. J Trauma 26 (7): 609–619
78. Hillner BE, Hollenberg JP et al. (1986) Postmenopausal estrogens in prevention of osteoporosis. Benefits virtually without risk of cardiovascular effects are considered. Am J Med 80 (6): 115–1127
79. Hirsch EO (1975) Sounding board: A problem and objective oriented approach to patient care evaluation. Engl J Med 292: 1348–1349
80. Hjalmarson AC (1987) Use of beta blockers in postinfarct prophylaxis: aspects on quality of life. Heart J 114: 245–250
81. Smith DW, Hogan AJ, Rohrer JE, Hogan AJ (1987) Activities of Daily Living as quantitative indicators of nursing effort. Med Care 25: 120–130
82. Horn SD, Horn RA (1986) Reliability and validity of the Severity of Illness Index. Med Care 24 (2): 159–178
83. Houston C, Pasanen W (1972) Patient's perception of hospital care. Hospitals JAHA 46: 70–74
84. Hunt SM (1984) Nottingham health profile. In: Wenger NK, Mattson ME (eds) Assessment of quality of life in clinical trials of cardiovascular therapies. Le Lacq, New York, pp 165–169
85. Hunt S (1988) Measuring health in clinical care and clinical trials. In: Teeling-Smith G (ed) Measuring health – a practical approach. Wiley, Chichester, pp 7–23
86. Hunt SM, McEwen J (1980) The development of a subjective indicator. Sociology Health Illnes
87. Hunt SM, McKenna SP et al. (1980) A quantitative approach to perceived health status: a validation study. J Epidemiol Community Health 34: 281–286
88. Hunt SM, McKenna SP et al. (1981) The Nottingham health profile: subjective health status and medical consultations. Soc Sci Med 15A: 221–229
89. Jachuck SJ, Brierley H et al. (1982) The effect of hypertensive drugs on the quality of life. J R Coll Gen Pract 32: 103–105
90. Jenkins CD, Stanton BA (19849 Quality of life as assessed in the recovery study. In: Wenger NK, Mattson ME (eds) Assessment of quality of life in clinical trials of cardiovascular therapies. Le Jacq, New York, pp 266–280
91. Jenkins CD, Stanton BA et al. (1983) Coronary artery bypass surgery. Physical, psychological, social and economic outcomes six month later. JAMA 250: 782–788
92. Jones MB (1977) Health status indexes: the trade-off between quantity and quality of life. Socioecon Plann Sci 11: 301–305
93. Julian DG (1987) Quality of life after myocardial infarction. Am Heart J 114: 241–244
94. Kann H-J (1975) Lebensqualität. Anmerkungen zur Wortgeschichte. Muttersprache 85: 50–52
95. Kaplan RM, Bush JW (1982) Health related quality of life measurement for evaluation research and policy analysis. Health Psychol 1: 61–80
96. Kaplan RM, Bush JW et al. (1976) Health status: types of validity and the index of well-being. Health Serv Res 11 (4): 478–507
97. Katz MM, Lyerly SB (1963) Methods for measuring adjustment and social behavior in the community. I. Rationale, description, discriminative validity and scale development. Psychol rep 13: 503–535
98. Katz S (1987) The science of quality of life. J Chronic Dis 40 (6): 459–463
99. Katz S, Akpom CA (1976) A measure of primary sociobiological functions. Int J Health Serv 6 (3): 493–508
100. Katz S, Downs TD et al. (1970) Progress in development of the index of ADL. Gerontologist 10: 20–30
101. Kaufhold G (1987) Zur Bedeutung des Typ-A-Verhaltensmusters für die Herzinfarktrehabilitation. In: Badura B, Kaufhold G, Lehmann H, Pfaff H, Schott T, Waltz M (eds) Leben mit dem Herzinfarkt. Eine sozialepidemiologische Studie. Springer, Berlin Heidelberg New York, pp 286–320

102. Kessner DM, Kalk CE et al. (1973) Assessing health quality – the case for tracers. N Engl J Med 288: 189–194
103. The Principal Investigators of CASIS and Their Associates (1981) The National Heart, Lung, and Blood Institute Coronary Artery Sugery Study (CASS). Circulation 63/Part II, I-1–I-81
104. Kleeberg UR (1984) Geriatrische Aspekte der Onkologie. Onkologie 7 (2): 63–66
105. Klonoff PS, Snow WG (1986) Quality of life in patients 2 to 4 years after closed head injury. Neurosurgery 19 (5): 735–743
106. Köhler T (1989) Über psychologische Faktoren in der Ätiologie der koronaren Herzkrankheit. Soz. Präventivmed [Suppl] 1: 13–14
107. Kornfeld DS, Heller SS et al. (1982) Psychological and behavioral responses after coronary artery bypass surgery. Circulation 66 [Suppl III]: III-24–III-28
108. Kornfeld DS, Heller SS et al. (1984) Psychologic and behavioral responses after coronary artery bypass graft surgery. In: Wenger NK, Mattson ME (eds) Assessment of quality of life in clinical trials of cardiovascular therapies. Le Jacq, New York, pp 233–238
109. Lane DA (1987) Utility, decision, and quality of life. J Chronic Dis 40: 585–591
110. Lembcke PA (1967) Evolution of the medical audit. J Am Med Assoc 199 (8): 543–550
111. Levine S, Croog SH (1984) What constitutes quality of life? A conceptualization of the dimensions of life quality in healthy populations and patients with cardiovascular disease. In: Wenger NK, Mattson ME (eds) Assessment of quality of life in clinical trials of cardiovascular therapies. Le Jacq, New York, pp 46–58
112. Levine MN, Guyatt GH et al. (1986) Quality of life in breast cancer: an instrument for clinical trials. Clin Invest Med 9: A47
113. Lomas J, Mohide A et al. (1987) Patient versus clinician item generation for quality-of-life measures: the case of language-disabled adults. Medical Care, 25, 764–769
114. MacKenzie CR, Charlson ME et al. (1986) Can the sickness impact profile measure change? An example of scale assessment. J Chronic Dis 39 (6): 429–438
115. MacCall S (1975) Quality of life. Soc Indicators Res 2: 229–248
116. McCullough LB (1984) Concepts of the quality of life: a philosophical analysis. In: Wenger NK, Mattson ME (eds) Assessment of quality of life in clinical trials of cardiovascular therapies. Le Jacq, New York, pp 25–36
117. McDowell I, Newell C (1987) Measuring health: a guide to rating scales and questionnaires. Oxford University Press, New York
118. McNair DML, Lorr M et al. (1971) Manual for the profile of mood states. Educational and industrial testing service, San Diego
119. McSweeny AJ (1984) Assessing the quality of life in patients with chronic obstructive pulmonary disease. In: Wenger NK, Mattson ME (eds) Assessment of quality of life in clinical trials of cardiovascular therapies. Le Jacq, New York, pp 250–265
120. Meenan RF, Gertman PM et al. (1980) Measuring health status in arthritis: the arthritis impact measurement scales. Arthritis Rheum 23: 146–152
121. Melzack R (1975) The McGill pain questionnaire: major properties and scoring methods. Pain 1: 277–299
122. Moberg DO, Brusek PM (1978) Spiritual well-being: a neglected subject in quality of life research. Soc Indicators Res 5: 303–323
123. Mumford E, Schlesinger HJ et al. (1982) The effect of psychological intervention on recovery from surgery and heart attacks: an analysis of the literature. Am J Public Health 72: 141–151
124. Myrtek M, Kreutel K et al. (1987) Lebenszufriedenheit und Rehabilitationsverlauf – eine Untersuchung an Herz-Kreislauf-Patienten. Rehabilitation 26 (1): 11–19
125. Najman JK, Levine S (1981) Evaluating the impact of medical care and technologies on the quality of life: a review and critique. Soc Sci Med 15F (2/3): 107–115
126. Oberman A (1984) Role of quality of life assessment in clinical trials. In: Wenger NK, Matson ME (eds) Assessment of quality of life in clinical trials of cardiovascular therapies. Le Jacq, New York, pp 67–84
127. Ochsner A (1976) Aging. J Am Geriatr Soc 24 (9): 285–293

128. Offe C (1974) "Lebensqualität" – eine Friedensformel sozialdemokratischer Innenpolitik? Leviathan 1–6
129. Olschewski H (1985) Möglichkeiten der Evaluierung und Analyse von Lebensqualität im Rahmen klinischer Studien. EDV Med Biol 16 (1): 8–16
130. Olsson G, Lubsen J et al. (1986) Quality of life after myocardial infarction: effect of long term metoprolol on mortality and morbidity. Br Med J 292: 1491–1493
131. Ott CR, Sivarajan ES et al. (1983) A controlled randomized study of early cardiac rehabilitation: the Sickness Impact Profile as an assessment tool. Heart Lung 12: 162–170
132. Pascoe GC, Attkinson CC et al. (1983) Comparison of indirect and direct approaches to measuring patient satisfaction. Eval Prog Plan 6: 359
133. Patrick D (1984) Quality of life measures for assessing disablement. In: Wenger NK, Mattson ME (eds) Assessment of quality of life in clinical trials of cardiovascular therapies. Le Jacq, New York, pp 220–232
134. Patrick DL, Bush JW et al. (1973) Toward an operational definition of health. J Health Soc Behav 14: 6–23
135. Patterson DR, Questad KA et al. (1987) Patient self-reports three months after sustaining a major burn. J Burn Care Rehabil 8 (4): 274–279
136. Perlman RA, Uhlmann RF (1986) Perceptions of quality of life in elderly patients with cardiovascular disease. Qual Life Cardivasc Care 2: 149
137. Pollard WE, Bobbitt R (1976) The sickness impact profile: reliability of a health status measure. Med Care 14 (2): 146–155
138. Potthoff P (1983) Anwendungsmöglichkeiten medizinischer Erfolgsmessung. Medizin Mensch Gesellschaft 8 (1): 10–17
139. Potthoff P (1988) Krankheitsbedingte Lebensbeeinträchtigungen in der Münchner Wohnbevölkerung. Öff Gesundheitswes 50 (6): 329–333
140. Priestman TJ, Baum M (1976) Evaluation of quality of life in patients receiving treatment for advanced breast cancer. Lancet I: 899–900
141. Questad KA, Patterson DR et al. (1988) Relating mental health and physical function at discharge to rehabilitation status at three months postburn. J Burn Care Rehabil 9 (1): 87–89
142. Rodgers WL, Converse PE (1975) Measures of perceived overall quality of life. Soc Indicators Res 2: 127-152
143. Rohde H, Rau E et al. (1984) Ergebnisse der Bestimmung des Lebensqualitätsindex nach Spitzer in der multizentrischen Magenkarzinom TNM-Studie (vorläufige Ergebnisse). In: Rohde H, Troidl H (eds) Das Magenkarzinom: Methoden klinischer Studien und therapeutische Ansätze. Symposium Köln 1983: Thieme, Stuttgart, pp 74–79
144. Sackett DR, Spitzer WO et al. (1974) The Burlington randomized trial of the nurse practioner: health outcomes of patients. Ann Intern Med 80: 137–142
145. Scheidt S (1987) Ischemic heart disease: a patient-specific therapeutic approach with emphasis on quality of life considerations. Am Heart J 114: 251–257
146. Schenk J (1989) Kardiologische Prävention – welchen Beitrag kann ein pharmazeutisches Unternehmen leisten? Soz Präventivmed [Suppl] 1: 30–31
147. Schuster B, Heim HE (1986) Lebensqualität von Karzinom-Patienten unter Chemo- und Radiotherapie. Prospektive Studie bei Gastrointestinal- und Bronchialkarzinom-Patienten. Onkologie 9: 172–180
148. Scott PJ (1976) Iatrogenic disease. NZ Med J 84: 405–408
149. Selby PJ, Chapman J-AE et al. (1984) The development of a method for assessing the quality of life of cancer patients. Br J Cancer 50 (1): 13–22
150. Shapiro S (1967) End result measurement of quality of medical care. Milbank Mem Fund Q 45: 7–30
151. Siara CS (1980) Komponenten der Wohlfahrt. Materialien zu Lebensbedingungen und Lebensqualität in der Bundesrepublik Deutschland. Campus, Frankfurt
152. Siegrist J (1987) Impaired quality of life as a risk factor in cardiovascular disease. J Chronic Dis 40 (6): 571–578

153. Siegrist J, Dittmann K et al. (1982) The social context of active distress in patients with early myocardial infarction. Soc Sci Med 16 (4): 443–453
154. Siegrist J, Matschinger H et al. (1987) Untreated hypertensives and their quality of life. J Hypertens 5 [Suppl 1]: 15–20
155. Siegrist J, Junge A (1988) Effects of antihypertensive therapy on quality of life. Mediographia 10: 38–40
156. Sigelman CK, Vengroff LP et al. (1979) Disability and the concept of life functions. Rehabil Couns Bull 23: 103–113
157. Smith HC, Frye RL et al. (1983) Does coronary bypass surgery have a favorable influence on the quality of life? Cardiovasc Clin 13: 253–264
158. Spitzer WO (1987) State of science 1986: quality of life and functional status as target variables for research. J Chronic Dis 40 (6): 465–471
159. Spitzer WO, Sackett DL et al. (1974) The Burlington randomized trial of the nurse practitioner. N Engl J Med 290: 251–256
160. Spitzer WO, Dobson AJ et al. (1981) Measuring the quality of life of cancer patients: a concise QL index for use by physicans. J Chronic Dis 34: 585–597
161. Starfield B (1974) Measurement of outcome: a proposed scheme. Milbank Mem Fund Q 52: 39–50
162. Stern MJ, Clearly P (1982) The National exercise and heart disease project: long-term psychosocial outcome. Arch Intern Med 142: 1093–1097
163. Stewart AL, Ware JE Jr et al. (1977) The meaning of health: understanding functional limitations. Med Care 15: 939–952
164. Stewart AL, Ware JE et al. (1978) Conceptualization and measurement of health for adults in the health insurance study: physical health in terms of functioning. Rand R-1978/2-HEW, Santa Monica
165. Storstein L (1987) How should changes in life-style be measured in cardiovascular disease? Am Heart J 114: 210–212
166. Sullivan D (1966) Conceptual problems in developing an index of health. DHWE-Publication no (HRA) 74-1017, Rockville
167. Sutherland HJ, Lockwood S et al (1989) Measuring satisfaction with health care: a comparison of single with paired rating strategies. Soc Sci Med 28 (1): 53–58
168. Swoboda H (1973) Die Qualität des Lebens: vom Wohlstand zum Wohlbefinden. DVA, Stuttgart
169. Taylor SH (1987) Drug therapy and quality of life in angina pectoris. Am Heart J 114: 234–240
170. Terleckyi NE (1975) Improvements in the quality of life. Estimates of possibilities in the United States, 1974–1983. National Planning Association, Washington
171. Till JE, McNeil BJ et al. (1984) Measurement of multiple components of quality of life. Cancer Treat Symp 1: 177–181
172. Troidl H, Kusche J et al. (1987) Quality of life: an important endpoint both in surgical practice and research. J Chronic Dis 40: 532–528
173. Van Dam FSAM, Somers R et al. (1981) Quality of life: some theoretical issues. J Clin Pharmacol 21 (1): 166S–168S
174. Walker SR, Rosser RM (eds) (1988) Quality of life: assessment and application. MTP Press, Lancaster
175. Wallwork J, Caine N (1985) A comparison of the quality of life of cardiac transplant patients and coronary artery graft patients before and after surgery. Quality Life Cardiovasc Care 1: 317–331
176. Ware J (1984) General Health Rating Index. In: Wenger NK, Mattson ME (eds) Assessment of quality of life in clinical trials of cardiovascular therapies. Le Jacq, New York, pp 184–188
177. Ware JE (1984) Methodological considerations in the selection of health status assessment procedures. In: Wenger NK, Mattson ME (eds) Assessment of quality of life in clinical trials of cardiovascular therapies. Le Jacq, New York, pp 87–111
178. Ware JE, Davies AR (1983) Behavioral consequences of consumer dissatisfaction with medical care. Eval Prog Plan 6: 291

179. Ware JE, Davies-Avery A et al. (1978) Conceptualization and measurement of health for adults in the health insurance study, vol V. General Health Perceptions. Rand Corporation: R-1987/5-HEW, Santa Monica
180. Weidman WH (1986) Cardiovascular risk modification in childhood: hyperlipidemia. Mayo Clin Proc 61 (11): 910–913
181. Welch AS, Welch BL (1987) The Rand Health Insurance Study: a summary critique. Med Care 25: 148
182. Wenger NK (1986) Quality of life: concept and approach to measurement. Adv Cardiol 33: 122–130
183. Wenger NK, Mattson ME et al. (1984a) Assessment of quality of life in clinical trials of cardiovascular therapies. Am J Cardiol 54 (7): 908–913
184. Wenger NK, Mattson ME (eds) (1984b) Assessment of quality of life in clinical trials of cardiovascular therapies. Le Jacq, New York
185. Wenzel H, Laaser U (1988) Lebensjahre und Lebensqualität. Ein Konzept zur Kostenbewertung der Hypertonie. In: Laaser U, Wilhelmsen L (eds) Epidemiologie und Prävention der Hypertonie. pmi, Frankfurt, pp 105–117
186. Williamson JW, Berdit M (1971a) The next step: quality assurance programs. Integrating health services research with continuing medical education. N Engl J Med [Suppl] 284: 98–103
187. Willamson JW (1971b) Evaluating quality of patient care – a strategy relating outcome and process assessment. JAMA 218: 564–569
188. Willamson JW (1978) Assessing and improving health care outcomes: the health accounting approach to quality assurance. Ballinger, Cambridge
189. Zacher HF (1973) Die Messung sozialen Glücks. Zur Problematik sozialer Indikatoren. Z Sozialreform 585–615
190. Zapf W (1972) Zur Messung der Lebensqualität. Z Soziol 1 (4): 353–376
191. Zapf M (1979) Lebensbedingungen und wahrgenommene Lebensqualität. In: Matthes J (ed) Sozialer Wandel in Westeuropa. Verhandlungen des 19. Deutschen Soziologentags. Campus, Frankfurt, pp 767–790
192. Zapf W (1984a) Lebensqualität in der Bundesrepublik. In: Glatzer W, Zapf W (eds) Lebensqualität in der Bundesrepublik. Objektive Lebensbedingungen und subjektives Wohlbefinden. Campus, Frankfurt, pp 391–401
193. Zapf W (1984b) Individuelle Wohlfahrt: Lebensbedingungen und wahrgenommene Lebensqualität. In: Glatzer W, Zapf W (eds) Lebensqualität in der Bundesrepublik. Objektive Lebensbedingungen und subjektives Wohlbefinden. Campus, Frankfurt, pp 13–26
194. Zyzanski SJ, Hulka BS et al. (1974) Scale for the measurement of 'satisfaction' with medical care: modifications in content format and scoring. Med Care 12 (7): 611–620

Measuring the Quality of Life in Patients with Heart Failure

Basic Principles, Methods, and Use in Clinical Studies

M. Bullinger and E. Pöppel*

Basic Principles of Measuring Quality of Life in the Field of Heart Disease

In order to illustrate the procedures for measuring quality of life in patients with heart failure, it is first necessary to discuss the issue of measurement and its aims.

The Issue

As described in various articles in this volume, the term "heart failure" refers to a functional disorder which is diagnosed primarily on the basis of clinical symptoms. It can be caused by various congenital or acquired dysfunctions or diseases – mostly by high blood pressure and coronary heart disease, but also by cardiomyopathy and valvular disease [7]. Patients with heart failure have a poor prognosis due to the frequently irreversible myocardial loss: mortality in New York Heart Association (NYHA) classes III and IV heart failure cases is 35% after 1 year and 80% after 3 years [27].

Depending on etiological factors and severity of the disease, treatment ranges from causal therapy and symptomatic procedures to heart surgery (coronary bypass, valvular surgery, heart transplantation). Symptomatic therapy can be divided roughly into medication that reduces the cardiac load (diuretics and vasodilators), cardiostimulants (digitalis, sympathomimetics), and intermediate preparations, whose use is still, however, of an experimental nature [7].

Clinically, heart failure is not a homogeneous disease entity, and the effects of the disease on the well-being and quality of life of patients with heart failure differ accordingly. For these patients, the disease manifests itself mainly as a reduction in and limitation of their physical performance due to inadequate performance of the cardiac pump. Dyspnea, fatigue, and general weakness develop progressively. Even simple daily activities become laborious with time, requiring planning and will power, and reach the point that even resting is possible only when seated and even then is accompanied by shortness of breath. Heart failure typically progresses in episodic fashion. Decompensa-

---

* In collaboration with M. Heinisch and M. Ludwig.

tion occurs repeatedly, with subsequent hospitalization. In addition to this massive limitation of physical activity, the patient suffers psychological stress due to the state of his health and the life-threatening nature of his disease, particularly when it takes a progressive course, as in other cardiovascular diseases [4, 24]. Furthermore, patients with severe heart failure are limited in their ability to function in everyday life (e.g., self care, work) and can experience changes in their relationships with other people (e.g., partner, family life, friends).

The way the patient experiences such limitations is not only dependent on the physical condition at that moment – interestingly enough, there is only minimal correlation in this regard between the patient's self-rating and the doctor's external assessment [57] – but also on the way the individual copes with the disease (i.e., on perception of, assessment of, and reaction to the disease; [25]). This, in turn, is also dependent on the feeling of being socially integrated (social support) and on socioeconomic circumstances [47]. However, not only the primary disease but also its treatment considerably affects the quality of life of patients with heart failure: for example, the use of medication in mild heart failure has different effects from those following heart transplantation in severe heart failure.

In the light of the heterogeneity of etiological factors, types of disease and possible types of therapy, the quality of life of patients with heart failure appears difficult to describe and even more difficult to measure. The problem increases when difficulties of definition, individual differences in the meaning of quality of life, and changes in the evaluative criteria for quality of life over time are considered [19]. Despite these difficulties, it is possible to measure the quality of life of patients with heart failure. Since medicine has increasingly begun to take quality of life into account, a consensus has been established over the use of the term. Although quality of life cannot be directly observed, it can be assessed on the basis of four components:

1. Physical condition
2. Psychological state
3. Social relationships
4. Ability to function in various spheres of life

There is also a unanimous opinion that quality of life should be assessed by the patient (self-rating), which does not exclude assessment by doctors, the medical team or relatives (external rating) [15, 28, 40, 42]. However, it is less clear how closely a means of measurement should be related to disease and therapy, whether disease-specific or nonspecific procedures should be used, and how extensively the assessment of quality of life should be pursued, i.e., whether global assessments or a differential evaluation of the components should be applied [26, 30]. The requirements of an instrument for measuring the quality of life of patients with heart failure do not depend, however, on the conceptual approach alone, but also on the aims of the measurement.

Aims

The measurement of quality of life can be pursued on the basis of descriptive, predictive, and evaluative approaches [37].

In the descriptive approach, description of the quality of life of a patient group compared to a control group is of interest (e.g., screening for the prevalence of feelings of impaired well-being in patients with heart failure and healthy people). A predictive approach would be to investigate the extent to which the quality of life of patients with heart failure suggests that a particular form of treatment is indicated (e.g., the use of quality-of-life-related prognostic factors in deciding for or against surgical intervention). An evaluative aim would be the use of quality of life measurement as an assessment criterion for therapeutic results (e.g., comparison of alternative treatments with regard to their effects on quality of life).

The demands made of a measuring instrument differ according to the approach to the problem: if the aim is descriptive, it should be possible to make generalizations about different groups, if the approach is predictive, it should differentiate clearly between different subgroups of heart failure patients, and, for evaluative purposes, it should react as sensitively as possible to therapy-induced differences in quality of life.

Assuming that measurement of quality of life in the field of heart failure is mainly concerned with evaluative approaches, i.e., those related to therapeutic research, the following aims of research into quality of life can be identified:

1. Documentation of the quality of life of patients with heart failure after treatment. Such cross-sectional studies relate to the patient's state after intervention. Here, the aim is to be able to make comparisons with the effects of alternative therapies from the point of view of health care economics, or to recognize the health care needs required by treated patients on the basis of psychosocial considerations. Investigations into the quality of life of patients after heart transplantation are an example of this [17].
2. Comparison of the quality of life of patients with heart failure before and after intervention. In such longitudinal studies, one or more groups of patients are observed over an extended period in order to obtain information on the effects of the intervention on quality of life. Investigations of quality of life before and after bypass or valvular surgery are an example of this (Bullinger and Naber, this volume).
3. Prospective, controlled comparison of the quality of life of patients with heart failure who were, by chance, given different alternative forms of treatment. This involves randomized clinical studies in which, from before the beginning of treatment to after its conclusion, the quality of life of all patient groups is assessed using the same method in several follow-up examinations. Investigations involving the controlled comparison of different approaches to medical treatment, e.g., of the effects of vasodilators, are an example of this [21].

For use in clinical therapeutic research, the procedures for measuring quality of life must fulfil quality criteria with regard to content, method, and practical application [1, 14, 35, 43, 55, 58].

Quality criteria for content relate to the adaptation of the measuring instrument to the clinical problem. The measuring instrument should cover those components of quality of life that are hypothetically influenced by the disease and therapy (e.g., sexuality). It should be suitable for repeated measurements and, at the same time, cover clinically meaningful periods in the assessment of quality of life (e.g., days, weeks). Furthermore, it should also be suited to the patient group (age, education, and life circumstances). Methodological criteria relate to the psychometric quality of the measuring instrument, i.e., its degree of standardization, its objectivity, its reliability, its validity, and especially its sensitivity to therapy-induced changes (responsiveness) [31]. The testing of these quality criteria is part of the development of a measuring instrument and is a protracted iterative process which requires the repeated testing of different forms of the instrument on different groups, with subsequent evaluation of the above-mentioned indices according to test theoretical principles. Practical criteria relate to applicability of the instrument in the study, i.e., the length and amount of time required to answer the questions, how easy the questions are to understand, their wording, and the number of response categories. They thus concern not only the patients' and doctors' acceptance of the questions, but also the possibility of constantly monitoring data input and of preventing information being omitted.

The subject and aims of research thus considerably influence the demands placed upon a measuring instrument. In the case of existing methods, this means that they must initially be tested for specific research questions in patients with heart failure.

Instruments for Measuring Quality of Life

Although research into quality of life is a new field, a large number of methods are already available for assessment by the patient or by medical personnel [8, 38, 51, 54]. These methods originate from the field of medicine, where they were developed for diagnosis and therapeutic research (e.g., in oncology or psychiatry), as well as from epidemiology and research into the health system, where they function as health indicators. Sociology has also examined quality of life, e.g., in the context of investigations on contentment with life, as has psychology, in particular clinical and medical psychology. Despite this, very few of the methods already developed are suitable for use in studies evaluating the quality of life [14]:

1. Most of the methods are of Anglo-American origin (particularly health indicators). They have rarely been translated and even more rarely tested psychometrically on groups of German patients.
2. Many methods, including German ones, are not suitable for the target population of patients with physical disease.

3. The components measured do not necessarily correspond to those areas that are important for quality of life.

Taking these limitations into consideration, those instruments whose use seems appropriate in studies on heart failure will be illustrated below. According to a taxonomy of quality of life measurements by Guyatt et al. [32], these consist largely of generic instruments, i.e. those which measure the different components of quality of life comprehensively and independently of the current state of the health of the individual, together with specific instruments which are orientated to the disease-, treatment-, or area-specific aspects of quality of life. The generic instruments relate to the integration of different components into a profile or index and to global assessments of quality of life. Specific instruments are mostly concerned with individual components or several areas specifically related to the disease and its treatment.

Studies providing an overview of research into quality of life in German-speaking patients with regard to generic and specific methods in disease in cardiology, oncology, and surgery are now available [10–12, 29, 33].

Generic Methods

The *Sickness Impact Profile* (SIP; [5]) measures functional and psychosocial aspects of the effect of illness on the patient in 12 categories using 136 behavior-related items. The SIP has been examined psychometrically in many studies and in different diseases with excellent results and, since its construction, has become the standard instrument of evaluative research. The responsiveness of the SIP, however, is still subject of discussion. There is a German version in existence, but it has not yet been subject to adequate psychometric testing (Potthoff, personal comment).

The *Nottingham Health Profile* (NHP; [34]) globally assesses self-rated state of health, in 38 questions and differentiates six different components, including limitation due to the state of health, globally and in six specific areas of life. There is a German version in existence which is currently being investigated psychometrically by our research group.

The *Psychological General Well-Being Index* (PGWB; [23]) assesses the subscales anxiety, depression, vitality, well-being, self-control, and health in 22 items. A German version was investigated, as a theoretical test, in 180 students [16]. Data are available from a total of over 200 healthy subjects and 300 cardiovascular patients (who had hypertension or coronary disease or who had undergone heart transplant, bypass, or valvular surgery). Analysis to date has shown that, instead of the six dimensions originally described, the German version has three dimensions: a stress factor, a well-being factor, and a health factor [29].

The *Affect Balance Scale* (ABS; [9]) is a short ten-item scale which translates the balance of positive and negative moods into a scale value. A German version was used by Badura and Waltz [4] in patients with myocardial infarction. The scale is easy to answer and evaluate and shows a high correlation with ratings of satisfaction with life.

The *Quality of Life Index* (QLI; [49]) is a short, easy to use instrument for external assessment (a self-assessment version is available) which considers five components of quality of life in three assessment categories – activity, everyday life, health, support, and prospects for the future – and summarizes them in an unweighted total score. Agreement between different assessors (interrating reliability) and correlation with other quality of life scales (convergent validity) are satisfactory. The strength of this instrument lies in the economical assessment (3 min to be filled in) of multiple aspects of quality of life for different patient groups; the weakness lies in the margin of interpretation and rough formulation of the categories. A German version is available from a gastric carcinoma study [45].

In the 11-stage *Anamnestic Comparative Self Assessment Scale* (ACSA; [6]) the patients assess their present quality of life in relation to the "best" (+5) and "worst" (−5) times of their life. Although the scale has scarcely been investigated psychometrically, it is often used because of the individual assessment of the patient's quality of life using only one scale value.

The so-called *ladder scales* follow a similar principle. The patient crosses off the stage between 0 and 10 which corresponds to his present quality of life [3].

The scales described up to this point measure disease-related generic quality of life, both in the form of profiles (SIP, NHP) and in terms of indices (QLI). Instruments that tend to focus on the area of psychological well-being (PGWB, ABS) have been illustrated, as well as a global rating (ACSA) as an example of a large number of similar single question scales.

The specific instruments discussed below initially relate to individual components as well as to disease-specific approaches.

Specific Methods

Although various questionnaires are available for assessing physical symptoms in clinical psychology, they are mostly unsuitable for clinical studies (e.g., symptom list BL; [53]). They were conceived as indicators of general physical condition and contain general vegetative symptoms which are neither disease- nor therapy-specific. Since the physical symptoms should specifically correspond to the clinical situation, new scales need to be constructed for this purpose. Our research group uses a flexible concept here in which relevant physical symptoms (e.g., those caused by heart transplantation) are identified in cooperation with the doctors and patients. The intensity of these complaints is measured via a five-point scale from "not at all" to "very strong" and retrospectively assessed for the past week by the patient.

In the area of the psychological state, a series of methods from psychiatry and clinical psychology are available, the psychometric characteristics of which have been tested for German-speaking patients [20]. This mostly involves lists of adjectives (e.g., I feel: irritated, depressed, tense) which either measure global well-being (using a total score, e.g., scale of well-being, Bf-S; [53]) or specific dimensions of mood, such as depression or anxiety. Our research group tested a scale of well-being which is in widespread Anglo-American use (Profile of Mood States, POMS; [39]) on different groups of healthy subjects and patients with chronic illness and found that the 35-question version had good psychometric characteristics [16, 29].

In contrast to the sparse attempts at measuring social well-being, a large number of scales are available from Anglo-American research in the field of geriatrics and rehabilitative medicine which assess the patient's functional competence, either by self-rating or by external assessment. These scales contain questions on the patient's self-care (eating, shopping, hygiene), on mobility (moving around inside and outside the house), and on other activities of daily life (e.g., tasks at home and at work; [38]).

An example of *disease-specific* procedures is provided by the Linear Analog Self-Assessment (LASA) scales which are self-assessment visual analog scales (VAS) consisting of lines 100 mm long with labelled poles (e.g., not at all versus very strong). Up to now, they have mainly been used in oncology. The best-investigated LASA scale originates from an investigation on the quality of life of 231 patients with breast cancer during treatment [46]. The authors reformulated the subscales of the Sickness Impact Profile (SIP) into LASA scales and based new items on clinical problems or adverse effects of the therapy. The 31-item instrument was tested for acceptability and psychometric criteria. The results showed that (with a modest time of 4 min required for completion) the modified SIP items were highly reliable. The disease-specific items were slightly less reliable. The following subclasses emerged for factor analysis: functional competence, general symptoms, psychological state, general and specific chemotherapeutic effects. Validity, determined by correlation with other instruments, was high. The instrument made it possible to detect differences between groups (discriminative validity) and therapy-related changes which were in agreement with clinical data (responsiveness).

The battery of questions which Croog et al. [22] used in their study should also be viewed as a disease-specific approach. This involved subscales from standardized questionnaires which were integrated into a comprehensive list of questions for hypertensive patients. This battery of questions was shortened by Siegrist and Junge [48] and investigated psychometrically in hypertensive patients in Germany.

In several German-language publications methods have been described specifically for use in patients suffering from hypertension and coronary heart disease [4, 24, 41, 44].

Alternative Approaches

Discussion of the use of disease-specific versus generic scales can be avoided by combining the two approaches, through so-called module models. Based on a module model of quality of life, which takes into account both generic and specific components, our research group has developed a four-page quality of life questionnaire which deals with physical complaints, psychological state, and functional competence and social relationships as well as assessing contentment and the importance of specific areas of life. This method is currently being applied in various clinical studies and investigated for psychometric properties [12, 14].

A module model also forms the basis of the questionnaire used by the quality of life working group of the European Organization for Research and Treatment of Cancer (EORTC) [2]. The questionnaire consists of a core instrument and an additional lung-cancer-specific scale. In the core instrument, the following components of quality of life (modules), which are generally relevant to malignant disease, are measured using 36 questions: functional condition, disease and therapy-related symptoms, psychological stress, social interaction, financial/economic problems, and general assessment of health and quality of life. The additional scale assesses symptoms and adverse effects of lung cancer treatment using 13 items. After translation into 11 languages (including German), the instrument is currently being tested for reliability, validity, and sensitivity in an international study on over 700 patients; in parallel with this, work is being done on additional scales for different diseases.

In the context of a decision-orientated approach to measuring quality of life, various methods have been developed (Standard Gamble, Time Trade Off; [42]). The Quality of Well-Being Scale (QWB; [36]) represents a combination of these health-economic and the psychometrically oriented methods. The QWB constitutes a new approach to measuring quality of life in so far as different areas of quality of life are assessed by interview, then summarized into a weighted index value for the biopsychosocial state of health. This index, which has now become the standard instrument for health economics, but is not yet available in German, is particularly suitable for cost-benefit analyses, for comparison between therapies, and for comparison across a range of diseases. An analysis of the extent to which this instrument can be used for German-speaking patients would be useful and is necessary.

Use in Clinical Studies

As it has been shown by the overview of instruments for measuring quality of life, there are a number of methods which could, in principle, be used in clinical studies. However, there are problems with regard to the selection of methods and their use in studies. In cardiology, the use of generic procedures, i.e., of health indicators in particular, has been primarily favored [50, 56]. While such methods are suitable for asymptomatic diseases such as hypertension (because the patients feel relatively "healthy"), their use in symptomatic populations (for whom, for example, heart surgery is indicated) appears very problematical [13]. In surgery, in particular, the use of methods specifically related to illness and its treatment is therefore strongly favored [58]. It is difficult to select an appropriate method because the available instruments are frequently not adapted to the clinical problem. A not insignificant number of researchers, therefore, develop their own instruments. This often takes place ad hoc without pilot testing or psychometric investigation, which is problematical because this, if anything, contributes to scientific confusion, rather than increased knowledge. The development of new methods is only to be recommended if this can be done with proper thoroughness and appropriate quality criteria with regard to content, method, and practical application [15]. If this is not the case, depending on the specific clinical approach, methods already in existence which fulfil such criteria should be chosen (e.g., German scales of well-being, translated and tested Anglo-American scales such as POMS, widespread international instruments such as the Spitzer QLIndex). When, in fact, such "standard" instruments are used with an adequate number of over 100 patients, a self-developed scale can be applied in parallel as well and its use explored in the study.

Even after careful deliberation of why and when to use which instrument, difficulties in organizing a study are common [14, 52]. Measurement of quality of life, like other measurements in the context of clinical studies, can fail as a result of relatively banal matters (e.g., forms for patients in investigations on the course of disease were mislaid, the doctor forgets to question patients before treatment, there is insufficient space and personnel for conducting the investigation). Interestingly enough, these factors are far more important than a refusal to take part or unsuitability of the patient for quality of life assessments, as is often wrongly assumed.

As it is not yet common practice in this country to question patients systematically and repeatedly about their quality of life in a clinical context – above all in the form of a research project – such problems are not surprising. However, they deserve early consideration during the planning phase of a research project: the measurement of quality of life requires the consent, motivation, and time of all concerned. Particularly when it is the main target criterion of clinical studies, no less effort should be put into its measurement than is the usual standard for somatic data.

In view of the fact that only one study to date has explicitly dealt with heart failure and quality of life [57], further steps are necessary for the adequate use of quality of life assessments in patients with heart failure:

1. Identification of specific clinical questions and the resulting demands made of instruments for measuring quality of life.
2. Critical evaluation of available methods, selection of suitable approaches, and clarification of the necessity for new developments.
3. Quality testing of the methods in German-speaking patients with heart failure.
4. Agreement about indications for quality of life assessments and about the use of scales appropriate to the objectives of specific types of study in the field of heart failure.
5. Development of guidelines regarding planning, execution, and statistical evaluation of the data.

Not least, the question should be considered how the information obtained on patients' quality of life should be used, and by whom. Although it is encouraging that medicine has become more sensitive to psychosocial aspects of disease and therapy, ethical implications should not be ignored. These arise because, however much it might be aimed at, research into quality of life is not free of value judgements: so who is to decide whether the poor quality of life experienced by study patients justifies giving up a certain treatment strategy? How are exclusions from therapy to be justified on the basis of quality of life related prognostic factors? May the application of a costly treatment be made dependent on the patient's expected fitness for work on health-economic grounds?

If research into quality of life is to contribute towards a more "humane" medicine, the interpretation of the concept by all involved should, at least, be discussed. Information on quality of life should not remain in the hands of those who are in control of health politics.

References

1. Aaronson NK (1989) Quality of life assessment in clinical trials: methodologic issues. Controlled Clin Trials 10: 1955–2085
2. Aaronson N, Bullinger M, Ahmedzai S (1988) A modular approach to quality of life assessment in cancer clinical trials. Recent results in cancer research 111: 231–248
3. Andrews FM, Candrall R (1976) The validity of measures of self-reported well-being. Soc Ind Res 3: 1–19
4. Badura P, Kaufhold G, Lehmann H, Pfaff H, Schott I, Waltz M (1987) Leben mit dem Herzinfarkt: Eine sozialepidemiologische Studie. Springer, Berlin Heidelberg New York
5. Bergner M, Bobbitt RA, Carter WB, Gilson BS (1981) The sickness impact profile: development and final revision of a health status measure. Med Care 19: 787–805
6. Bernheim J, Avonts G, De Schampheleire D, Gelissen I, Sauer AM (1985) Anamnestic comparative self-assessment (ACSA) to correlate medical and psychosocial variables with subjective quality of life of cancer patients. Proceedings of the American Society for Clinical Oncology 4: 248–257
7. Bertel O (1983) Neuere Aspekte der medikamentösen Behandlung von herzinsuffizienten Patienten. Schweiz Med Wochenschr 113: 739–745
8. Biefang S (1985) Messung psychosozialer Erfolgs- und Prädiktorvariablen. Habilitationsschrift, University of Ulm
9. Bradburn NM (1964) The structure of psychological well-being. Aldine, Chicago NORC 1969, Chicago
10. Bullinger M (1989) Forschungsinstrumente zur Erfassung der Lebensqualität bei Krebs. In: Verres R, Hasenbring M (eds) Psychosoziale Onkologie. Springer, Heidelberg New York (Jahrbuch der medizinischen Psychologie, vol 3)
11. Bullinger M (1991) Quality of life: definition, conceptualization and implications – a methodologist's view. Theoretical surgery (in press)
12. Bullinger M (1990) Quality of life research in West Germany: the Munich experience. In: Schumaker S, Furberg C, Czajkowski S, Schron E (eds) Quality of life and cardiovascular disease. Springer, New York
13. Bullinger M (1991) Lebensqualität unter antihypertensiver Therapie: konzeptuelle, methodische und praktische Aufgaben. In: Bullinger M, Ludwig M, Steinbüchel N (eds) Lebensqualität bei kardiovaskulären Erkrankungen. Grundlagen, Erhebungsmethoden und Ergebnisse. Hogrefe, Stuttgart pp 4–13

14. Bullinger M, Hasford J (1991) Evaluating quality of life measures for German clinical trials. Controlled Clin Trials (in press)
15. Bullinger M, Pöppel E (1988) Lebensqualität in der Medizin: Schlagwort oder Forschungsansatz? Dtsch Ärztebl 85: 679–680
16. Bullinger M, Heinisch M, Luwig M, Geier S (1990) Skalen zur Erfassung des Wohlbefindens: Psychometrische Analysen zum „Profile of Mood States" (POMS) und zum „Psychological General Well-Being Index" (PGWI). Z Diff Diagn Psychol 11/1: 53–61
17. Bullinger M, Angermann C, Kemkes M (1991) Lebensqualität nach Herztransplantation. In: Bullinger M, Ludwig M, Steinbüchel N (eds) Lebensqualität bei kardiovaskulären Erkrankungen. Grundlagen, Erhebungsmethoden und Ergebnisse. Hogrefe, Stuttgart, pp 153–163
18. Bullinger M, Ludwig M, Steinbüchel N (1991) Lebensqualität bei kardiovaskulären Erkrankungen. Grundlagen, Erhebungsmethoden und Ergebnisse. Hogrefe, Stuttgart
19. Calman KC (1987) Definitions and dimensions of quality of life. In: Aaronson N, Beckmann J (eds) Quality of life of cancer patients. Raven, New York
20. CIPS (1983) Internationale Skalen für Psychiatrie. Beltz, Weinheim
21. Cohn JN, Archibald DG, Ziesche S (1986) Effect of vasodilator therapy on mortality in chronic congestive heart failure. Results of a Veterans Administration cooperative study. N Engl J Med 314: 1547–1552
22. Croog SH, Levine S, Testa MA, Brown B, Bullpitt CJ, Jenkins CD, Klerman G, Williams GH (1986) The effects of antihypertensive therapy on the quality of life. N Engl J Med 314: 1657–1664
23. DuPuy HJ (1984) The psychological general well-being (PGWB) index. In: Wenger NK, Mattson ME,Furberg CD, Ellinson J (eds) Assessment of quality of life in clinical trials of cardiovascular therapies. LeJacq, New York
24. Fahrenberg J, Myrtek M, Wilk D, Kreutel K (1986) Multimodale Erfassung der Lebenszufriedenheit: Eine Untersuchung an Herz-Kreislaufpatienten. Psychother Psychosom Med Psychol 36: 347–354
25. Feifel H, Strack S, Nagy VT (1987) Coping strategies and associated features of medically ill patients. Psychosom Med 49: 616–625
26. Feinstein AR (1986) Clinimetric perspectives. J Chronic Dis 40: 635–640
27. Franciosa JA, Willen N, Ziesche S, Cohn JN (1983) Survival in men with severe chronic left ventricular failure due to either coronary heart disease or idiopathic dilated cardiomyopathy. Am J Cardiol 51: 831–836
28. Furberg CD (1985) Assessment of quality of life. In: Friedman L, Furberg C, Demets D (eds) Fundamentals of clinical trials. 2nd Ed. PSG, Littletown
29. Gross M (1991) Psychometrische Eigenschaften zweier Fragebögen zur Erfassung der psychischen Dimension der Lebensqualität. In: Bullinger M, Ludwig M, Steinbüchel N (eds) Lebensqualität bei kardiovaskulären Erkrankungen. Grundlagen, Erhebungsmethoden und Ergebnisse. Hogrefe, Stuttgart, pp 91–109
30. Guyatt GH, Bombardier C, Tugwell PX (1986) Measuring disease-specific quality of life in clinical trials. Can Med Assoc J 134: 889–895
31. Guyatt GH, Walter S, Norman G (1987) Measuring change over time – assessing the usefulness of evaluative instruments. J Chronic Dis 40: 171–178
32. Guyatt GH, Veldhuyzen Van Zanten SJO, Feeny DH, Patrick DL (1989) Measuring quality of life in clinical trials: a taxonomy and review. Can Med Assoc J 140: 1441–1448
33. Heinisch M (1991) Meßinstrumente zur Erfassung der Lebensqualität am Beispiel der Hypertonie – eine kritische Übersicht. In: Bullinger M, Ludwig M, Steinbüchel N (eds) Lebensqualität bei kardiovaskulären Erkrankungen. Grundlagen, Erhebungsmethoden und Ergebnisse. Hogrefe, Stuttgart pp 54–63
34. Hunt S (1984) The Nottingham health profile. In: Wenger NK, Mattson ME, Furberg CD, Ellinson J (eds) Assessment of quality of life in clinical trials of cardiovascular therapies. LeJacq, New York
35. Hunt S (1988) Measuring health in clinical care and clinical trials. In: Teeling-Smith G (ed) Measuring health – a practical approach. Wiley, Chichester

36. Kaplan RM, Bush JW, Berry CC (1976) Health status: types of validity and the index of well-being. Health Serv, Res, 11: 478–507
37. Kirshner B, Guyatt G (1985) A methodological framework for assessing health indices. J Chronic Dis 38: 27–36
38. McDowell I, Newell C (1987) Measuring health: a guide to rating scales and questionnaires. Oxford University Press, New York
39. McNair D, Lorr M, Droppleman LF (1971) Manual for the profile of mood states. Educational and Industrial Testing Service, San Diego
40. Najman JM, Levine S (1981) Evaluating the impact of medical care and technology on the quality of life: a review and critique. Soc Sci Med 15F: 107–115
41. Otten H, Schmeider R, Rüddel H (1987) Disparate effects of initial antihypertensive therapy on well-being. J Hypertens [Suppl] 5 (1): 537–540
42. Patrick DL, Erickson P (1988) Assessing health-related quality of life for clinical decision-making. In: Walker SR, Rosner RM (1988) Quality of life assessment and application. MTP, Lancaster
43. Pocock SJ, Geller NL, Tsiatis AA (1987) The analysis of multiple endpoints in clinical trials. Biometrics 43: 487–498
44. Potthoff P (1988) Krankheitsbedingte Lebensbeeinträchtigung in der Münchner Wohnbevölkerung. Öff. Gesundheitsw 50 (6) 329–333
45. Rohde H, Rau E, Gebbensleben B (1984) Ergebnisse der Bestimmung des Lebensqualitätsindex nach Spitzer in der multizentrischen Magenkarzinom TNM-Studie. In: Rohde H, Troidl H (eds) Das Magenkarzinom. Thieme, Stuttgart
46. Selby PJ, Chapman JA, Etazadi-Amoli J, Dalley D, Boyd NF (1984) The development of a method for assessing the quality of life of cancer patients. Br J Cancer 50: 13–22
47. Siegrist J (1987) Impaired quality of life as a risk factor in cardiovascular disease. J Chronic Dis 40: 571–578
48. Siegrist J, Junge A (1988) Effect of antihypertensive therapy on quality of life. Medicographia 10: 38–40
49. Spitzer WO, Dobson AJ, Hall J, Chesterman E, Levi J, Shepherd R, Battista RN, Catchlove BR (1981) Measuring the quality of life of cancer patients. A concise QL-index for use by physicians. J Chronic Dis 34: 585–597
50. Storstein L (1987) How should changes in life-style be measured in cardiovascular disease? Am Heart J 114: 210–212
51. Teeling-Smith G (1988) Measuring health – a practical approach. Wiley, Chichester
52. Van Dam FSAM, Aaronson N (1987) Practical problems in conducting cancer psychological research. In: Aaronson NK et al. (eds) Quality of life of cancer patients. Raven, New York
53. Von Zerssen D (1976) Klinische Befindlichkeitsskalen Bf-S. Beltz, Weinheim
54. Walker SR, Rosner RM (1988) Quality of life: Assessment and Application. MTP, Lancaster
55. Ware JE (1987) Standards for validating health measures: definition and content. J Chronic Dis 40: 473–480
56. Wenger NK, Mattson ME, Furberg CD, Ellinson J (1984) Assessment of quality of life in clinical trials of cardiovascular therapies. LeJacq, New York
57. Wiklund I, Lindvall K, Swedberg K, Zupkis RV (1987) Self-assessment of quality of life in severe heart failure. An instrument for clinical use. Scand Psychol 28: 220–225
58. Wood-Dauphinee S, Troidl H (1986) Endpoints for clinical studies: conventional and innovative variables. In: Troidl H, Spitzer WO, McPeek B, Mulder DS, McKneally MF (1986) Principles and practice of research. Strategies for surgical investigators. Springer, Berlin Heidelberg New York
59. Wood-Dauphinee S, Troidl H (1989) Assessing quality of life in surgical studies. Theor Surg 4: 35–44

Quality of Life after Bypass and Valvular Surgery: Results of a Prospective Study

M. Bullinger and D. Naber

Quality of Life as a Criterion for Evaluation in Heart Surgery

In the light of the increasing availability and frequency of open-heart surgery, the question of how to evaluate its outcome arises. In heart surgery, this question is discussed with particular regard to bypass and valvular surgery. The numerous medical studies primarily target classical indicators of therapeutic success (peri- and postoperative mortality, morbidity, and complications) but also look at the incidence of psychiatric abnormalities [19, 21, 32, 46]. There have also been some studies on disturbances of neuropsychological function after heart surgery [1, 31, 48, 50]. A target criterion which is particularly well documented and important for cost-benefit estimations in health care is the return to work after surgery [39, 51, 53].

As a result of the rather unsatisfactory findings on postoperative employment status and the one-sided preference given to somatic indicators of therapeutic success, there have been an increasing number of recent studies which place emphasis on the quality of life of cardiac patients. Cross-sectional studies on the outcome of bypass grafts and valvular surgery have become available, but these were often conducted with a small number of cases [2, 5] or using a retrospective approach [29, 49]. There are, however, a number of methodologically acceptable prospective observation studies in groups of more than 100 patients which assess quality of life before and at one, at the least, point in time after surgery [14, 22, 25, 33, 35, 40–42].

Randomized clinical studies which compare two or more therapeutic alternatives and which also consider the patient's quality of life have been rare to date. One exception here is the Coronary Artery Surgery Study (CASS) [15], which prospectively investigated over 5 years the quality of life in 780 patients who, over a period exceeding 5 years, were treated with either drugs or surgery. In spite of its adequate design, however, this study is an example of the conceptual and methodological problems involved in the term "quality of life". Quality of life was mainly evaluated by means of somatic complaints (e.g., chest pain), activities of daily life (e.g., self-sufficiency), physical exercise (e.g., mobility), and employment status. These variables, however, were assessed externally by the doctor using a very simple classification scheme. Although physical symptoms and functional capacity are considered to be components of the quality of day-to-day life, the failure of the CASS study to consider the psychological and social indicators of quality of life, the absence of self-rating by the patients, and recourse to an ad hoc means of assessment rather than the application of standardized and validated procedures are apparent.

In accordance with the recommendations of review articles on this subject, the quality of life of cardiac patients should be assessed in close relationship with the clinical problem, largely from the point of view of the patient and with regard to physical, psychological, social, and functional aspects. Moreover, it is also desirable to include external evaluations by doctors and health care professionals as well as by relatives, and to include neuropsychological methods for measuring function (e.g., mental performance capacity). Last but not least, the psychosocial factors which influence quality of life deserve more attention. In assessing both quality of life and the factors influencing it, measuring instruments which are standardized, psychometrically well-tested, and widely used should be employed, so that results can be compared beyond the scope of individual studies [9, 10, 57]. Studies which follow such methodological recommendations are, however, the exception. Moreover, they differ considerably, not only in the instruments used but also in the questions asked, patient populations, and study design.

Based on these considerations, our own study, which will be discussed below, was directed towards obtaining a comprehensive picture of the patients' condition before and after bypass graft or valvular surgery. Great importance was attached to the interdisciplinary nature of the approach (cardiology, clinical psychology, psychiatry, neuropsychology), the quality of life components (physical, psychological, social, functional, and mental), the responders (patients, relatives, doctors), the degree of standardization of the methods (psychometrically tested questionnaires, well-researched systems of clinical documentation, classical neuropsychological tests), the inclusion of basic psychosocial data and clinical information, and the use of a longitudinal observation design (before surgery and up to one year after surgery).

In the literature on quality of life, emphasis to date has been placed on controversial questions of improvement [58] or worsening [30] of psychological well-being and the effect of preoperative management of the disease on the physical and psychosocial course following operation [8, 16, 24, 26]. The questions on self-rated quality of life up to three months after surgery which will be discussed below by way of example covered:

1. Changes in the quality of life before and after surgery.
2. Relationships between postoperative quality of life and cognitive illness processing.
3. Differences in postoperative quality of life related to clinical and sociodemographic characteristics.

Methodology

Design

In cooperation with the German Heart Center, Munich, a prospective observation study was conducted. This is still in progress and documents the quality of life of patients undergoing heart surgery over a period from 2 weeks before surgery to 12 months thereafter. Inclusion criteria were age (18–65 years), either aortic valve replacement or coronary artery bypass graft as indications, and written consent to participate in the quality of life study. Exclusion criteria included emergency operations, private health insurance, and insufficient knowledge of the German language. The recruiting period for the study was 6 months.

Variables

In the course of the study, data on the following groups of variables were assessed: (1) basic psychosocial data; (2) quality of life from the patient's point of view; (3) quality of life from the point of view of the patient's relatives; (4) external psychiatric evaluation of psychopathological symptoms; (5) tests of neuropsychological function; and (6) clinical data (see Table 1).

Basic psychosocial data consisted of personality traits as measured by the Freiburg Personality Inventory (FPI; [23]) and *anxiety* using the State Trait Anxiety Inventory (STAI; [52]). In a questionnaire translated from American English, the sense of internal (self) versus external (others, fate) *control* in relation to the course of the disease was as assessed (SKF, Selbstkontrollfragebogen; Self-control questionnaire; Klauer 1987, unpublished work). Moreover, six strategies for *coping with the disease* were surveyed using a questionnaire (FEKB, Fragebogen zur Erfassung der Krankheitsbewältigung; Questionnaire for the assessment of coping with illness [45]): avoidance and wishful thinking, depressive response, rational problem solving, religious response and search for meaning, and *Carpe diem* and enjoyment of life, and social withdrawal. The extent of social support was assessed using a list of questions translated from American English with the total score in this providing a measure of the feeling of social integration [18]. In addition, a number of open and closed questions on the current perception of disease and surgery were asked, aided in part by using 100-mm-long visual analog scales (VAS). The *patient's living situation* was assessed using a questionnaire on sociodemographic data and information on the social network. This information was collected by mail 14 days before surgery; single questions (e.g., occupation) were reassessed in the course of the study. Some questionnaires (FPI, FEKB) were presented on several occasions.

Quality of life from the patient's point of view was measured using standardized measures as well as newly and self-designed questionnaires. In the *psychological realm,* the *Psychological General Well-being Index* (PGWB; [20]), translated into German and psychometrically tested by us, was applied [11].

Table 1. Variables analyzed and times of measurement

Variables	Preoperative		Postoperative					
	Week −2 (at home)	Day −1 (in hospital)	Days (in hospital)			Months (at home)		
	1	2	3	4	5	6	7	8
Basic psychosocial data								
Personality (FPI)	×						×	
Anxiety (STAI)	×							
Social support (SS)	×							×
Self-control (SK)	×							×
Coping (FEKB)	×							×
Perception of surgery (VAS)	×					×	×	×
Life situation	×					×	×	×
Quality of life (self-rating)								
Psychological well-being (PGWB)	×					×	×	×
Symptoms	×					×	×	×
Everyday life	×					×	×	×
Feelings (POMS)	×					×	×	×
Assessments	×					×	×	×
Global rating	×					×	×	×
Quality of life (relatives)								
Feelings (POMS)	×					×	×	×
Perception of surgery (VAS)	×					×	×	×
Personal stress	×					×	×	×
Psychiatric symptoms								
AMP scores		×	×	×	×			
Feelings (POMS)		×	×	×	×			
Neuropsychological functions								
Attentiveness test D2		×			×			
Repetition of numbers test (HAWIE)		×			×			
Number symbols test (HAWIE)		×			×			
Color-word interference test		×			×			
Trail making test		×			×			
Benton test		×			×			
Word list		×			×			
Clinical data								
Preoperative data (including diagnosis, waiting time)	×	×						
Perioperative data (including duration, complications)			×					
Postoperative data (including medication, NYHA status)			×	×	×	(×	×	×)

By posing 22 questions, it assesses the psychological state during the past month on the basis of the dimensions anxiety, depression, well-being, self-control, state of health and vitality. A psychometrically-tested German version of *Profile of Mood States* (POMS; [43]) with 35 questions was used to assess mood scales for depression, fatigue, vigor and irritability. To assess the *physical component* of quality of life, a list of symptoms consisting of 12 questions was developed in cooperation with heart surgeons especially for patients who had undergone heart surgery. The total score for questions relating to symptoms reflects the stress caused by the symptoms of the primary disease, the medication, and after-effects of surgery. In the area of *functional capacity,* we used both the American *Beth-Israel Scale* for assessing physical functions as well as our own *questionnaire on daily life,* which uses 42 questions to assess activity in various areas of daily life, particularly in social areas. A low total score reflects severe impairment in daily life. With regard to the *assessments* of quality of life, *satisfaction* with and importance of 15 areas of life were analyzed. Total scores for the individual scales and their product were calculated. In addition, the global rating of quality of life was indicated using a five-point scale ranging from 1 (excellent) to 5 (miserable). The evaluations of quality of life from the *patient's point of view* were conducted by mail 2 weeks before and 3, 6, and 12 months after surgery. At the same points in time, the *patient's relatives* also received a questionnaire to analyze their appraisal of the patient's psychological state (POMS), his general state (VAS scales) and the relatives' own stress induced by the patient's state of health.

Psychiatric symptoms were documented by the physician with the *AMP system* [4]. Using this standard psychiatric instrument, the degree of severity of various psychopathological symptoms (e.g., paranoid-hallucinatory symptoms, depression, mania) can be assessed by a trained observer [27]. The AMP was documented in hospital the day before surgery and also 4, 7, and 10 days afterwards. At the same time, the patients filled out the *POMS scale.*

Neuropsychological functions were measured at the German Heart Center the day before and 10 days after surgery using several standardized neuropsychological tests to measure attention, concentration, learning, memory, and psychometric coordination. The tests consisted of the *Attention Stress Test D2* [7], which assesses concentration and speed of reaction, two tasks taken from the *Hamburg-Wechsler intelligence test* (HAWIE), namely, repetition of numbers and number symbols tests [56], which measure various aspects of memory, ability to concentrate, and speed of perception (coupled with visuomotor competence in the number symbols test). The *Color-Word Interference Test* [55], which measures cognitive functions in perception and naming of objects and reaction speed, was also used. The *Trail Making Test* (Part B; [47]) assesses concentration and psychomotor response, the *Benton Test* [6] assesses powers of visuospatial imagination and observation, and a *word list* developed at the Institute of Medical Psychology by Dr. J. Ilmberger assesses learning and memory.

Clinical data were available from the internal records of the German Heart Center and these consisted of *preoperative* information (case history, diagnosis, indication, waiting time), *perioperative* features (diagnosis, type of surgery, anesthesia, duration of extracorporeal circulation, complications) as well as data on the inpatient *postoperative* course (parameters of cardiovascular function, complications, and medication).

Assessment

During data input samples were monitored and tested for plausibility. Missing values (maximum of 20% per scale) were handled by mean substitution. The statistical assessment was directed towards descriptive, exploratory, and deductive analyses and was performed using the SPSS-PC program. Parametric methods for testing differences between the groups and/or over the period of time (*t* test, variance analyses) were used only if the specific requirements were fulfilled (Kolmogoroff-Smirnow test). Otherwise, nonparametric tests were used (Wilcoxon, Mann-Whitney). The Pearson correlation coefficient was used as a measure of the relationship between variables. The following questions were discussed:

1. Do the pre- and postoperative quality of life indicators differ?
2. What is postoperative quality of life dependent upon?
3. Do the defined patient subgroups differ in postoperative quality of life?

The following results pertain to the patients' self-rated quality of life at two assessment times (before and after surgery) in relation to psychological, clinical, and sociodemographic baseline data.

Results

Patient Sampling

During the recruitment phase 172 patients were registered for surgery at the German Heart Center. According to the hospital's information, they fulfilled the inclusion criteria and were informed about the study in a letter and also by a follow-up telephone call. Six patients in this group were excluded because they did not fulfil the inclusion criteria, five patients withdrew from surgery, and for nine patients surgery was postponed by more than 6 months for organizational or health reasons. Six patients refused to participate, another 10 were willing to be available for the study only during their stay in hospital. Thus, informed consent was given by 135 subjects, 20 patients did not return the questionnaire, and one questionnaire was not evaluable.

At the first time of measurement, 2 weeks before the operation, data on 114 patients (75 men, 39 women) were available. Their average age was 57.3 ± 10.8 years. Bypass surgery was planned in 55 patients, valvular surgery in 49, and both forms of surgery in 10. The majority of patients (75%) were married and living with their partner. Regarding education, 74 of the patients had completed lower compulsory schooling, 23 had attended a nonclassical secondary school ("Realschule"), and 15 had a higher education ("Abitur"). With regard to vocational education, completion of a vocational degree ($n = 48$) was most common, followed by qualifications from technical college ($n = 14$) and polytechnic or university degrees ($n = 13$). Only 22 patients had no school qualifications.

Before surgery, 20 patients had been in full-time employment (17.9%), 26 were on sick leave (22.8%), 28 had taken early retirement (25%), 23 had reached the age of retirement (20.5%) and 14 were doing housework. Their former or present position was predominantly in the civil service ($n = 49$; 43%); 29 patients were manual workers (25.4%), five were office staff, and five were self-employed (4.4%).

Differences in Quality of Life Indicators Before and After Surgery

The POMS scale of well-being (with its four subscales), the index of psychological well-being PGWB (with its six subscales), the total scores from the questionnaires on symptoms, daily life, functional capacity, and contentment, and the global rating of quality of life (five-point scale) were used as quality of life indicators.

In the global quality of life rating, an improvement was found in comparing the state before the operation with that after the operation. Before surgery, 38.7% considered their quality of life to be "quite good"; afterwards, 61.4% gave this rating. By analogy, the percentage of "average" ratings fell from 42.2% before to 23.8% after surgery (Fig. 1). In the Wilcoxon paired comparison test, the difference between the times of measurement proved to be significant ($Z = -2.98$, $n = 91$, $p = 0.0029$).

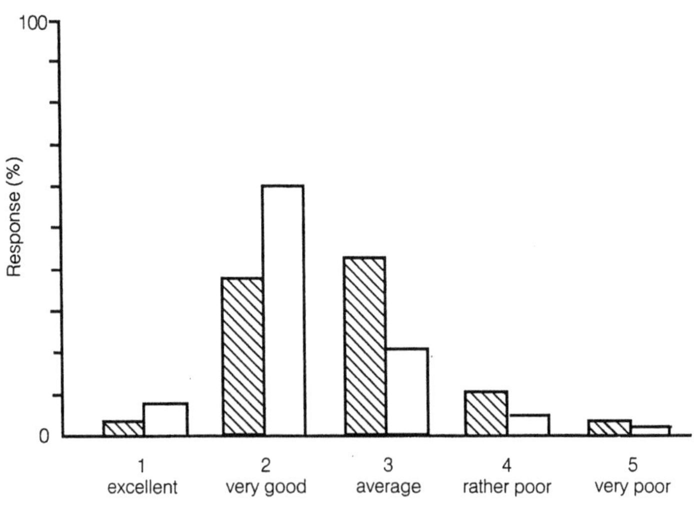

Fig. 1. Quality of life global rating (five-point scale)

Table 2. Quality of life indicators: differences over the period of time from 4 days before the operation to 3 months afterwards

Variable	Range[a]	Preoperative		Postoperative		Statistical test[b]	Degree(s) of freedom	Level of significance
		Mean	SD	Mean	SD			
Feelings (POMS)								
Depression	*0–56	23.3	(7.2)	20.1	(8.3)	Z = –4.41	86	0.0001
Fatigue	*0–28	13.3	(4.2)	11.4	(3.6)	t = 4.2	86	0.0001
Vigor	0–28*	13.4	(4.7)	15.05	(4.4)	t = –3.32	81	0.001
Irritability	*0–28	9.7	(2.9)	8.7	(2.8)	Z = –2.50	88	0.01
Well-being (PGWB)								
Anxiety	*0–25	15.9	(3.1)	16.9	(3.1)	Z = –2.2	89	0.028
Depression	*0–15	11.2	(2.9)	12.1	(2.8)	t = –3.45	93	0.001
Well-being	0–20*	10.2	(2.0)	10.8	(2.1)	Z = –2.8	90	0.0047
Self-control	0–15*	8.3	(1.2)	7.8	(1.3)	Z = –3.2	90	0.001
Health	0–15*	7.8	(1.6)	7.9	(1.7)	Z = –0.72	90	n.s.
Vitality	0–20*	9.4	(2.3)	10.6	(1.7)	Z = –413	91	0.001
Symptom score	*0–96	47.3	(11.8)	39.4	(10.4)	Z = –4.72	65	0.0001
Everyday life	0–168*	157.3	(23.6)	161.0	(27.1)	t = –0.43	67	n.s.
Functional capacity	*0–36	18.3	(5.6)	18.0	(6.9)	Z = –0.96	85	n.s.
Contentment (total)	0–80*	42.1	(10.7)	45.0	(10.7)	+ = 2.91	81	0.005
Quality of life (global)	*1– 5	2.7	(0.8)	2.3	(0.7)	Z = –2.98	91	0.009

a best result;
b $t = t$ test; Z = Wilcoxon paired comparison test.

Significant improvements of this type after surgery, when compared to the preoperative state, were also revealed by the psychological indicators of quality of life. The POMS subscales thus showed less depression, fatigue, and irritability as well as increased vigor after the operation.

In the PGWB, the subscales anxiety, depression, well-being, and vitality improved in a similar fashion. After their operation, the patients reported greater satisfaction with life and less stress from their symptoms. However, no significant changes were demonstrated in the PGWB subscale health, which measures worries about health, and there was a worsening in the subscale self-control, which reflects the feeling of mental integrity. There were no demonstrable differences in scales for daily life and for physical capabilities (Table 2).

The Relationship between Postoperative Quality of Life and Preoperative Variables

There is a clear and significant relationship between pre- and postoperative indicators of quality of life. The patient's state after surgery, however, is not only dependent on the respective preoperative status, but also on the amount of anxiety, social support, coping strategies, and feelings of control. Anxiety is

Table 3. Correlation of quality of life indicators

	Psychosocial Predictors							
	Pre-operative status (day – 14)	Anxiety (STAI)	Social support (SS)	Denial (FEKB)	Depressive response (FEKB)	Social with-drawal (FEKB)	Internal Control (SK)	External Control (SK)
Feelings (POMS)								
Depression	.58**	.57**	–	.32*	.42**	–.29*	–	–
Fatigue	.43**	.44**	–.29*	–	–	–	–	–
Vigor	.50**	–.43**	–	–.27*	–.32*	–.39*	.31*	–
Irritability	.46**	.58**	–	–	.26*	.27*	–	–
Well-being (PGWB)								
Anxiety	.42**		.30*	–	–	–	–	–
Depression	.57**		–	.36**	.46**	.31*		.36*
Well-being	.57**		.32*	–.30	–.41	–.31	.29*	–.27*
Self-control	–		–	–	–	–	–	–
Health	.56**		–	–	–	–.24		
Vitality	.27*		–	–	.29*	.29*		
Symptom score	.58**	.54**	–30*	.34*	.52**	.26*	–	–
Everyday life	.63**	–.52**	.33	–.41*	–.41**	–.46**		–.39*
Functional capacity	.57**	–.27*	–.28*	–	–	.26*	–	–
Contentment (total)	.58**	–.51**	–.28*	–	–.32*	.36*	–	–
Quality of life (global)	.33**	.39*	–.18	.26*	.33*	.35*	–.26*	–

* $p \leq 0.05$; ** $p \leq 0.01$

thus correlated with a deterioration in mental health (POMS, PGWB), greater stress from symptoms, impairment in daily life, limited functional capacity, and reduced contentment and quality of life. The coping strategies of denial, depressive response, and social withdrawal also appear detrimental to quality of life. On the other hand, evidence points to the protective function of social support as well as internal control, e.g., the conviction that one can influence one's own state of health (Table 3).

Group Differences in the Postoperative Quality of Life

In connection with the changes in quality of life due to the operation in the group as a whole, the interesting question arises as to whether specific subgroups of patients experience difference in the extent of the changes in their quality of life. To this end, the differences between post- and preoperative values for the individual quality of life indicators were calculated for each patient and tested for differences in relation to type of operation (bypass vr. valvular), postoperative psychopathological abnormalities (yes vr. no), sex (male vs. female), life situation (alone vs. with a partner), level of education (lower or secondary or high school) and employment (full-time, sick-leave, early retirement, or old-age pension).

The findings here indicated less satisfaction in the valvular surgery patients ($F=4.2$, $p=0.04$), less vitality in patients with psychiatric abnormalities ($F=2.41$, $p=0.05$), increased stress due to symptoms in women ($F=5.32$, $p=0.02$), and reduced self-control in men ($F=4.30$, $p=0.411$). Patients who lived alone considered that their quality of life had deteriorated ($F=5.45$, $p=0.02$). Patients with less education had increased scores for anxiety ($F=3.61$, $p=0.01$) and for depression in the PGWB ($F=3.20$, $p=0.27$). Patients who were not in employment reported increased limitation of their physical capabilities postoperatively ($F=2.80$, $p=0.03$).

Significance of the Postoperative Evaluation of Quality of Life

The values for the correlation between the five-point global rating of quality of life and the individual indicators 3 months after surgery are generally high ($r=0.40$–0.75). Based on the patients' own experiences, the overall assessment of quality of life appears to be particularly dependent upon the extent of freedom from depression, contentment with life, and ability to carry out the tasks of everyday life. It seems to depend less on the self-rated physical state.

Compared with data from our own studies [13], the overall rating of the quality of life from excellent (1) to miserable (5) of heart surgery patients scored postoperatively was somewhat higher (2.3) than the quality of life of patients operated on for renal cell carcinoma (2.4, $n=101$), but was lower than that of heart transplant patients (2.0, $n=41$), patients with coronary heart disease (1.8, $n=111$), and age-matched healthy subjects (1.2, $n=66$).

Discussion

The results of the study suggest that the self-rated quality of life of heart surgery patients is improved 3 months after treatment, when compared to 2 weeks before it, particularly in the psychological area. Similar positive postoperative assessments resulted from our own studies in heart transplant patients [13]. In recently published prospective studies of coronary bypass grafts, the majority of findings also reveal improved psychological well-being [22, 25, 35, 40, 41].

Despite the reduction in the burden of subjective symptoms, and contrary to the literature (e.g., [54]), however, our own study revealed no significant changes in relation to physical capabilities in daily life and the subjective state of health. The published investigations, however, are based on longer follow-up periods of at least 6 months to over 5 years. Clinical data which are not yet available should be included in order to determine whether our own results reflect the limitations in the state of health caused by the recent surgery, which may not differ from the disease-related limitations present before the operation.

The deterioration in self-control could reflect a loss of mental integrity in a subgroup of patients: during their hospitalization, over 35% of the patients had psychological abnormalities, which included acute depressive moods and reversible exogenous psychoses. The experience with one's own psychological state immediately after surgery could thus contribute to a negative assessment of power over one's own feelings. In an earlier investigation, not only was a comparable percentage of psychiatric abnormalities found, but there was also a connection with the patients' attempts at self-control [46].

In this light, the relationships between quality of life and cognitive illness processing before surgery and the postoperative state of the patients appear interesting. In keeping with the literature, our own results indicate a relationship, particularly between pre- and postoperative well-being [2, 29, 42]. In this context, connections between anxiety about the preoperative situation (with regard to the operation) and postoperative course are especially evident [16, 24].

In the 1970s, in particular, personality traits were taken into consideration as generic factors relevant to the therapeutic result, albeit without success [38]. More recently, on the other hand, coping strategies were considered as predictors, with variable results. Whilst Bruce and coworkers [8], for example, found that coping with the disease had no influence on the postoperative state in 100 patients, Meffert et al. [44] discussed the detrimental effect of denial as a coping strategy, and Folks and coworkers [26], in their prospective investigation on 121 patients, concluded that denial facilitates adaptation to the postoperative condition. Our own results in relation to self-rated quality of life in the 3-month follow-up suggest that three coping strategies have a maladaptive effect: denial, depressive response, and social withdrawal. On the other hand, our results point to the protective value of internal locus of control and social support. Although this has been found in psychocardiological studies, it has not yet been investigated to any notable extent in cardiac surgery.

Analyses of the dependence of quality of life on selected clinical and sociodemographic variables are not conclusive. Differences with regard to sex, educational level, type of surgery, or psychiatric symptoms show significant results for only a few indicators and are to be used to generate hypotheses.

In all, the results of this study should not be overinterpreted, as they reflect only part of the data and the initial analyses. Clinical data have not yet been included, and relationships between psychiatric, neuropsychological and relatives' data still need to be tested. The long-term course has not yet been considered and detailed analyses at the subscale level are still lacking. Furthermore, problems typical of longitudinal observational surveys of quality of life also exist, namely problems with the selection bias, study drop-outs, and missing values in the questionnaires. These difficulties, however, do not exclude recommendations as to the method and content of further quality of life investigations.

Methodologically, it is recommended that quality of life be integrated into clinical studies in which one or more patient group is selected according to clinical criteria, allocated to a treatment (randomized, if applicable), and investigated over time. In addition to information about the psychosocial effects of the intervention, this procedure guarantees the homogeneity of patient groups and therapy, as well as continuous standardized documentation [9]. With regard to the instruments for measuring quality of life, the results indicate the sensitivity of scales of well-being and suggest, for further studies, that the armamentarium be reduced because of the correlation between different methods [17].

As to the contents, further investigations should place emphasis on the consequences of surgery as experienced by the patients, and the psychosocial predictors and therapeutic implications thereof. Although the quality of life of patients appears to improve after surgery, the interindividual differences in indicators of quality of life should be taken into account. For some of the patients, as well as their relatives [28], surgery can represent a stressful event [34]. Studies of the quality of life help to find new approaches, to facilitate adaptation to surgery not only beforehand but also afterwards [3, 36, 37].

References

1. Aaberg T, Aalund P, Kihlgren M (1983) Intellectual function late after open-heart surgery. Ann Thorac Surg 36: 680–683
2. Adrian J, Crankshaw DP, Tiller JWG, Stanley RO (1988) Affective, cognitive and subjective changes in patients undergoing cardiac surgery – a preliminary report. Anesth Intensive Care 16: 144–149
3. Anderson EA (1987) Preoperative preparation for cardiac surgery facilitates recovery, reduces psychological distress and reduces the incidence of acute postoperative hypertension. J Consult Clin Psychol 55: 513–520
4. Arbeitsgemeinschaft für Methodik und Dokumentation in der Psychiatrie (1971) Das AMP-System. Manual zur Dokumentation psychiatrischer Befunde. Springer, Berlin Heidelberg New York
5. Bass C (1984) Psychosocial outcome after coronary artery by-pass surgery. Br J Psychiatry 145: 526–532
6. Benton AL (1981) Der Benton Test. Huber, Bern
7. Brickenkamp R (1981) Test D 2, Aufmerksamkeits-Belastungs-Test, Handanweisung. Hogrefe, Göttingen
8. Bruce EH, Bruce RA, Hossak KF, Kusumi F (1983) Psychological coping strategies and cardiac capacity before and after coronary artery bypass surgery. Int J Psychiatry 13: 69–84
9. Bullinger M, Hasford J (1990) Testing and evaluating quality of life measures for German clinical trials. Controlled Clin Trials (in press)
10. Bullinger M, Pöppel E (1988) Lebensqualität in der Medizin: Schlagwort oder Forschungsansatz? Dtsch Ärztebl 85: 679–680
11. Bullinger M, Heinisch M, Ludwig M, Geier S (1990) Studien zur Erfassung des Wohlbefindens – psychometrische Analysen zum "Profile of Mood States" (POMS) und zum "Psychological General Well-Being Index" (PGWB). Z Diff Diagn Psychol 11: 53–61
12. Bullinger M and the Quality of Life Study Group (1990) Quality of life research in West Germany – the Munich experience. In: Shumaker S, Furberg C, Czajkowski S, Schron E (eds) Quality of life and cardiovascular disease. Springer, New York (in press)
13. Bullinger M, Angermann C, Spes C, Zellner M (1991) Lebensqualität nach Herztransplantation. In: Bullinger M, Ludwig M, Steinbüchel N (eds) Lebensqualität bei kardiovaskulären Erkrankungen. Hogrefe, Stuttgart, pp 153–163

14. Bunzel B, Eckersberger F (1987) Veränderung der Lebensqualität nach aortokoronarem Bypass und Klappenersatz: Ein Gradmesser des subjektiven Operationserfolges. Thorac Cardiovasc Surg 35: 242–247
15. CASS Principal Investigators (1983) Coronary artery surgery study (CASS) – a randomized trial of coronary artery bypass surgery. Quality of life of patients randomly assigned to treatment groups. Circulation 68: 951–960
16. Davies-Osterkamp S, Moehlen K (1978) Post-operative course after cardiac surgery as correlated with pre-operative fear and its control. Med Psychol 4: 247–260
17. Deyo RA, Inui TS (1984) Toward clinical applications of health status measures: sensitivity of scales to clinically important changes. Health Serv Res 19: 275–289
18. Donald CA, Ware JE (1982) The quantification of social contacts and social resources. Rand Corporation, Santa Monica (R-2937–HHS)
19. Dubin WR, Field HL, Gastfriend DR (1979) Postcardiotomy delirium: a critical review. J Thorac Cardiovasc Surg 77: 586–594
20. Du Puy HJ (1984) The Psychological General Well-being (PGWB) Index. In: Wenger NK, Mattson ME, Furberg CD, Ellinson J (eds) Assessment of quality of life in clinical trials of cardiovascular therapies. LeJacq, New York, pp 170–183
21. Egerton N, Kay JJ (1964) Psychological disturbance associated with open-heart surgery. Br J Psychiatry 110: 433–439
22. Erikson J (1988) Psychosomatic aspects of coronary artery bypass graft surgery. A prospective study of 101 male patients. Acta Psychiatr Scan 77: Suppl, 340: 1–112
23. Fahrenberg J, Hampel R, Selg H (1985) Die revidierte Form des Freiburger Persönlichkeitsinventars FPI-R. Diagnostica 31: 1–21
24. Flemming B, Dahme B, Goetze P, Huse-Kleinstoll G, Kalmar P, Meffert J, Speidel H (1978) Some psychological predictors for psychosis after open-heart surgery. Thoraxchirurgie 26: 458–462
25. Folks DG, Blake DJ, Fleece L, Sokol RS, Freeman AM (1986) Quality of life six months after coronary artery bypass surgery: a preliminary report. South Med J 79: 397–399
26. Folks DG, Freeman AM, Sokol RS, Thurstin AH (1988) Denial: Predictor of outcome following coronary bypass surgery. Int J Psychiatry Med 18: 57–66
27. Gebhardt R, Pietzker A, Freudenthal K, Langer C (1981) Die Bildung von Syndromen im AMP-System. Arch Psychiatr Nervenkr 231: 93–109
28. Goldsmidt T, Brooks N, Sethia B, et al. (1984) Coronary artery bypass surgery – impact upon a patient's will. A pilot study. Thorac Cardiovasc Surg 32: 337–340
29. Gregori AN, Hetzer R, Schwarz B, Mayer B, Buser K, Lichtlen PR, Borst HG (1983) Veränderung der "Lebensqualität" nach Koronarrevaskularisation. Z Kardiol 72: 12–17
30. Gundlee MJ, Reeves BR, Tate S, Raft D, McLauren LP (1980) Psychosocial outcome after coronary artery surgery. Am J Psychiatry 137: 1591–1594
31. Hammeke TA, Hastings JE (1988) Neuropsychologic alterations after cardiac operation. J Thorac Cardiovasc Surg 96: 326–331
32. Heller SS, Kornfeld D (1986) Psychiatric aspects of cardiac surgery. Adv Psychosom Med 15: 124–139
33. Heller SS, Frank KA, Kornfeld DS, Malm JR, Bowman FO (1974) Psychological outcome following open-heart surgery. Arch Intern Med 134: 908–914
34. Holroyd KA, Lazarus RS (1988) Stress, coping and somatic adaptation. In: Goldberg L, Bresnitz S (eds) Handbook of stress: theoretical and clinical aspects. MacMillan, New York, pp 21–35
35. Jenkins CD, Stanton B-A, Savageau JA, Denlinger P, Klein MD (1983) Coronary artery bypass surgery. Physical, psychological, social and economic outcomes six months later. JAMA 250: 782–788
36. Jordan J, Overbeck, Joos W (1983) Psychische Bewältigungsmechanismen bei offenen Herzoperationen in Abhängigkeit von der Persönlichkeitsstruktur des Patienten. Die Möglichkeit der Zusammenarbeit zwischen Psychosomatik und Kardiologie. Z Psychosom Med Psychoana Med 29: 380–403

37. Kendall PC (1982) Stressful medical procedures: cognitive behavioural strategies for stress management and prevention. In: Meichenbaum D, Jahremko ME (eds) Stress reduction and prevention. Plenum, New York, pp 159–191
38. Kilpatrick DG, Miller WC, Allain AN, Huggins MB, Lee WH (1975) The use of psychological test data to predict open-heart surgery outcome: a prospective study. Psychosom Med 37 (1): 62–73
39. Kinchla J, Weiss T (1985) Psychological and social outcomes following coronary artery bypass surgery. J Cardiopulm Rehabil 5: 274–283
40. Kornfeld DS, Heller SS, Frank K, Wilson SN, Malm JR (1982) Psychological and behavioral responses after coronary artery bypass surgery. Circulation 66: 24–28
41. Magni G, Unger HP, Valfre C, Polesel E, Cesari F, Rizzardo R, Paruzzolo P, Galluci V (1987) Psychosocial outcome one year after heart surgery. A prospective study. Arch Intern Med 147: 473–477
42. Mayou R, Bryant B (1987) Quality of life after coronary artery surgery. Q J Med New Series 62: 239–248
43. McNair D, Lorr M, Droppelman LF (1971) EITS manual for the profile of mood states. Educational and Industrial Testing Service, San Diego
44. Meffert HJ, Dahme B, Flemming B, Goetze P, Huse-Kleinstoll G, Rodewald G, Speidel H (1979) Open-heart surgery from the psychological point of view and resulting therapeutic considerations. Psychother Psychosom 23: 148–156
45. Muthny FA (1989) Krankheitsverarbeitung bei chronisch körperlich Kranken. Prax Psychother Psychsom 34: 64–72
46. Naber D, Bullinger M (1985) Neuroendocrine and psychological variables relating to post-operative psychosis after open-heart surgery. Psychoneuroendocrinology 10: 315–324
47. Reitan RM (1958) Validity of the trail making test as an indicator of organic brain damage. Percept Mot Skill 8: 271–276
48. Savageau JS, Stanton BA, Jenkins CD, Klein MD (1982) Neuropsychological dysfunction following elective cardiac operation. I. Early assessment. J Thorac Cardiovasc Surg 84: 585–594
49. Schoenberg B, Zuercher M, Baur HR (1985) Lebensqualität nach Herzklappenersatz. Schweiz Med Wochenschr 115: 239–241
50. Shaw PJ, Bates D, Cartlidge NE, French JM, Heavisidie D, Julian D, Shaw D (1986) Early intellectual dysfunction following coronary bypass surgery. Q J Med 225: 59–68
51. Smith HC, La von Hammes N, Gupta S, Vliestra RE, Elveback L (1983) Employment status after coronary artery bypass surgery. Circulation 65: 120–125
52. Spielberger CD, Gorsuch RL, Lushene RE (1979) STAI manual for the state-trait-anxiety-inventory. Palo Alto, Consulting Psychologists Press
53. Stanton BA, Zyzanski SJ, Jenkins CD, Klein MD (1982) Recovery after major heart surgery: medical, psychological, and work outcomes. In: Becker R et al. (eds) Psychopathological and neurological dysfunctions following open-heart surgery. Proceedings. Springer, Berlin Heidelberg New York, pp 217–226
54. Stanton BA, Jenkins CD, Savageau JA, Thurer RL (1983) Functional benefits following coronary artery bypass graft surgery. Ann Thorac Surg 37: 286–290
55. Stroop JR (1935) Studies of interference in serial verbal reactions. J Exp Psychol 18: 643–662
56. Wechsler D (1982) Handanweisung zum Hamburger-Wechsler-Intelligenztest. Huber, Bern
57. Wenger NK, Mattson ME, Furberg CD, Elinson (1984) Assessment of quality of life in clinical trials of cardiovascular therapies. Am J Cardiol 54: 908–913
58. Zyzanski SI, Stanton BA, Jenkins CD, Klein MD (1981) Medical and psychosocial outcomes in survivors of major heart surgery. Psychosom Res 23: 213–221

Guidelines for Improving Prognosis, Stress Tolerance, and Quality of Life in Patients with Heart Failure: Cardiological and Psychological Aspects

G. F. Hauf and W. Langosch

Cardiological Aspects

In patients with congestive heart failure, what factors determine prognosis, and what factors are responsible for subjective well-being and thus ultimately (at least in part) for quality of life?

Prognosis

Despite considerable advances in diagnosis and drug therapy, it has not been possible to demonstrate any significant alteration in survival rates in recent decades. Only in the most recent studies has it been possible to demonstrate some improvement in prognosis, due to altered drug schedules and the introduction of new drug classes as well as the possibility of heart transplantation in appropriately selected patient groups, [6, 8, 40]. However, it is certainly still necessary to base our conclusion with respect to survival rate on data such as those collected in the Framingham Study. The 1-year and 5-year survival rates of male patients with congestive heart failure were 79% and 38%, the corresponding values for females being 86% and 57% [30]. Comparable rates are cited in other studies, with 1-year mortality rates between 30% and 60%, depending on the patient groups examined, with heart failure secondary to coronary heart disease or primary cardiomyopathy [29].

A knowledge of factors potentially influencing prognosis is also important for the management of an individual patient. First of all, it is important that the cause of myocardial damage should be elucidated. Existing severe myocardial damage may be entirely reversible when it is due to an acute inflammatory process, to nutritional or toxic factors, or even to correctable volume or pressure overload of the ventricle. Of course, even these conditions may in individual cases progress to a chronic irreversible state.

If patient groups whose heart failure is due to chronic ischemic heart disease are compared to groups without ischemic myocardial damage, patients in the former group appear to have a much worse prognosis [6, 15, 30].

Hemodynamic parameters such as the magnitude of the mean arterial blood pressure, left ventricular filling pressure, systemic vascular resistance, and cardiac index have also been put forward as prognostically important factors. However, a significant correlation to the survival rate has only been demonstrable in with respect to the left ventricular end-diastolic pressure [15].

Further parameters such as the ejection fraction or oxygen uptake (VO_2) also have prognostic value [7]. Even biochemical data such as increased plasma norepinephrine levels or lowered serum sodium concentrations have also turned out to be prognostically decisive [17, 39, 42]. Cardiac dysrhythmia must also be cited as another prognostically significant factor. Appropriate studies have revealed complex ventricular arrhythmias in approx. 35%–95% and additional self-limiting ventricular tachycardia in approx. 25%–80% in patients with congestive heart failure. Correlations have also been shown between ventricular function and the biochemical data mentioned above, but also as a possible side effect component of drug therapy [2, 10, 29, 38].

Stress Tolerance

What factors are responsible for the limitation of everyday stress tolerance in the presence of chronic heart failure? Studies of this question have revealed that in the chronic stage of myocardial damage there is no correlation between maximum stress tolerance or maximum oxygen uptake (VO_2 max) and, e.g., pulmonary capillary pressure or ejection fraction. It does appear, however, that maximum stress tolerance depends on the value of the cardiac index and on oxygen transport capacity [16, 26, 31, 46]. A further factor limiting physical stress tolerance appears to be the state of the peripheral working musculature. It has been shown that in chronic heart failure there is an alteration in the structure and metabolism of working muscle. This manifests itself partly in increased tone with increased wall rigidity of the precapillary vessels, perhaps due to an increased intracellular sodium content, and partly in a reduction of the capillary density. In addition, at the cellular level, there is both a reduction and a structural alteration in the mitochondria of the myocytes. A reduction in the anaerobic/aerobic threshold and a shift in ambient pH are assumed to be due to these factors. It is postulated that peripheral chemoreceptors give feed sack to the respiratory center, giving rise to the symptom of dyspnea and also to that of premature peripheral exhaustion [12, 28, 29, 41, 47, 50].

Drug-Related Improvement in Prognosis

Few data are yet available on the question of possible improvements in prognosis due to drug therapy. At the present time, the essential data of relevance here are a Veterans Administration Cooperative Study [6, 7] and the Scandinavian Consensus Study [8].

In the Veterans Administration Cooperative Study, 642 male patients with congestive heart failure, different NYHA classes, were studied for an average period of 2.3 years. They all received basic therapy with digitalis and diuretics, and were randomized double-blind to test the efficacy of either prazosin (20 mg/day), or a combination of hydralazine (300 mg/day) with ISDN (160 mg/day), against placebo. Only the hydralazine-ISDN treatment group showed a significant reduction in mortality and a simultaneous improvement in left ventricular function. In this group the 1-year mortality was reduced by 38% and the 2-year mortality by 25%.

In the Consensus Study, 253 patients initially with NYHA class IV were studied over an average period of 6 months. In a double-blind randomized trial, in which conventional therapy with digitalis/diuretics and in some cases vasodilators was continued, it was shown that additional treatment with an ACE inhibitor (enalapril 40 mg/day) also causes a significant improvement in prognosis, with a 40% reduction in the 6-month mortality rate, and a 30% reduction in the 1-year mortality rate, together with a significant improvement in NYHA class.

Drug-Related Improvement in Physical Stress Tolerance

In contrast to the limited information on the effect of drug therapy on prognosis, information is available from a larger number of studies on the efficacy of individual drug groups with regard to physical stress tolerance. Relevant findings are available from a number of studies, with different designs, some lasting over 1 year, involving, e.g., digitalis, nitrates, ACE inhibitors, and newer positive inotropes, e.g., dobutamine, amrinone, milrinone.

These studies showed increases in stress tolerance of up to 50%; however, they generally did not show a regular relationship with significant improvements in NYHA classification or with ejection fraction [3, 4, 9, 11, 23, 24, 25, 40, 45].

Information to Patients

In case of congestive heart failure, the first step should be to confirm the diagnosis and determine the etiology.

As pointed out above, this question is of crucial importance in assesssing prognosis. A snapshot of the clinical picture at one time can generally not differentiate between the reversible situation in an inflammatory process or the presence of toxic, particularly nutritional, factors (e.g., ethanol). At this point, the patient and perhaps his relatives should be fully informed of the seriousness of the present situation, but no over-hasty conclusions about prognosis should be expressed.

As soon as diagnostic measures, and observation of the course of disease, have generated clear doubts about the reversibility of the clinical situation, the prognosis and also the possibilities of intervention should be discussed in detail with the patient, and if possible this discussion should include his closest relatives. As the bottom line, the patient must be clearly made aware both of the severity of his condition and also of the essential importance of his personal commitment to overcoming it. Stabilization of the patient's clinical condition, or at least a delay in its progression, undoubtedly depends not only on appropriate drug therapy in combination with the essential degree of patient compliance, but also on the continued observance of strict dietary measures including fluid restriction and appropriate limits on physical activity. This last point in particular frequently involves major changes in professional activity and also in spare-time occupations. The family environment is particularly essential in overcoming this problem.

The schedule of drug therapy must be carefully selected and discussed on a broad basis of mutual trust. The patient must be emphatically warned against any uncontrolled increase in physcial activity, and even against himself testing out his own performance limits, if and when there is any marked improvement in his condition with time. Unfortunately, everyday experience repeatedly demonstrates that patients have great problems in continuing to limit their physical activities in the case of symptom reduction. The consequence is often a renewed or first attack of myocardial decompensation, with problems relating not only to further drug therapy, but also to psychological aspects.

In addition, it is essential that the possible drug side effects should be carefully discussed with the patient. This should be combined with strict instructions that, in the event of any suspicion of such complications, he should immediately consult his physician and should not himself undertake changes in drug regimen. Well-informed patients will undoubtedly be more willing to tolerate the transient headaches associated with vasodilator therapy than unprepared patients who are faced with an unknown symptom that causes them considerable discomfort. In this connection, patients should also be warned that cardiac arrhythmias may be aggravated during drug treatment, particularly with diuretics, perhaps in combination with digitalis; and that they should therefore immediately consult their doctor if they experience suspicious symptoms such as increased dizziness or presyncopal attacks [38].

Depending on the further course of the disease, long-term schedules should already be considered, which may include the possibility of heart transplantation. In such cases, a good basis of trust between patient and doctor is absolutely decisive in order to enable careful preparation to this dramatic event, not only for the patient but also for his family, especially also in relation to postoperative problems and long-term supervision.

Subsequent Stress Tolerance

These facts should not of course lead to the conclusion that patients with significant congestive heart failure should be subjected to total inactivity, if possibly even on tranquillizers. Instead, an attempt should be made to devise an individual training schedule, based both on the patient's psychosocial situation and on his ventricular and circulatory function. This schedule should ideally include measures such as relaxation training, stress-mastery training, and also exercise training. As pointed out above, physical performance is partially limited by the state of the peripheral working musculature. This can be corrected, at least partially, both by drug therapy such as long-term ACE inhibitor treatment, and also by graduated physical training. It has been suggested that these measures cause diversion of blood circulation toward the active working muscle, as well as an increase in capillary density, and in addition an increase in mitochondrial numbers, which is ultimately responsible for an improvement in O_2 utilization [5, 12, 16, 28, 29, 44, 47, 48, 49, 50].

It is recommended to start such an exercise training based on hemodynamic data, obtained at rest and during exercise by means of a right heart catheter examination. Subsequent monitoring should be carried out by echocardiography, at intervals selected on an individual patient basis, to exclude any initially subjectively undetected worsening of the myocardial situation, so that correction of the exercise therapy plan can be carried out in good time.

At our hospital, a special exercise training schedule for patients with very severe myocardial damage, usually patients waiting for heart transplantation, was started a few months ago. Before being entered into this exercise therapy group, patients undergo individual cycle ergometry stress testing in sitting position. Their work capacity is tested under close medical supervision, with limits set primarily by their own symptoms; this involves graduated increases in effort at 1-min intervals, starting with 12.5 W and with subsequent steps increasing by 12.5-W intervals. In addition, the maximal muscle strength is tested also in the sitting position on the cycle ergometer. This exercise starts with a stress of 25 W, graduated increases of 25 W each time at 10-s intervals, once again with limitations set by the patient's symptoms. These studies are supplemented by laboratory tests including lactate, catecholamines, and if possible also measurement of ventilatory volumes.

On the basis of these data, patients then undergo daily 15-min interval ergometer training, involving 20–30 s high intensity training, to approx. 50% of maximal muscle strength and 60–80 s low intensity training at 12.5 W. At individually determined intervals of 7–14 days, patients again undergo testing of their symptom-limited stress tolerance, and a control echocardiogram. On this basis the exercise therapy schedule may be newly adjusted. This ergometer training is supplemented by gymnastic exercises, also carried out in groups, aimed particularly at exercising the arm and spinal musculature. Patients also undergo individually adapted physiotherapy, which may include climbing stairs or walking in the open air.

Results to date have shown that these measures can achieve a clear improvement in symptom-limited physical stress tolerance; this is primarily explained by correction of the pathologically altered peripheral parameters [43].

In patients with less marked myocardial damage, it would be possible to carry out a similar program as a group exercise in the patient's own home town. For this purpose it would be essential to have close cooperation between the cardiological center or the local cardiologist, the general practitioner, and the exercise therapist.

On the basis of what has been said above, diagnosis of severe congestive heart failure should not necessarily be followed by the termination of the patient's professional activity. In individual cases it should be discussed to continue the patient's former job in combination with mild physical activity. Of course this will often require the cooperation of the employer, and understanding and willingness to help of the patients' colleagues.

Psychological Aspects

Chronic congestive heart failure often goes together with the need for the patient to make many changes in his way of life, either because he feels unwell or because of his physicians' recommendations. The way in which individual patients concretely come to terms with their disease and the changes in well-being that accompany it depends crucially on the way they assess the limitations resulting from their disease, and their own practical options.

Adaptive Requirements for the Patient with Heart Failure

The primary assessment process relates to the significance that the patient attributes to his disease as far as his own well-being is concerned, i.e., whether his primary emotion is one of harm and loss, threat, or challenge. The extent of his experience of threat, however, is a consequence of the secondary evaluation, in which the patient assesses the appropriateness of his currently available options and resources for coping with his disease. Depending on the results of the primary and secondary assessment processes, different strategies may be selected to cope with the disorder [14]. The most essential adaptive tasks that face the patient with heart failure are the following (cf. [33]):

- Coping with a multiplicity of adverse symptoms (dyspnea, exhaustion, etc.).
- Coping with the side effects of drug treatment (headache, feeling faint, etc.).
- Maintaining good relations with the doctor.
- Maintaining emotional stability and an appropriate level of self-esteem despite uncertainty about the further course of the disease and fear of renewed myocardial decompensation.
- Coping in advance with illness-related limitations on physical performance and stress tolerance and acceptance of imposed limitations, even when he feels well.
- Maintenance of good relations with relatives, friends, and acquaintances.

It is precisely because the patient with heart failure goes through repeated periods in which he feels almost no impairment in his health, that it is important for his physician always to view the patient's chosen strategies for coping with his illness as the outcome of an effort to cope with the demands and burdens to which he is subject at the present time.

It is an excessively one-sided view to regard the patient's efforts to cope with his disorder as an adaptive process only from the viewpoint of long-term physical health; since if subjective well-being and the quality of social functioning at any particular time are included in the assessment, the overall judgment may vary very widely. Lazarus [21] accordingly emphasizes that the functional process of coping with the disorder must not be viewed separately from the circumstances in which it takes place. If we wish to alter the patient's efforts to cope with his disorder, we must first understand how the coping and assessment process has taken place in the particular patient concerned, and what factors were involved in determining it. It is therefore often not sufficient to give the patient an objective explanation of the adverse consequences for his health that were to be expected if he overstepped the performance guidelines laid down by his doctor; this will not always prevent him from attempting such behavior, since he has not yet integrated his new knowledge into his present motivation-related and emotional pattern.

Role of Personal Factors in Coping Strategies

The psychological state of a cardiac patient only depends on his cardiac findings to the extent that the latter affect his well-being, or give rise to anxiety and insecurity due to uncertainty about the further course of the disease. Thus it has been shown in coronary patients that neither the presence of impaired left ventricular function, nor coronary status, is closely correlated with psychological features such as emotional lability, social anxiety, lack of self-confidence, or depression [18].

The form taken by the patient's efforts to cope depends predominantly on the following factors:

- Subjective illness-related beliefs about control, i.e., the patient's subjective expectation as to how far the course of his illness is influenced by his own behavior, or is a matter of chance and destiny, removed from his own control.
- The patient's subjective attributions of causation, i.e., his awareness of the cause of his disease and its further course. If stress is laid on causative factors that accentuate inactive behavior, e.g., lack of movement, this increases the danger of the patient's overstepping his performance limits, while if emphasis is laid on factors such as a hectic working environment or general stress, this tends to promote passivity and self-sparing behavior (cf. [13]).
- The habitual orientation of the patient's perception in stress situations, involving an opposition between the tendency to concentrate attention on stress-provoking elements, and the tendency to withdraw attention from these aspects. Both strategies are inappropriate if they are applied rigidly, i.e., without consideration of the situational factors in the individual case.
- The patient's individual capacity to efficiently use all resources available to him, i.e., his ability, when in a stressful situation, to make use of various types of available support such as information, practical offers of help, or relief of the emotional strain.
- The patient's habitual level of anxiety – since a chronic tendency toward anxiety leads to higher anxiety levels in stressful situations.
- The patient's subjective life plans, interests, and values. If short-term and long-term goals are not in harmony, the patient's motivation in carrying through alterations in his behavior pattern will also be weak and unstable.

The large number of factors affecting coping strategies explains why the simple imparting of information is not always sufficient to induce the patient to adopt the behavior pattern that his doctor would wish.

Types of Coping Strategies

Among the factors listed by Beutel [1] as general stresses of chronic disorders, the most important for the patient with heart failure are the following: the essential irreversibility and progression of the disease process, the unpredictability of its course, reduced physical capacity, the threat to physical integrity, long-term dependence on medical specialists, manifold losses in the patient's personal and social life, and limited prospects for the future. How can the patient with heart failure attempt to cope with these factors?

These stresses – aggravated by the often short interval between the occurrence of the first symptoms and the development of, e.g., pulmonary embolism, and by the condition of pulmonary embolism itself, which is subjectively experienced as life-threatening result in the adaptive demands of the disease mentioned above. The choice of coping strategies depends essentially on the significance that the patient gives to his illness, his view of its causation, and his conviction about its control.

Lipowski [27] distinguishes the following types of significance that can be ascribed to an illness:

- A challenge to be overcome.
- An enemy who provokes anxiety and hostility, and must be fought against with all the strength at the patient's command.
- An unjust punishment, provoking aggression and rage.
- A just punishment, which is accepted on the basis of subjectively perceived feelings of guilt.
- A weakness, experienced as a subjective insult and a personal failure, sometimes associated with a feeling of shame.
- An irreparable loss and damage, experienced as something so overpowering that it results in depression or hostility.

The predominant significance attached to the illness determines the coping strategy chosen. For example, if the illness is experienced as a weakness of which the patient is ashamed, he will be very hesitant about obtaining help from others; if on the other hand feelings of anxiety about his own life or about the future of others occupy the foreground of his attention, he will be very ready to accept reassurance and support from others. If he sees the illness as a challenge, whose further course he can help to influence, he will seek for problem-solving approaches, while emotion-oriented coping efforts will be preferred if his predominant feeling is that he can only influence the further course of his illness to a small extent or not at all.

The different coping strategies are subdivided by Lazarus et al. [22] into problem-oriented strategies aimed at alterations in environmental pressures and barriers, and emotion-oriented strategies aimed only at modifying awareness of the problem situation. Problem-oriented coping strategies include confrontative coping, self-control, efforts to achieve social support, and planned problem solving, while distancing, flight/avoidance, and positive reevaluation are classified as palliative approaches. Moos [32] distinguishes three categories of coping strategies, which are also important for patients with heart failure:

1. Emotion-oriented strategies such as affective regulation, aimed at maintaining hope and controlling one's own emotions; emotional relief, e.g., in the form of open expression of rage, annoyance, despair, anxiety; or in the form of acting-out behavior, e.g., violent criticism of the doctor and/or nursing personnel, uncontrolled consumption of nicotine or alcohol, or resigned acceptance.
2. Problem-oriented strategies such as a search for information relevant to the further course of the disease; acceptance of the offered practical and emotional support; establishment of concrete intermediate goals, e.g., relating to increases in physical performance capacity; opening up possibilities for satisfaction of personal needs as a compensation for illness-related limitations on previous use of leisure time, e.g., by forming an outpatient cardiac group.
3. Evaluation-oriented strategies such as avoidance, denial, trivialization (all aimed at diminishing the existential threat posed by the illness), redefinition of the illness, laying selective stress on potentially good or useful aspects of it; concentration of attention on some isolated aspect of the illness, e.g., present physical well-being; and reference back to previous positive experiences in coping with illness, e.g., physical training.

In addition, the illness can be exploited as a technique for getting attention, care, or consideration from others [1].

In general it can be assumed that wishful thinking, in the form of flight/avoidance strategies, represents an unfavorable coping technique, since it is incompatible with a realistic pattern of behavior appropriate to the illness, while problem-oriented behavior combined with redefinition favors acceptance of the illness-related limitations of performance, preserving emotional stability and restoring an adequate feeling of self-esteem. Folkman et al. [14] have shown that problem-oriented coping strategies depend greatly on the individual situation of the patient, including his social environment, while emotion-oriented strategies are relatively stable and thus more strongly dependent on the patient's personality. In treating the patient with heart failure, therefore, it is necessary to take account both of his life circumstances and of his individual prejudices in the evaluation of his illness (e.g., his style of perception, attributions, anxiety level).

Psychological Measures in Patients with Heart Failure

The overriding aim of psychological measures applied to patients with heart failure must be to provide support in carrying out the difficult task of reconciling their previous life and future plans with their illness [35], or integrating their functional limitations and the changes in life-style dictated by their disease into their working and general life environment [34]. A possible though global criterion of successful adaptation might be that of "handicap limitation" – i.e., living with the recognition of one's real restrictions. The opposite poles to this attitude are, on the one hand, a total denial of illness, which except in the first phase of the illness rules out a constructive analysis of it, and on the other hand, a generalized disease concept, marked by hypochondriacal self-observation and subjective invalidity. The next question is how to approach the goal of "handicap limitation".

Beutel [1] emphasizes the following special characteristics of psychological measures in chronic invalids:

- Motivation to accept psychological treatment is limited, mainly because of the preexisting burden of drug treatment, but also due to prejudices against psychotherapy.
- An important limitation is set by the chronic patients' perception of their limited life expectancy; this leads them to concentrate on present problems, resulting in increased attention being paid to issues such as forced dependency, increasing physical limitations, or loss of previous opportunities for fulfilment of needs.
- The psychotherapist must be clearly aware that psychotherapeutic efforts are in principle secondary to drug treatment.

It follows from the above that psychotherapeutic measures in patients with heart failure will generally be of a supportive nature and aimed at specific problems. It is important to promote the patient's coping opportunities in such a way as to expand his options for self-help and help from his social environment, in particular his partner.

Lazarus [21] is undoubtedly right in pointing out that, in chronic patients, treatment strategies can only be successful if due account is taken of the patient's life problems, temporal planning and life-style. He recommends that the therapeutic approach should be narrowly aimed at the patient's actual life situation and prospects; this assumes that psychotherapeutic intervention has to be based on adequately detailed knowledge of the following factors: the patient's previous efforts to cope with the adaptive demands posed by his heart failure; the personal factors involved in his coping strategy; the quantity and quality of help available to the patient, and actually accepted by him, from others; the significance he himself attaches to his disease; and the situational factors limiting his problem-oriented coping strategy. It is only against the background of such knowledge that an assessment can be made of what has, e.g., led a patient to deviate from dietary measures recommended by his doctor, or to show unsatisfactory compliance in limiting his physical activity. The cause may lie in information that he subjectively rates as inadequate; denial of the illness or of its seriousness; loss of previously highly valued opportunities for need satisfaction, without corresponding opportunities for compensation; a feeling that the cause of the illness is scarcely open to influence by his own behavior; reference back to earlier experience showing that physical training has beneficial effects on his well-being; shame at having to expose his physical weakness to others; the setting of unrealistic goals, leading to such great discrepancies between his present and desired physical state that he gives up the attempt to attain his goal in frustration; situational complications at his workplace; etc. Therapeutic measures must be selected in the light of the dysfunctional coping behavior that has been worked out as being crucial in each individual case.

Measures relating to individual cases can be supplemented by group measures aimed at providing the patient with a differentiated repertoire of coping strategies, and increasing his flexibility in coping with everyday burdens. This sort of stress-mastery training was described, e.g., by Langosch [19]. An exercise training program matched to the patient's physical performance capacity can also contribute to restoring patients' confidence in the dependability of their body, particularly in those who have a tendency to regard their body with some anxiety [20]. It should always be remembered that it is not physical conditioning as such, but the acquisition of positive body experiences that is psychologically stabilizing. From the psychological view-point, exercise training should therefore focus on the development of a positive body image and increased body awareness.

The practice of relaxation techniques, such as progressive muscular relaxation, can favor the acquisition of positive body experiences, and the use of such techniques thus appears indicated in patients with heart failure. If relaxation training is to continue to produce positive psychological effects, however, it must be carried out in such a way that the patient is ultimately capable of using relaxation in his everyday life as a technique to improve his mastery of problem situations. In the light of Oest's reports of success in the treatment of panic attacks [37], the procedure recommended by him [36] appears to be suitable here.

Since the emotional support offered by the patient's partner is one of the coping strategies in principle available to him, effort should be directed toward ensuring that the patient is able to accept the help offered to him, and in addition it should be determined whether he is receiving the type of help required. Group discussions may make it easier for the patient to accept help and may make him better aware of the type of help he requires. Supplementary talks with patients and their partners can contribute to ensuring that the partner also acquires a clearer idea of what types of behavior are useful.

Psychological measures in patients with heart failure comprise – particularly in the initial phase – a combination of individual and group therapy. During the further course of the illness, there should always be a return to individual treatment if the patient experiences psychological crises, which may express themselves, e.g., in poor compliance; appropriate treatment of the psychological and behavioral problems cannot be successful without detailed analysis of the patient's coping process and related factors to date.

References

1. Beutel M (1988) Bewältigungsprozesse bei chronischen Erkrankungen. Edition Medizin, VCH, Weinheim
2. Bigger JT (1987) Why patients with congestive heart failure die: arrhythmias and sudden cardiac death. Circulation 75 [Suppl IV]: 28–35

3. Captopril-Digoxin Multicenter Research Group (1988) Comparative effects of therapy with captopril and digoxin in patients with mild to moderate heart failure. J Am Med Assoc 259: 539–544
4. Chatterjee K et al. (Captopril Multicenter Research Group I) (1985) A cooperative multicenter study of captopril in congestive heart failure: hemodynamic effects and long-term response. Am Heart J 110: 439–447
5. Cohen-Solal A (VO2 French Study Group) (1989) Improving exercise tolerance in patients with chronic heart failure: should we treat the heart or the periphery? Eur Heart J 10: 866–871 (editorial)
6. Cohn JN, Archibald DG, Ziesche S et al. (1986) Effect of vasodilator therapy on mortality in chronic congestive heart failure. Results of a Veterans Administration Cooperative Study. N Engl J Med 314: 1547–1552
7. Cohn JN, Archibald DG, Francis GS et al. (1987) Veterans administration cooperative study on vasodilator therapy of heart failure: influence of prerandomization variables on the reduction of mortality by treatment with hydralazine and isosorbide dinitrate. Circulation 75 [Suppl IV]: 49–54
8. Consensus Trial Study Group (1987) Effects of enalapril on mortality in severe congestive heart failure. Results of the cooperative North Scandinavian enalapril survival study. N Engl J Med 316: 1429–1435
9. Creager MA, Massie BM, Faxon DP, Friedman SD, Kramer BL, Weiner DA, Ryan TJ, Topic N, Melidosian CD (1985) Acute and long-term effects of enalapril on the cardiovascular response to exercise and exercise tolerance in patients with congestive heart failure. J Am Coll Cardiol 6: 163–170
10. Dargie HJ, Cleland JGF, Leckie BJ, Inglis CG, East BW, Ford I (1987) Relation of arrhythmias and electrolyte abnormalities to survival in patients with severe chronic heart failure. Circulation 75 [Suppl IV]: 98–107
11. DiBianco R, Shabetai R, Kostuk W, Moran J, Schlant RC, Wright R (Milrinone Multicenter Trial Group) (1989) A comparison of oral milrinone, digoxin, and their combination in the treatment of patients with chronic heart failure. N Engl J Med 320: 677–683
12. Drexler H, Banhardt U, Meinertz T, Wollschläger H, Lehmann M, Just H (1989) Contrasting peripheral short-term and long-term effects of converting enzyme inhibition in patients with congestive heart failure. A double-blind, placebo-controlled trial. Circulation 79: 491–502
13. Faller H (1989) Das Krankheitsbild von Herz-Kreislaufkranken. Ein Gruppen- und Methodenvergleich. In: Bischoff C, Zenz H (eds) Patientenkonzepte von Körper und Krankheit. Huber, Bern, pp 571–579
14. Folkman S, Lazarus RS, Gruen RJ, De Longis A (1986) Appraisal, coping, health status and psychological symptoms. J Pers Soc Psychol 50: 571–579
15. Franciosa JA, Wilen M, Ziesche S, Cohn JN (1983) Survival in men with severe chronic left ventricular failure due to either coronary heart disease or idiopathic dilated cardiomyopathy. Am J Cardiol 51: 831–836
16. Franciosa JA, Leddy CL, Wilen M, Schwartz DE (1984) Relation between hemodynamic and ventilatory responses in determining exercise capacity in severe congestive heart failure. Am J Cardiol 53: 127–134
17. Hirsch AT, Dzau VJ, Craeger MA (1987) Baroreceptor function in congestive heart failure: effect on neurohumoral activation and regional vascular resistance. Circulation 75 [Suppl IV]: 36–48
18. Langosch W (1989a) Psychosomatik der koronaren Herzkrankheiten. Edition Medizin, VCH, Weinheim
19. Langosch W (1989b) Welche psychosozialen Aspekte sind in der Herzinfarkt-Nachsorge besonders wichtig? In: Halhuber MJ (ed) Umfassende Herzinfarkt-Nachsorge in Klinik and Praxis. Huber, Bern
20. Langosch W (1989c) Zur Bedeutung des körperlichen Trainings von Herzpatienten – aus der Sicht des Psychologen. In: Halhuber MJ, Krasemann EO, Mücke F (eds) Ratgeber zur Betreuung von Herzgruppen. Echo, Cologne

21. Lazarus RS (1989) Bewältigung krankheitsbedingter Belastungen. In: Kaplan A, Branckaerts J, Erben R, Milz H (eds) Gesundheitsförderung und chronische Krankheit. Bundeszentrale für Gesundheitliche Aufklärung, Cologne, pp 172–180
22. Lazarus RS, de Longis A, Folkmons, Green R (1985) Stress and adaptational outcomes. The problem of unconfounded measures. Am Psychol 40: 770–779
23. Leier CV, Huss P, Lewis RP, Unverferth DV (1982) Drug-induced conditioning in congestive heart failure. Circulation 65: 1382–1387
24. Leier CV, Dalpiaz K, Huss P, Hermiller JB, Magorien RD, Bashore TM, Unverferth DV (1983) Amrinone therapy for congestive heart failure in outpatients with idiopathic dilated cardiomyopathy. Am J Cardiol 52: 304–308
25. Leier CV, Huss P, Magorien RD, Unverferth DV (1983) Improved exercise capacity and differing arterial and venous tolerance during chronic isosorbide dinitrate therapy for congestive heart failure. Circulation 67: 817–822
26. Lipkin DP, Canepa-Anson R, Stephens MR, Poole-Wilson PA (1986) Factors determining symptoms in heart failure: comparison of fast and slow exercise tests. Br Heart J 55: 439–445
27. Lipowski JZ (1983) Psychosocial reactions to physical illness. Canad Med Assoc J 128: 1069–1072
28. Mancini DM, Davis L, Wexler JP, Chadwick B, Le Jemtel TH (1987) Dependence of enhanced maximal exercise performance in increased peak skeletal muscle perfusion during long-term captopril therapy in heart failure. J Am Coll Cardiol 10: 845–850
29. Massie BM, Conway M (1987) Survival of patients with congestive heart failure: past, present, and future prospects. Circulation 75 [Suppl IV]: 11–19
30. McKee PA, Castelli WP, McNamara PM, Kannel WB (1971) The natural history of congestive heart failure: the Framingham Study. N Engl J Med 285: 1441–1446
31. Miyagi Y, Nakajima H, Sasayama S (1989) Relation of exercise capacity to resting arterial oxygen transport in left ventricular dysfunction. Am J Cardiol 63: 457–460
32. Moos RH (1986) Coping with life crisis. An integrated approach. Plenum, New York
33. Moos RH, Tsu VD (1977) The crisis of physical illness: an overview. In: Moos RH (ed) Coping with physical illness. Plenum, New York
34. Naschold F, Novak P (1980) Bedingung für eine Systemanalyse des Gesundheitswesens: Integrale Erklärung von Krankheit in der heutigen Gesellschaft. In: Schönbäck W (ed) Gesundheit im gesellschaftlichen Konflikt. Urban and Schwarzenberg, Munich
35. Nerenz DR, Leventhal H (1983) Self-regulation theory in chronic illness. In: Burish TG, Bradley LA (eds) Coping with chronic disease. Research and Applications. Academic, New York
36. Oest LG (1987) Applied relaxation: description of a coping technique and review of controlled studies. Behav Res Ther 25: 397–409
37. Oest LG (1988) Applied relaxation vs. progressive relaxation in the treatment of panic disorder. Behav Res Their 26: 13–22
38. Packer M (1985) Sudden unexpected death in patients with congestive heart failure: a second frontier. Circulation 72: 681–685
39. Packer M, Lee WH, Kessler PD, Gottlieb SS, Bernstein JL, Kukin ML (1987) Role of neurohumoral mechanisms in determining survival in patients with severe chronic heart failure. Circulation 75 [Suppl IV]: 80–92
40. Pfeffer MA, Lamas GA, Vaughan DE, Parisi AF, Braunwald E (1988) Effect of captopril on progressive ventricular dilatation after anterior myocardial infarction. N Engl J Med 319: 80–86
41. Poole-Wilson PA, Buller NP, Lipkin DP (1988) Regional blood flow, muscle strength and skeletal muscle histology in severe congestive heart failure. Am J Cardiol 62: 49E–52E
42. Rector TS, Olivari MT, Levine TB, Francis GS, Cohn JN (1987) Predicting survival for an individual with congestive heart failure using the plasma norepinephrine concentration. Am Heart J 114: 148–152
43. Samek L, Hauf GF, Roskamm H (1990) Interval training in candidates for heart transplantation. Eur Heart J 11 (Abstract supplement)

44. Sullivan MJ, Higginbotham MB, Cobb FR (1988) Exercise training in patients with severe left ventricular dysfunction. Hemodynamic and metabolic effects. Circulation 78: 506–515
45. Sullivan MJ, Atwood JE, Myers J, Feuer J, Hall P, Kellerman B, Forbes S, Froelicher V (1989) Increased exercise capacity after digoxin administration in patients with heart failure. J Am Coll Cardiol 13: 1138–1143
46. Sullivan MJ, Higginbotham MB, Cobb FR (1989) Exercise training in patients with chronic heart failure delays ventilatory anaerobic threshold and improves submaximal exercise performance. Circulation 79: 324–329
47. Sullivan MJ, Knight JD, Higginbotham MB, Cobb FR (1989) Relation between central and peripheral hemodynamics during exercise in patients with chronic heart failure. Muscle blood flow is reduced with maintenance of arterial perfusion pressure. Circulation 80: 769–781
48. Terjung RL, Mathien GM, Erney TP, Ogilvie RW (1988) Peripheral adaptations to low blood flow in muscle during exercise. Am J Cardiol 62: 15E–19E
49. Wenger NK (1989) Quality of life: Can it and should it be assessed in patients with heart failure? Cardiology 76: 391–399
50. Zelis R, Sinoway LI, Musch TI, Davis D, Just H (1988) Regional blood flow in congestive heart failure: concept of compensatory mechanisms with short and long time constants. Am J Cardiol 62: 2E–8E

Quality of Life of Infants and Children with Cardiac Disease and Cardiac Failure Both Before and After Cardiac Surgery

E. W. Keck

Patients who require treatment by the pediatric cardiologist for cardiac failure are generally very young: neonates and infants. In pediatric practice, "big" children, in other words pre-school and school-age children suffering from the classical features of cardiac failure, are much less common. How does this shift toward the beginning of life and the early years come about? To answer this, we must investigate the causes of cardiac failure in children and their main symptoms. They are: (a) congenital cardiac malformations, (b) disturbances of rhythm, (c) diseases of the myocardium, and (d) cardiac valve dysfunctions.

The signs and symptoms of cardiac failure in infants are: failure to thrive, lack of weight gain; tachypnea, dyspnea, tachycardia; increased sweating, pallid gray skin color; and weak pulse, not always easy to feel.

The signs of disease develop after birth, sometimes after only a few hours on the first day, sometimes after a few days or weeks.

If we group the congenital cardiac and vascular malformations together into three main types, they are as follows:

1. Stenosis or atresia of heart valves or vessels (aorta or pulmonary) (Fig. 1 a, b)

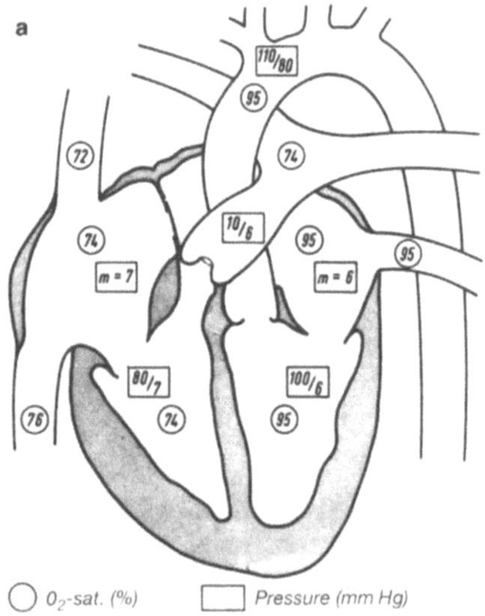

Fig. 1 a.

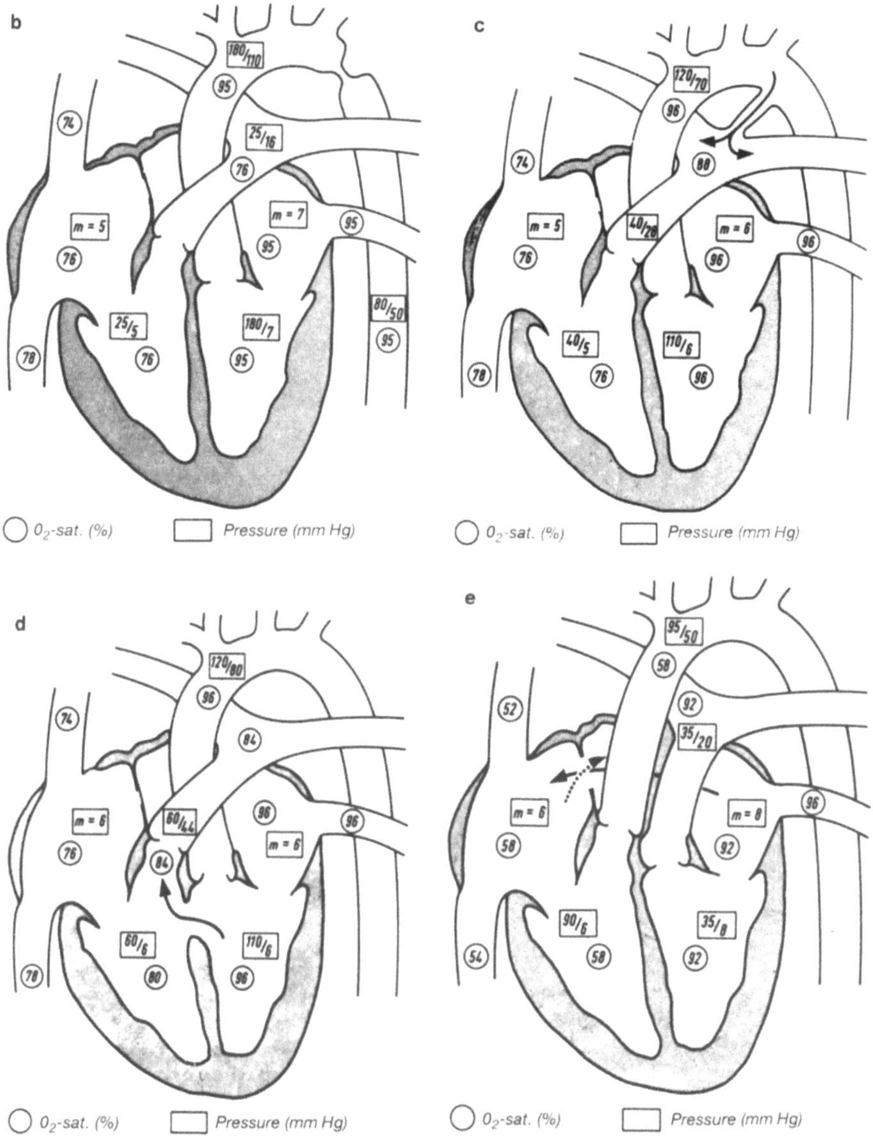

Fig. 1a–e. Stenotic or atresic heart valves or vessels. **a** Stenosis of the pulmonary valve. **b** Coarctation of the aorta. **c, d** defects which lead to connections between the systemic and the pulmonary circulations. **c** Persisting ductus arteriosus. **d** Ventricular septal defect. **e** Malpositioning of the great cardiac arteries. Transposition of the great arteries

2. Defects which create connections between the systemic and pulmonary circulations (Fig. 1c, d)
3. Errors in the origin or termination of the great cardiac arteries or veins (Fig. 1e)

Before birth, the pulmonary circulation is at rest, and the fetal paths of circulation, ductus arteriosus, and foramen ovale, are open, gaseous exchange and supply of nutrients occur in the placenta, which is supplied by the systemic circulation. Thus the embryo, and later the fetus, grows without major disturbances and develops normally. A *neonate* that appears to be healthy and without

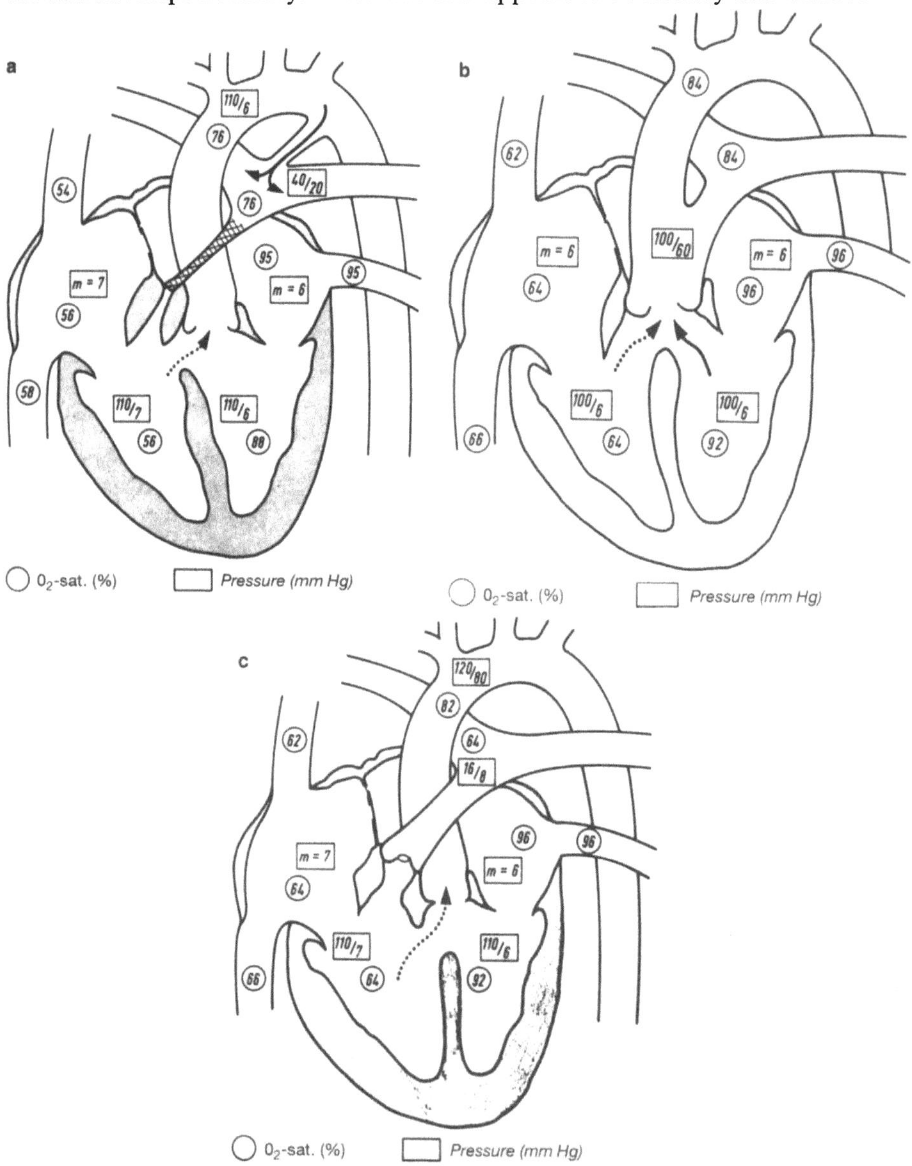

Fig. 2 a–c. Inadequate perfusion of the lungs with dependence upon an open ductus arteriosus. **a** Pulmonary atresia with VSD and PDA. Connection of the lungs to the systemic circulation. **b** Truncus arteriosus. Reduced perfusion of the pulmonary circulation with cyanosis. **c** Fallot's tetralogy any sign of disease at first is brought into the world. Opening up of the pulmonary circulation and closure of the ductus or foramen ovale then lead to the unmasking of abnormalities of the heart and blood vessels:

1. Inadequate perfusion of the lungs (in pulmonary stenosis or atresia) with dependence upon an open ductus arteriosus (Fig. 2 a–c).
2. Connection of the lungs and the systemic circulation [ventricular septal defect (VSD), truncus, patent ductus arteriosus, and others].
3. Parallel diversion of the circulation [transposition of great arteries (TGA) (Fig. 1 e)].

The heart and circulation are not prepared and not trained for these hemodynamic errors. The left ventricle, which is still weak, is not able to cope with the amount of blood from the overflowing lungs, or cannot overcome barriers such as aortic stenosis or coarctation of the aorta, so that it fails. Increased pre- and afterload are present. Back pressure develops in the left atrium and pulmonary veins and is expressed as dyspnea, increased fluid in the lungs, and pulmonary edema. Or hypoxia and, with it, the clearly recognized symptom "cyanosis" dominate the developing clinical picture: reduced pulmonary perfusion and origin of the aorta from the right ventricle (ventricle serving the pulmonary circulation). Hypoxia, acidosis, and stasis in the great veins of the heart develop, with hepatomegaly and edema.

How do these signs of disease impair or affect the life of the neonate or infant; in other words, its quality of life? Normal life is thought of as existing in obscurity: sleep, feeding and growing, gradual development of mind and body, maturation of the lungs and of the CNS, the brain with its awakening functions. Rapid exhaustion when feeding, breathlessness, restlessness when awake, and the sleep of exhaustion perceptibly disturb normal development. The reserves in illnesses which are at first mild, such as colds (rhinitis), are rapidly exhausted and the blood-engorged lungs become predisposed to infection, i.e., bronchitis and bronchopneumonia. Dilated branches of the pulmonary artery and pooling of blood in the peripheral vascular tree hamper ventilation and gaseous exchange.

Since infants depend for their life upon the complete support and care of the mother, the *mother's life,* her quality of life, and that of the family are fundamentally changed. The sight of the rapidly breathing, pale cyanotic child, whose feeding times are laborious, greatly disturbs the mother and, indeed, both parents. Frequently, small feeds with inadequate liquid intake given day and night disturb family life and destroy the peace in which the developing infant needs to be cocooned.

Inadequate weight gain, less than 80 g/week, or even a cessation of weight gain, are reasons for major concern. Some of these symptoms will not immediately lead to the diagnosis of cardiac disease or cardiac failure being suspected. Only a thorough examination, taking note of murmurs (which are sometimes absent or atypical), with assessment of the quality of the peripheral arterial pulse and the skin color, reveals the presence of cardiac disease. Further diagnosis with ECG, phonocardiography, X-rays, and echocardiography leads to the diagnosis, which only in exceptional cases requires confirmation by cardiac catheterization and angiocardiography. These invasive methods of investigation provide finer and more exact information about hemodynamic and morphological details. They require admission to hospital and thus mean a further major encroachment upon the life of the infant, its mother, and its family.

Treatment

The treatment of cardiac failure and errors of cardiac function includes general measures, medical treatment, and palliative or corrective surgery.

Apprehension, fear, and hope, which fill older children and adults when they learn that they have cardiac disease and require surgery, are transferred wholly to their parents in the case of neonates and infants. Informative discussion about the nature of the cardiac abnormality, the type and risk of diagnostic interventions, the necessity for and possibility of surgical measures, the prognosis with or without invasive treatment – all these are included in discussions between the doctor and parents. Here, the pediatric cardiologist may observe the needs and fears, the stresses and deep attacks upon the life of the mother, the two parents, and the whole family. His ability to empathize, his readiness to clarify things in discussion, and his leadership are put to the test as much as his scientific knowledge and skills.

Conservative treatment of cardiac failure can only be carried out with the help of the parents and with nursing staff in the hospital. The so-called general measures such as bedding, administration of food as many small meals or by nasogastric tube, administration of oxygen, and avoidance of disturbances can only be attained and prove successful if parents and nurses approach them with understanding and consistency. Medical treatment with digitalis, diuretics, possibly with afterload reduction or catecholamines, requires dosage and administration of minute quantities of medication. While parenteral administration is confined to hospital, oral use and, particularly, long-term treatment is also carried out at home. The recognizable response to treatment then leads to a reduction in breathlessness, in respiratory and heart rates, to better staying power in drinking and eating, and, after initial weight loss due to correction of edema and hypervolemia, to growth and "real" weight gain. The improvement in the quality of life of infants is apparent, and it affects the parents and the rest of the family. By careful monitoring, e.g., by the pediatric practitioner in attendance, with review of the doses of medication and their adaptation to body weight and monitoring of signs of cardiac failure, the initial success can be maintained.

The favorable natural history of the most common cardiac abnormalities makes the first months of life and the first year the most difficult. A spontaneous reduction in the size of the defect (in VSD), adaptation of pulmonary vascular resistance to too high a flow volume (Q pulm) or the supply of the lungs by collateral vessels (in pulmonary atresia) and the strengthening of the ventricles in aortic or pulmonary valve stenosis, lead to a reduction in cardiac failure and, in spite of a possible delay, allow the child to develop.

Unfortunately the natural history and prognoses of other cardiac abnormalities are less favorable. Very early, often highly risky, interventions are then required to preserve the life of the neonate or to alleviate the severe symptoms of failure. A few examples are as follows:

1. Connection of the circulations in TGA by dilatation of the holes in the atrial septum with balloon atrioseptostomy (BAS)
2. Attachment of the pulmonary artery to the aorta or its branches in high-grade pulmonary stenosis or atresia
3. Correction of coarctation of the aorta and, with it, improvement of renal blood flow and blood flow in the lower half of the body
4. Opening up of a high-grade "critical" aortic valve stenosis
5. Narrowing of the pulmonary artery, "banding" in anomalies that lead to flooding of the lungs and to early development of an increase in obstructive pulmonary resistance and vessel disease (in DORV, DILV, large and multiple VSD)

Preschool Children

Cardiac Surgery

The optimal age for surgery in the case of many simpler congenital malformations of the heart is in the nursery and preschool years, in other words, in the 4th–6th years of life. Several favorable conditions occur together at that time:

1. The cardiac anomaly has proved to be constant; i.e., a ventricular septal defect, for example, shows, by its spontaneous history over several years, that it is not becoming smaller or closing.

2. The anatomical relationships are favorable to intracardiac operations on the heart-lung machine because of the patient's size and weight. The greater difficulties and higher risks of surgery in infancy are no longer present; any further improvement as a result of increase in body size is not to be expected.

3. The children, at this age – surrounded by the care of parents, nurses, and doctors - do not have the fear of surgery or the anxieties of older children and adults. The risks and possible complications which this kind of surgery brings with it are hidden from them, and the postoperative complaints and discomforts during the - usually short – period of intensive treatment are tolerated patiently at this age. The most important thing for these small patients is to ensure the presence of the mother – or both parents – and a speedy return to a position where they have greater freedom, e.g., their favorite, well-loved tricycle. It is far from unusual for these children to be sitting on their tricycles and riding from room to room 4 days after the closure of an atrial or ventricular septal defect, or after correction of a coarctation of the aorta or aortic stenosis. The quality of their lives rapidly returns to normal. This phase is some-

what delayed in children with preoperative cyanosis, i. e., in Fallot's tetralogy and similar anomalies.

It is, however, precisely these patients whose life is markedly, decisively altered, for the 4- to 7-year-olds with atrial or ventricular septal defects and coarctation of the aorta are free of subjective symptoms prior to the operation; they are almost always capable of performing as they should at that age – although the term "performance" is scarcely appropriate. They can lead an unhindered, normal life, playing and romping at playschool or nursery. They do not suffer from cardiac failure and require digitalis or diuretics. *Why,* one may ask, *do they require surgery at all?*

The Apparently Healthy Child With Cardiac Anomalies

It is precisely this question that is asked by parents! Again, it is the care-laden parents whose life is changed – at least in the period of waiting and the anxious period during and immediately after cardiac or vascular surgery. It is they whose usual trains of thought and daily routines are affected by awareness of the impaired "wholeness" of their child, and by concern about and fear of hospital admission and cardiac surgery; indeed, of deeply invasive surgery. What is more, this is a child who appears healthy, lives without pain or symptoms, and is growing. And again, the parents and families are more greatly affected by the existence of a congenital malformation of the heart than the unconcerned patient himself, unless he is prevented by groundless and unnecessary fear and solicitude from going about freely and without hindrance. It is enough for the children to pause, to take a break from their games of their own accord when signs of stress-insufficiency – generally dyspnea and fatigue – occur.

Ill-considered advice from someone else: "You shouldn't run so fast, you shouldn't climb that, bathe or swim for so long . . ." etc. is unnecessary! This overprotection only deepens the awareness of having heart disease in an unhelpful fashion and also leads to children becoming recalcitrant, continually wanting to make exceptions, and hiding their reluctance to carry out necessary activities behind their cardiac anomaly. These are the types of damage which cannot be remedied by successful operation!

Indications for Heart Surgery

The indication for surgery at the optimal age mentioned above is knowledge of the spontaneous course, the natural history of congenital malformations of the heart. A hemodynamically relevant deficiency, e.g., ASD and VSD with large left-to-right shunt (i. e., when at least 40%–50% of the pulmonary circulation is derived from blood flowing through the defect) (Fig. 3a,b), or stenosis of valves or vessels with pressure gradients in excess of 50 mmHg, or with clinical cyanosis, leads to a reduction in life expectancy. Careful investigation by Campbell shows that most anomalies of this type halve life expectancy, i.e., half of these patients die in the 4th decade of life. Cardiac failure

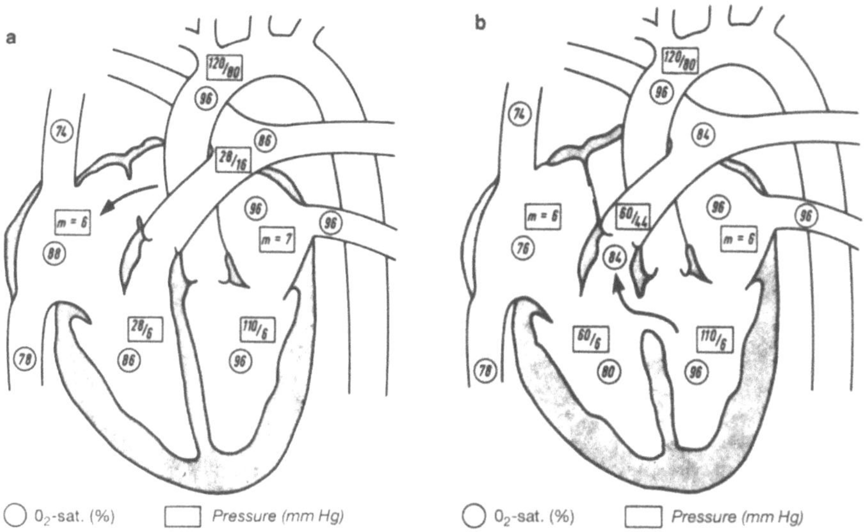

Fig. 3. **a** Atrial septal defect *(ASD)* with large left-to-right shunt. **b** Ventricular septal defect (VSD) with large left-to-right shunt

and supervention of disturbances of rhythm, lead to an early and decisive reduction in physical performance. Pregnancy and giving birth and the daily stresses of housework and occupation, become tiring, difficult to tolerate, or even impossible; all this at an age at which healthy people are in the middle of their careers and watching their families grow up. The onset of overt symptoms should no longer determine the time of operation. Knowledge of the history without treatment modifies the indications toward a corrective intervention. Very different are the symptoms before and the changes brought about by the operation in children with frank cyanosis, as has already been indicated. Cardiac anomalies with reduced perfusion of the lungs and a right-to-left shunt, such as those present in Fallot's tetralogy, affect growth, development, and stamina when playing or walking to a marked degree. The hypoxia and the chronic lack of oxygen lead to obvious changes. The children show delayed development and learn to sit and walk late. They are, as one generally expresses it, a little retarded. This should not be taken to include their mental development. Parents are concerned that the reduced oxygen content may injure the brain and have an adverse effect upon the mental performance of their children. This is not the case. There are, however, children whose symptoms are so great that they make attendance at nursery or school difficult, if not impossible. They are not able to travel to school by foot or, later, by bicycle. They have to be driven. Some are taught at home, being unable to endure 4 h at nursery or 5 h at school; and all this places restrictions upon their lives. In some, hypoxia appears to produce a change in mood. Small children cry often and are discontented, querulous, and unhappy. It is a great joy to see how these children change for the better when palliative, or – better still – corrective surgery alleviates or cures the cyanosis. The external appearance, with blueness of the extremities and lips, with clubbing of the nails and drumstick fingers, changes, and the child becomes happy, energetic, and strong. It acquires an appetite and grows. The parents think they have a new child, and say so; and it is not unusual for the anniversary of the operation to be celebrated as another birthday. These experiences give pediatric cardiologists, surgeons, and all their colleagues new hope and strength in their work. Dangers that the child is unaware of, and which the parents do not fully believe in, such as the development of a brain abscess due to hypoxia, polycythemia, and connections between the arteries serving the brain and venous return form the body, are now averted. Long-term studies of physical development, on the subcutaneous fat and muscle, reveal the effect of this type of heart abnormality and its correction [1].

Urgent and Life-Threatening Indications

The occurrence of attacks of hypoxemia renders operative intervention necessary at a time far removed from the optimal age for a selective operation, namely in late infancy and at the start of the toddler period, i.e., early in the 2nd year of life. A state and attacks of hypoxemia are threatening events. Unprepared parents believe that they are going to lose their child. At first the child screams and its breathing is very labored, it then becomes more cyanotic and apathetic, and may roll its eyes, have a fit, and lose consciousness. Infants with Fallot's tetralogy, and often those very children who show little sign of cyanosis at rest, may get these attacks.

The cause is a reduction in pulmonary perfusion due to functional narrowing of the infundibulum of the right ventricle (RVOTO) and an increase in the right-to-left shunt through the VSD. Polycythemia with relatively low Hb content is responsible for a predisposition toward hypoxic attacks.

It is extremely important – nay, essential – to tell the parents of these children about the possibility, manifestations and course, and cause and treatment of such attacks at the time of diagnosis, during a thorough discussion of the case. The parents must not experience their child's first attack without preparation. We give them instruction – using their own child – in emergency treatment (Fig. 4) which mimics squatting in older children. In this way panic, the telephone and the emergency doctor with his blue light – and thus a delay in or the omission of more immediate measures – are avoided.

The marked extent to which these cardiac anomalies change and affect the quality of life of these young patients and that of their entire family is well known.

Palliative surgery, which involves connecting the great pulmonary arteries to the systemic circulation, in other words Blalock and Taussig's anastomosis, was and is the salvation of these children. We have had 44 years' experience of helping "blue babies" with this operation. But, in some children, even restoration of the hemoglobin with oral iron may cause these attacks to disappear. We – and others – have experienced good results with the administration of β-

Fig. 4. An 18 months old child with tetralogy of Fallot. A hypoxic attack is treated with this knee-chest position which resembles squatting of all children. (Drawing by Pascal Keck)

blockers. Dociton (Propanolol) can widen the infundibulum and the outflow channel of the right ventricle, and can thus alleviate the occurrence of attacks of hypoxemia. This conservative treatment may spare some patient's palliative surgery, so that they can undergo major corrective surgery for their Fallot's tetralogy in the 3 or 4th year of life.

This procedure requires good anatomical relations, i.e., the pulmonary artery and its branches must be well developed. Such niceties can only be recognized by angiocardiography and thus necessitate cardiac catheterization.

School-Age Children

So far, we have been discussing the effect of congenital malformations of the heart upon the lives of *infants* and, thus, upon the lives of their families, and the problems and situations that arise as a result of the symptoms and treatment of *pre-school children.* Let us now turn to *schoolchildren.* This is a large expanse of time, 8–12 important years in the life of a human being! The 6- to 14-year-olds are more aware of the symptoms of their disease. Interruptions in schooling, being excused from gymnastics and school games, together with hospital admissions for diagnosis or treatment, place a different burden upon the child itself. The parents must care for their other children, and therefore cannot always remain with the child in hospital. This leads to tearful partings and homesickness. Both these painful experiences have a deeper effect and are more painful for most children than the physical pain caused by cardiac catheterization or an operation. Since admission of the mother has become possible and can be used liberally, this picture has changed. The majority of parents make use of this possibility. They take turns to remain with their child, while the other partner cares for the brothers and sisters at home. Much anxiety is dispelled this way. After surgery, the child, still intubated and ventilated, wakes to see its mother and father: a great comfort. Part of the nursing care can be taken over by the parents: care of the mouth, skin, and nappy area. The sight of their own child with an endotracheal tube, intravascular lines, drains, urinary catheter, fresh operation wound, and visual and auditory alarms on the monitors, together with the other children, who are often severely ill, in the intensive care ward – all this is an experience and a burden that one never experiences in everyday life outside the hospital. Not all parents are mature enough to face these experiences and situations, particularly when lack of sleep and separation from their family and home are added. Thus the whole period of hospital admission is a time of new experiences, renunciation, and sacrifice, which binds parents and children even closer to one another. It is a period of profound renewal in the life and quality of life of the entire family.

When all therapeutic measures, cardiac surgery, and treatment of cardiac failure or disturbance of rhythm have run their course successfully, there is a rapid return to normal life, school, and the family. It is a source of joy, but also of wonder, to see how the children react at their first outpatient visit, some 4 months after surgery. As soon as they know that they can rely upon the examination being painless, and that the visit will have a fixed time limit, the reunion with the hospital and its doctors and nurses is freed from tension and there is no recognizable stress or fear.

The Child Who is Gravely Ill

The picture is quite different for children who have to be readmitted for intensive diagnosis and treatment. Tears and worries are common, and the parents are also worried and anxious. Does the doctor always make sure that readmission to hospital is absolutely necessary? Has he given even a passing thought to the crucial changes that this means for a family, for the parents, and their situation at work?

Nevertheless, all these trials and tribulations are endurable. They may leave scars on the soul and on the skin. But the death of a child is far more painful and is never completely overcome or forgotten. Some malformations of the heart are incompatible with life. The early death of a neonate usually occurs during treatment in an intensive care ward. Close contact between mother and child has scarcely developed; separation soon after birth was unavoidable (Fig. 5). For some malformations, the diagnosis can be made using methods that do not involve shedding blood – that are not invasive – such as echocardiography in hypoplastic left heart syndrome. Diagnostic operations and unsuccessful attempts at treatment should be avoided. The loss of a child, often their first, is painful for the parents. It can, however, be overcome by hope for a new healthy child, particularly when they are young.

The anxious questions and concerns which occur before another birth, the knowledge of the risk of recurrence, which come into counselling, require another paper. These are different and novel facets when we consider the term: quality of our human life!

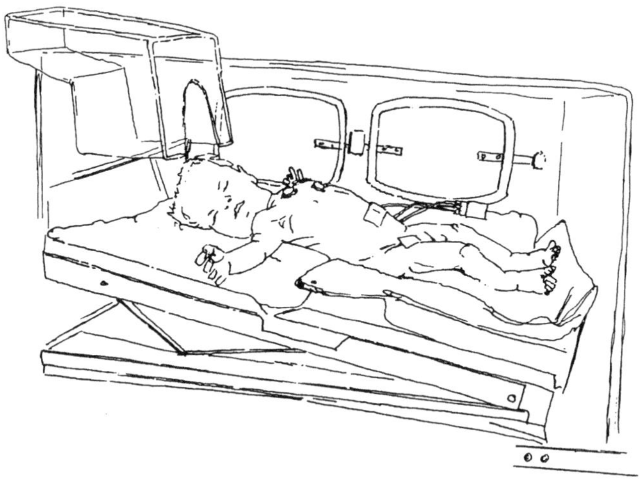

Fig. 5. A newborn child in an incubator. This is how the mother sees her baby. A close contact is almost impossible. But oxygen and temperatur can be controlled and altered. (Drawing by Pascal Keck)

The death of an older child impinges upon the life of parents and brothers and sisters in quite a different way. There is no cardiology or heart surgery without mortality, without loss of some patients. Several admissions for invasive diagnostic techniques and operations, with thorough discussion and difficult decisions, have passed and have bound doctors and parents closer to one another. The doctors and nurses have also become more closely attached to the sick child. And now the child will not return home from the cardiac surgery unit or it dies after days, sometimes weeks, of treatment in the intensive care unit. Failure of all efforts, release from a heart condition that was too difficult to treat? The course of events and opinions about it are very different, but the pain and loss of the cherished and much-loved child are always hard and always the same for the parents.

Older Children and Young Adults

Older schoolchildren and young adults are the third large group of patients whose life and problems we wish to discuss. They include three very different types of heart patient:

1. Young people who underwent successful surgery in childhood and who are now symptom free and able to live without restrictions or treatment
2. Patients whose cardiac abnormality was improved by medical or surgical treatment, but whose lives are restricted by late consequences and residual effects and who require treatment
3. Inoperable patients who suffer more markedly from the restrictions and symptoms which their cardiac disease brings as their bodies develop, and as their lives become increasingly demanding

We must also mention a large number of young people with a mild abnormality of the heart which has no consequences for their way of life or life expectancy, and which exists only as an undercurrent in the consciousness of the person affected. These young people have problems similar to or even easier than group (1) above.

It is very important that those who have been treated successfully and those with minimal heart disease view themselves as whole and healthy. It can now be seen how important it was whether or not they were encouraged to lead a normal life by their family when they were younger children at the time of their operation. Unnecessary restrictions and special treatment imposed by parents now prove to have been errors of behavior which are difficult to correct. Any effort to participate and compete, e.g., in sport, within the school and outside, in school examinations or in training and in study, can only develop if the brake was not put on in childhood. One of the most important pieces of medical advice runs thus: despite heart disease, as normal a life as possible! That there are certain restrictions in energetic sports – according to the type and severity of the heart disease – is self-evident. Advice should be given on an individual basis.

Exemption from military service for those who have undergone cardiac surgery already represents a significantly different treatment in comparison with healthy subjects. Only a proportion of young men greet it as a welcome measure; some wish to serve, but others are glad to begin a continuous period of training, or to start their studies. Exemptions from military service and knowledge of the heart disease or operation they have survived may, however, also be a hindrance in seeking a training post and a job. Competition is severe, and unemployment among young people is significant. Rejection when seeking work also threatens the now healthy and successfully treated former heart patient. It is here that the doctor must intervene and use convincing arguments. The effect of a certificate of reduced capacity for work, carelessly given, is hard to shake off. Parents want to do something good for their child's future, which may be transformed into the opposite and become a hindrance. Early, forceful explanation and warning by the doctor is essential. It is not easy to talk someone out of a supposed financial advantage. Happily, the majority of young people aim for normality and wish neither for special consideration nor for exceptional treatment. In many, the large thoracotomy scar is the only thing that reminds them of the heart disease and the operation.

The Scar

This is never a problem for young men, but may be for girls, if a great deal of it can be seen at the neckline or when wearing a bathing costume. In general, those who are asked about it do not make any severe complaints. They themselves are used to their scar, and they expect their friends to like and accept them, scar and all.

Own Children

With all the effort of those who have been cured to lead a normal life at work and in the family, one point of view should not be forgotten: the offspring. Those who are themselves affected and their – perhaps future – marriage partner must know that the possibility of having a child with heart disease exists. The word "risk" should, perhaps, be avoided. Any parent-to-be risks having a child with a congenital defect. In the case of cardiac and vascular malformations, 8 out of 1000 neonates have this type of defect (0.8%). Until recently, a risk of reappearance in the next generation of about 4% was quoted for the most important and most common abnormalities. A very careful investigation has now shown that the figure is much higher, namely 14%. Mild defects and those which close spontaneously (small ventricular septal defects) were, however, included. But the mental strain of transmitting to the next generation and to one's own child a defect which one has had to endure oneself is not easily removed and requires understanding on the part of one's partner.

Late Sequelae and Problems

When the patient has survived heart disease, but obvious symptoms thereof are still present and serve as a daily reminder that he or she is not completely well, the problems mentioned above are much more prominent and self-evident. Complicated congenital malformations of the heart, such as Fallot's tetralogy, atrioventricular septal defects (AVSD), and transposition of the great arteries (TGA), cannot always be completely corrected. Residual defects, impaired perfusion of the lungs and disturbances of rhythm, cardiac failure in response to a load and the necessity for continuous medication hinder the day-to-day life at home, on the way to work and during the working day. The patient has to give up many things: he requires medical attention and care. He will be fortunate if he is employed in a job, freely chosen by himself, and is thus able to earn his living. Whether it is possible to find an understanding and considerate companion in life as well is another question. But one is surprised, time and time again, at the extent to which young people overcome a handicap and how, despite the external signs of their defect and diminished physical performance, a new family is started. The care required to keep a sick person occupied and to give him permanent employment will seem very restrictive if not offputting to many employers. Paid leave for sickness or early retirement does not correspond to the requirements of an employer in a society based on performance. It is therefore important that the doctor counsels the family and patient in good time, so that he will strive for a good education which prepares him for jobs not requiring hard physical activity. A completed education, and one that is as long as possible, will open the way to occupations that are compatible with a restricted way of life for many young patients with heart disease.

Life Expectancy

Although there are some preliminary studies on working life and performance in so-called normal life with family, job, and leisure pursuits, there is not yet any conclusive information about life expectancy for those who have had heart surgery. The first intracardiac operations for septal defects and complex malformations took place in the years 1950–1960, in other words, some 35 years ago. The risks and perioperative complications were, at the time, higher than they are today. Operative techniques and surgery results have been improved. Patients operated upon in childhood and now in the 5th and 6th decades of life are still rare. Symptoms of cardiac failure in old age, late postoperative complications as a result of irreversible myocardial changes, of reduced coronary artery perfusion, or of disturbances of rhythm and conductive tissue defects are probably more common than in men of a similar age who have not had a congenital cardiac malformation or other form of heart disease in childhood. Thus, we still do not know whether coronary heart disease, which is among the most common causes of death in our society, is more common in those with hypertrophied and dilated hearts as a result of hemodynamic errors due to congenital disease – or perhaps less common? Will the coronary vessels, which were displaced during anatomical correction of transposition of the great arteries (TGA), succumb to late narrowing and occlusive diseases, which lead to ischemia and infarction? This type of operation has only been carried out in large numbers during the past 10 years.

There are thus still many unanswered questions about these three important areas of human life:

1. Individual physical health and performance
2. Life expectancy of individuals
3. The incidence of congenital heart disease in subsequent generations.

Moreover, these three important facets of life are called into question at a much earlier point in time in young people with inoperable heart disease.

Inoperable Patients

Are there still cardiac malformations that cannot be operated upon in 1991? The answer is "yes." Palliative surgery, heart, and heart-lung transplants may, it is true, improve the quality of life and lead to an increased life expectancy, but any talk of a cure or full health is out of the question.

The close relationship between heart and lungs, the functional unit of the pulmonary circulation, right atrium and right ventricle, pulmonary arteries and all their branches, the fine pulmonary vascular bed, its veins and the left atrium, is changed and disturbed in many heart diseases. Thus, in pulmonary atresia, perfusion of the lungs via branches of the systemic circulation is secured by a persisting ductus arteriosus or by collaterals from the bronchial arteries and other main branches of the aorta. Artificial connections – anastomoses (e.g., Blalock-Taussig) – may improve or take over responsibility for lung perfusion. But too great a number of arterial connections between lungs and systemic circulation, particularly those involving systemic pressure, make corrective surgery impossible. This is also true of diseases in which the lungs are connected with the high-pressure section of the systemic circulation via the pulmonary trunk; in other words, in malformations which lead to pulmonary hypertension, to obstructive disease of the pulmonary vessels (Eisenmenger): large atrioventricular septal defects (AVSD), ventricular septal defects (VSD), truncus, single ventricle, transposition with VSD (TGA with VSD) are some of these. If they are not operated upon successfully in the early years – in other words, before irreversible changes in pulmonary vessels have occurred – or treated palliatively by artificial narrowing of the pulmonary trunk, the Banding operation, later surgery is impossible.

Pulmonary hypertension and pulmonary sclerosis may, however, result in a fairly prolonged span of life with few symptoms. The initial increase in blood flow to the lungs, with dyspnea and left heart failure, is then throttled back, and the right ventricle becomes strong and hypertrophied and shows no obvious signs of right heart failure. Patients with this circulatory conformation can live into the 4th decade. When marked cyanosis, polycythemia and high blood viscosity develop, when hepatomegaly and edema occur, or when performance in general and at work is restricted, the desire is awakened for more comprehensive treatment and more thorough improvement. These are the patients who may benefit from a heart-lung transplant. But they must know, and the doctor must prepare them for the fact, that rejection and infection have to be monitored and prevented by close medical care and continuous medication. A free and healthy life without medicaments that have intrusive side effects is not possible. Progressive changes in the lungs, pulmonary fibrosis with restricted lung function, are seen in the majority of patients and cloud the hopes of a prolonged life with few symptoms. A second transplant is a major undertaking which not very many people are able to face, either physically or psychologically.

References

1. Baum D, Stern MP (1977) Adipose hypocellularity in cyanotic congenital heart disease. Circulation 55: 916
2. Bentovim A (1983) Psychiatric and intellectual assessment. In: Anderson RH, Macartney FJ, Shineborne EA, Thynan M (eds) Paediatric cardiology, vol 5. Churchill Livingstone, Edinburgh, pp 309–316
3. Campbell M (1968) Natural history of persistent ductus arteriosus. Br Heart J 30: 4–13
4. Campbell M (1970) Natural history of atrial septal defect. Br Heart J 32: 820–826
5. Campbell M (1970) Natural history of coarctation of the aorta. Br Heart J 32: 633–640
6. Corone P, Vasseur F, Chareire ML (1986) Relationship of problems of young adults with congenital heart disease. In: Link. (An information bulletin about psycho-social problems in congenital heart disease) R. Verhaaren. Univ Hosp Gent, pp 32–33
7. Dennis NR (1983) Genetic conselling. In: Anderson RH, Macartney FJ, Shinebourne EA, Thynan M (eds) Paediatric cardiology, vol 5. Churchill Livingstone, Edinburgh, pp 328–331

8. Engle MA, Goodwin JE (1981) Fate of apparently successfully treated survivors of congenital heart disease (introduction). In: Godman MJ (ed) Paediatric cardiology. Churchill Livingstone, Edinburgh, pp 546–552
9. Ferencz C (1981) Some observations on psychosocial influences. In: Godman M (ed) Pediatric cardiology. Churchill Livingstone, New York, pp 561–565
10. Ferencz C, Wiegmann FL Jr, Dunning RE (1980) Medical knowledge of young persons with heart disease. J Sch Health 29: 133–136
11. Hilgenberg F (1989) Psychosoziale Aspekte bei der Betreuung chronisch herzkranker Kinder. In: Singer H (ed) Herzerkrankungen im Kindes- und Jugendalter. Perimed-Fachbuch-Verlagsgesellschaft, Erlangen (Kinderheilkunde und Jugendmedizin, vol 4): 115–121
12. Kaplan S (1977) Long-term results after surgical treatment of congenital heart disease. Mod Concepts Cardiovasc Dis 46: 115–121
13. Keck EW (1984) Angeborene Herzfehler: operiert – geheilt? Postoperative Probleme herzoperierter Jugendlicher und junger Erwachsener. Herz/Kreisl 16: 213, 269–276
14. Keck EW (1985) 'Cured' congenital heart disease. In: Julian DG, Wenger NK (eds) Cardiac problems of the adolescent and young adult. Butterworths, London, Boston, pp 28–44
15. Keck EW (1989) Befunde und Probleme nach Herzoperationen bei Kindern und Jugendlichen. In: Singer H (ed) Herzerkrankungen im Kindes- und Jugendalter. Perimed-Fachbuch-Verlagsgesellschaft, Erlangen (Kinderheilkunde und Jugendmedizin, vol 4): 87–114
16. Manning JA (1981) Insurability and employability problems of the cardiac survivor. In: Godman MJ (ed) Paediatric cardiology. Churchill Livingstone, Edinburgh, pp 569–572
17. Moss AJ (1979) What every primary physician should know about the postoperative cardiac patient. Pediatrics 63: 320–330
18. Pernot C (1982) Schooling of the cardiac child. In: Link (ed.) An information bulletin about psycho-social problems in congenital heart disaease. R. Verhaaren, Univ Hosp Gent, vol 1: 13–16
19. Rautenburg HW (1986) Herzfehler im Kindes- und Jugendalter. Katamnestische Langzeitstudie über 25 Jahre. Enke, Stuttgart
20. Silverman D (1983) Family adaption to congenital heart disease: adjusting to physical and moral realities. In: Anderson RH, Macartney FJ, Shinebourne EA, Thynan M (eds) Paediatric cardiology, vol 5. Churchill Livingstone, Edinburgh, pp 317–326
21. Sommerville W (1981) Licence to drive. In: Godman MJ (ed) Paediatric cardiology. Churchill Livingstone, Edinburgh, pp 566–568
22. Somerville J (1983) Assessment after operation of congenital heart disease (introduction). In: Anderson RH, Macartney FJ, Shinebourne EA, Thynan M (eds) Paediatric cardiology, vol 5. Churchill Livingstone, Edinburgh, pp 297–299
23. Ternestedt Östman B-M (1989) Hjärtopererad som barn – Hur gick det sedan? En deskriptiv studie av 32 hjärtopererade personer. (Heart operation as a child – What about the future? A descriptive study of 32 individuals.) Graphic System, Göteborg (Socialmedicinsk tidsskriftsserie nr 50)
24. Whittemore R, Hobbins JC, Engle MA (1982) Pregnancy and its outcome in women with and without surgical treatment of congenital heart disease. Am J Cardiol 50: 641–651

Health Care and Economic Aspects of the Quality of Life of Heart Failure Patients

W. Pohl and J.-M. Graf v. d. Schulenburg

Nature of the Problem

An improvement in medical care, with its consequent reduction in mortality, may increase life expectation; but it also leads to an increase in the number of persons who are chronically ill, and to questions about their quality of life. Studies have shown that heart disease patients, in particular, experience a distinct decrease in their quality of life [10]. It follows that information regarding the quality of life and the burdens created by illness and disability in the lifespan preceding death is required.

Furthermore, funding is necessary for research on diseases – such as heart failure – the mortality which is increasing. However, funding is also necessary for research on and the provision of new preventive measures [26]. If sufficient funds were available to satisfy all the needs of society in general, and of the health care system in particular, then an economic analysis in this sector would be unnecessary. This condition, however, is not fulfilled and the above remarks apply to every form of organization in the health care system and to every country – with the exception of "the land of milk and honey" [1].

The inexhaustible demands of the health sector on the one hand, and the lack of available means on the other, make economic action necessary. Every measure in health care is linked to the consumption of social resources (work, capital, and energy), which consequently cannot be usefully applied to other areas of the health care system or of society. The costs of one type of health care are thus always balanced by the lost benefits of other measures which could not be carried out. It is the mission of health care economics to draw attention to this situation, and to develop approaches and methods for comparing benefits and opportunity costs (i. e., benefits lost because of the alternative use of resources). While doctors concentrate correctly on the possibilities for treating their patients, the economic perspective widens to include the alternative use of limited resources.

In order to illustrate this point of view, a basic summary of the health care from an economic point of view will be provided in the next section. The third section deals with a more medical delineation of the problems of heart failure and quality of life. In the fourth section, the economic problems will be illustrated on the basis of cost-benefit analyses.

Importance of Economic Considerations in Health Care

Intensified by the cost explosion in health care, and by ever decreasing financial resources, questions about the financing of applied resources and also about their alternative uses have increasingly gained in significance. Through the use of such economic criteria within the health care system, a new discipline, health care economics, has developed.

A comprehensive, generally recognized definition or delineation of the concept has not been forthcoming to date [14]. Health care economics should not only be seen as a mean of formulating an economic policy for reducing costs; rather more than ever before, the aim is to describe, explain, and make prognoses about the mechanics of disease and health by means of economic analysis and its methods as well as to select those instruments and analytical methods from the multitude of economic approaches which are best able to deal with the particular characteristics of the health care system [1]. It is a further aim of health care economics to develop a special theory to explain the specific relationship between input and output in the health care system [15].

Because of the problem of limited resources, it is the task of economics to analyze the efficiency of production and the distribution of limited resources [14]. Health indicators as well as mortality and morbidity figures point toward the targets and confirm the success with which the efficiency of health systems is measured [15]. Since quality of life is receiving more and more attention, additional criteria as to the quality of life will gain in significance in the future. The benefit of new therapeutic methods often lies mainly in the fact that the quality of life of the patient is improved, and the wishes of patients to undergo a treatment improving quality of life are increasing.

In order to solve the problem of limited resources, economists put forward optimality criteria [14]. According to the principles of economics, as an expression of the general principle of rationality, the maximum realization of a goal should be achieved within the funding available, or a specific goal should be attained with a minimum of resources [16].

Furthermore, resources that have, for example, been used within the health care system for the treatment of heart failure can no longer be used for treating other illnesses, nor for other purposes such as the construction of recreational parks. What is utilized in producing one benefit will be lost to the production of other benefits (alternative or opportunity-cost concept). Cost-benefit analysis is one of the methods of investigation employed here.

Economic theory shows that the use of financial resources in the health care system is optimal when the marginal yield of the resources used is distributed ceteris paribus in all areas or when a changed expense structure does not lead to an increase in the total yield [15]. Nevertheless, problems arise in making goals operational and in the analysis of effects, without which efficiency control in the health care system is hardly possible [15].

In this sense, health care economics which strives to have practical relevance cannot be "pure" economics [1]. In addition to the interdisciplinary cooperation between doctors, psychologists, and sociologists, health care economics has to take into account special aspects of the health care system.

Benefits provided by the health care system are predominantly services governed by the "uno-actu principle" [19]. As a rule, production and consumption coincide. Consequences of the uno-actu principle include the impossibility of storing the product (e.g., doctors' services), the availability of spare capacity and the extent to which the limit can be exceeded, and an increased uncertainty in the supply of health care services due to strong personal preferences which render quality control more difficult.

In addition, health care services are concerned with the future; i.e., the benefit of consumption today manifests itself at a later date. In combination with future benefits, the aspect of meritorious behavior has to be considered. Since, under certain circumstances, the initial demand for health care benefits is insufficient because of inadequate estimation of future demands (Böhm-Bawerk), an additional demand will develop on the part of some third party, the state or special institutions, which will be called upon because of their greater expertise in the subject matter [19]. The degree of such encroachments upon the sanctity of the individual remains controversial and worthy of discussion.

Moreover, health care services are, for the most part, a collective benefit to which the principle of nonexclusion applies, so that everyone expects to be entitled to a free ride. In order to discourage such expectations, and also to avoid suboptimal care, membership of a health insurance plan, for example, is compulsory. This in turn leads to inefficiency – disproportionately high utilization and unnecessary treatment.

Finally, the price of health care is essentially nonmarket or noncompetitive, having been laid down by the administration [17].

Although the peculiarities of the health care system have received attention from economists, the application of economic criteria to the health care system has sometimes been severely criticized, since a health care economy based on averages can be accused of inhumanity and placed on a level with euthanasia [1]. This criticism is the result of a misunderstanding. It is neither the aim nor the mission of health economies to determine how much an individual or a society should spend on medical care, but to try and ensure the effective and efficient use and distribution of the necessary resources available. The consequences for human beings of health care economic analysis must be regarded as ethical and moral problems [1].

A further criticism at the methodological level is directed at the use of the economic approach to the health care system. Interdisciplinary cooperation between economists and doctors is necessary.

This also seems sensible in view of the availability of insufficient data. In comparison with other countries, information on the state of health of the population, the resources of the health care system, the utilization of services, and the costs and financing of the health care system is inadequate [13]. In view of this paucity of data, it is not possible, for example, to estimate the economic costs of heart failure according to the ICD codes. It should be pointed out here that research into health care economics is seldom impeded by the fact that data are missing, but rather that data already compiled are difficult to obtain, have been poorly handled, and have been inadequately commented upon, even though their collection was financed with public monies. In addition to this there is, in the majority of cases, an adequate link between available data records and new data.

Before we give special consideration to the economic aspects of the quality of life of patients with heart failure, we should discuss more fully the relationship between heart failure and the quality of life.

Heart Failure and Quality of Life

Heart failure generally means inadequate functioning of the heart considered as a pump. The diseased heart no longer has the power to pump the blood returned to it by the veins into the peripheral circulation [25]. Heart failure is caused by multiple factors. The causal factors are, among others, ischemic heart disease, idiopathic cardiomyopathy, hypertension, dysfunctions in hormone and electrolyte metabolism, and arrhythmias [28]. Acute heart failure occurs, above all, as a result of myocardial infarction [20] (on the criteria for heart failure [28]).

While the incidence of myocardial infarction has slightly decreased during the last few years, the frequency and significance of heart failure have increased. Moreover, the risk for patients with heart failure of sudden cardiac death is markedly greater than for patients with coronary heart disease, or for patients in the 1st year after a myocardial infarction. In other words, the prognosis for patients with heart failure is worse than for those with coronary heart disease.

Investigations have shown that the probability of death occurring within 5 years from the time the diagnosis was made is over 50%, and in patients with therapy-resistant heart failure death within 1 year is to be expected in more than 50% of cases [28]. The significance of heart failure is also clearly illustrated by statistics on the cause of death. In 1987, 79754 people died in West Germany of a myocardial infarction, and in the same year heart failure was diagnosed as the cause of death in 50136 cases [29]. However, it is questionable whether the statistics on the cause of death provide reliable data. Comparison of the diagnosis on death certificates with autopsy findings has revealed a high rate of error. Despite this, it should be stated that the frequency and significance of heart failure have increased.

Following the recommendations of the New York Heart Association (NYHA), heart failure has been divided into four stages – among other things, for the assessment of therapeutic measures. These stages are applied in order to estimate the degree of severity based on stress tolerance (Fig. 1).

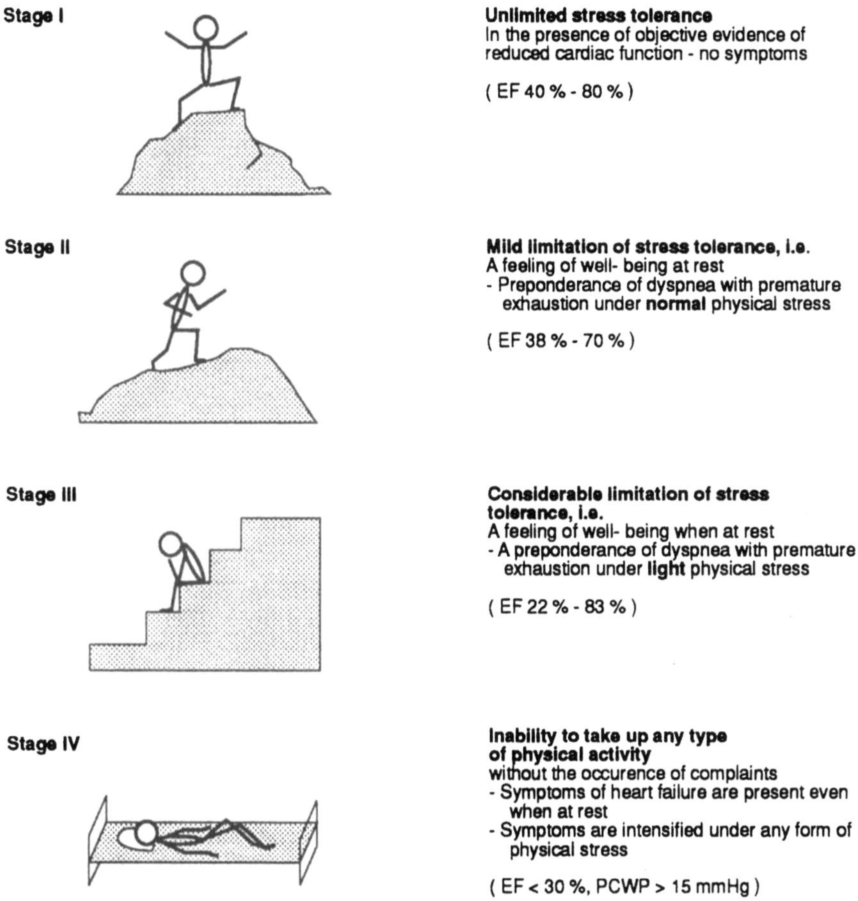

Fig. 1. Classification and severity of heart failure. *EE*, ejection fraction; *PCWP*, pulmonary capillary wedge pressure

Although, in stage I, reduced heart function can be objectively demonstrated, there are no clinical symptoms. In stage II, there is a mild reduction in stress tolerance and symptoms appear under normal physical stress, while in stage III even slight physical stress produces symptoms, and a considerable limitation of stress tolerance is present. As can be seen in Fig. 1, stage IV is characterized by massive symptoms [28]. Interactions between somatic, psychological, and social factors have to be considered [2]. Reductions in stress tolerance, physical complaints, and the immediate fear of death can influence the quality of life for these patients. Studies have shown that the quality of life has to be considered in the evaluation of therapeutic success [5].

At this point, the question must be asked, what is meant by "quality of life"? In order to remain within the scope of this article, an extensive discussion of the concept of quality of life will be dispensed with and only a short overview will be given. Research on health indicators examines questions of the definition and measurement of quality of life [3]. Brusis and Biefang [5] found that the concept of quality of life is "indeterminate and nebulous," and that every definition of the term must be made in terms of some standard.

A description of the quality of life is impossible without taking the patient's subjective impressions into account, but even the determination of the features for measuring quality of life makes normative decisions essential. Quality of life generally encompasses physical, psychological, and social health. Beyond this, the socioeconomic situation should also be considered, as heart diseases are often linked to loss of job, change of profession, early retirement, and a decrease in professional and socioeconomic status. Furthermore, it should be borne in mind that the definition of quality of life is dependent on several factors, as well as on the individual's cultural circle, religion, and level of education. Moreover, as was noted by Inglehart, the definition of quality of life also varies with age [21].

However, a multitude of standard questions still remains. One aim of the treatment of heart failure is to decrease the clinical degree of severity from NYHA III or NYHA IV to NYHA II or NYHA III, for example. Nevertheless, the question remains, to what "normal condition" should patients be brought through medical treatment, when should individuals be perceived as being "healthy," when should they be considered to be sick, and to what degree does purely medical treatment improve satisfaction with life and the quality of life?

Studies have shown that heart patients experience a decrease in quality of life [10] and a loss of satisfaction with life. Physical complaints and emotional instability increase [24]. These investigations were mostly carried out on patients with myocardial infarction and chronic coronary ischemia [10, 24]. Owing to the fact that acute heart failure can be the result of a myocardial infarction, and that chronic coronary ischemia is one of the causes of heart failure, a certain relevance of the investigation for heart failure can be assumed. This has also been confirmed by physicians. Furthermore, the quality of life of heart patients can be considerably impaired by professional, family, and sexual problems as well as by the danger of a drop in social status and insecurity with regard to the course of the illness, fear of a cardiac infarction, or sudden cardiac death [10]. In addition, awareness of the apparent cause of the disease may influence the quality of life. One investigation into cardiovascular patients underlines the special (and also economic) significance of satisfaction with life for the ability to work [24]. Medical and clinical factors alone do not suffice to determine this. It is rather that the differences in satisfaction with life have a negative effect on the ability to work, and thus increase a reliance upon the health care system which is not medically justifiable [10]. The question therefore arises, to what extent does an improvement in quality of life bring about a saving of economic resources?

Cost-Benefit Analyses

Overview of the Techniques Used in Cost-Benefit Studies

As a result of illnesses, claims are made on health care services. This ties up economic resources that would normally be used for other purposes. In addition to these direct costs, indirect costs are incurred as a result of absenteeism from work due to illness, and of disability and premature death. All goods and services produced by the economy could be produced with more leisure time, or the range of products available could be broadened, if these costs did not exist [23]. If one wishes to measure effectiveness and efficiency, then input and output must be compared with each other. Effectiveness (productivity) is considered to be the degree to which the goal of the health care system is attained, and efficiency (i. e. economic efficiency) is understood as a comparison of the attainment of health care goals with the funds used. Cost-benefit analyses represent one method for measuring effectiveness and efficiency. Behind this lies an operational system which has been developed by fiscal economists for the evaluation of public projects and project alternatives, and which is regarded as an aid to decision-making for public investment and aims at a better supply of public goods to the population [12].

Particularly because of the data situation, problems of practical application arise, since there are no statistics regarding the level of apportionment of costs for various diseases, or according to ICD codes. Hence, costs for the treatment of heart failure cannot be determined separately. Not without reason, researchers in the field of health services [11] demand an improvement in the state istics, because in the current situation it is difficult to assess the available instruments.

Cost-benefit analyses should not be used to anticipate political decisions. They may, however, serve as an aid for making political decisions [17]. The level of knowledge of the decision-makers should be improved, and the situation should be made clearer [4]. Cost considerations are one aspect of the fight against disease [17].

Three methodological procedures are available: (a) cost-benefit analysis (CBA), (b) utility analysis (UA), and (c) cost-effectiveness analysis (CEA). In the case of utility analysis, monetarily measurable parameters and the effects of a measure are stated in single utility values. Weighting depends on subjective approximations [4]. The CBA attempts to measure all costs and benefits in monetary parameters, and the alternative with the best cost-benefit ratio is preferred [4]. Table 1 provides a summary of the costs which have to be considered in the CBA and disease-cost studies. The general empirically measurable consumption of resources is regarded as the direct cost – treatment, care, rehabilitation, and the like. The loss of real net output due to inability to work, disability, and premature death is regarded as the indirect cost [18].

In order to ascertain the costs, two concepts are available: the prevalence approach and the incidence approach. In the prevalence approach, the presence of disease and premature death are included in the calculation,

Table 1. Direct, indirect, and psychosocial costs [19] of illness/disability/death

People directly affected

Direct Costs
1. Basic costs
   Prevention
   Treatment
   Rehabilitation
   Care
2. Additional costs
   Diet, the purchase of health literature, transportation to the doctor, monitoring of food, etc.

Indirect Costs
1. Indirect costs as a result of morbidity
   Employment disability, decreased fulfilment of functions, etc.
   Change of job, lost opportunities for professional advancement, etc.
2. Indirect costs as a result of mortality
   Loss of resources through premature death

Psychosocial Costs
1. Psychosocial costs as unmeasured economic costs
   For example, decreased productivity without employment disability, change of job
2. Psychosocial costs in the strict sense
   Decrease in feeling of well-being, fear of illness, suffering, etc.

People indirectly affected

Direct Costs
1. Basic costs are not incurred; however, insofar as the psychosocial effects on people who are not initially affected by illness/disability/death lead to illness, disability, or death, then transfer to the group of persons directly affected may follow
2. Additional costs
   See also under People directly affected.

Indirect Costs
Indirect costs as a result of morbidity and mortality
The time required to care for the sick and dying, etc.

Psychosocial Costs
1. Psychosocial costs as unmeasured economic costs
   For example, change of job, without the person affected himself being sick
2. Psychosocial costs in the strict sense
   Decrease in feeling of well-being, fear of illness, suffering, etc.

whereas the incidence approach includes the expense of an increase in disease during the period of observation [17].

The traditional way used for ascertaining indirect costs is the human capital approach. The income which can be attained by those who are employed is used as a comparable measure of loss caused by mortality and morbidity in terms of net value [7]. The weaknesses of the CBA also lie in the difficulty of evaluating the benefits monetarily [9]. Although considerable progress was made at a very early stage in the development of methods for determining and evaluating non-monetary costs and benefits [8], the use of these methods in the health care system frequently remains very unsatisfactory.

In CEA, effectiveness is determined for effects which do not permit monetary evaluation and is used in relation to the collected aims. All values that have a determinable monetary dimension are stated in monetary units, as is the case in CBA. In CBA, psychosocial costs are not considered and are treated as intangible effects. Such a disregard for intangible effects causes problems in the final estimation of the cost-benefit relationship, if this measure is intended to obtain intangible effects; for example, an improvement of the quality of life [6]. Moreover, CEA permits multidimensional observations to be made. When considering the economic health care aspect of the quality of life of patients with heart failure, cost-effectiveness analysis shows comparative advantages. The procedure for CEA is illustrated below.

Illustration of Cost-Effectiveness Analysis

Firstly, the problem has to be defined. The situation which is to be changed has to be described within the framework of a condition analysis. In connection with the quality of life, of patients with heart failure, high rates of mortality and morbidity, insufficient satisfaction with life, and the quality of life (as well as fears about premature death) can be pointed out which indirectly create high economic costs. This situation may be seen by the decision-makers as one worth changing. Furthermore, a goal system and a preference structure should be drawn up. In our special case, the primary goals would be improved satisfaction with life and the quality of life, reduced fear, pain and suffering, reduced morbidity and mortality, increased life expectancy, and improved acceptance of the disease. Concrete, subordinate goals may be derived from these primary goals [6]. Some subordinate goals would include the improving of self-support activities, improved mobility in and outside of the home, independent management of the household, and an improvement in sexual, marital, family, and professional activities. However, the goals must be rendered operational, and the fact that there are several goals makes it necessary to set down a preference structure. In order to investigate the attainment of goals, indicators have to be determined which have a mutually comparable weighting. The goals have to be weighted with respect to one another on each hierarchical level, and with respect to their significance for the next higher goal [6]. It should be borne in mind that, when choosing a system of goals and when determining a preference structure, subjective value judgments come into play. In order to obtain concrete measurable results, the contributions to goals (manifestations of the attainment of goals) have to be evaluated, for example, with a point system of 0–10 points. 0 points describes the attainment of goals that are only just acceptable, and a worthwhile manifestation within the realistic range receives 10 points [6]. Within an evaluation interval, evaluation functions and rankings can be defined. What is important is to expose the subjective value judgments inherent in an evaluation of this nature.

Satisfaction with life and quality of life can be determined by using differential patient polls. One such method is the Rosser-Matrix [22], in which patients

Table 2. Rosser's classification of illness states [29]

Disability		Distress
I.	No disability	A. No distress
II.	Slight social disability	B. Mild
III.	Severe social disability and/or slight impairment of performance at work	C. Moderate
	Able to do all housework except very heavy tasks	D. Severe
IV.	Choice of work or performance at work very severely limited Housewives and old people able to do light housework only but able to go out shopping	
V.	Unable to undertake any paid employment Unable to continue any education Old people confined to home except for escorted outings and short walks and unable to go shopping Housewives able only to perform a few simple tasks	
VI.	Confined to chair or to wheelchair or able to move around in the house only with support from an assistant	
VII.	Confined to bed	
VIII.	Unconscious	

are categorized into a matrix on the basis of the criteria listed in Table 2 [27]. Values were allocated to the fields of the matrix by various people – doctors, nurses, nursing staff, and patients; their means were standardized between 0 (death) and 1 (without impairment) as weighting factors for the evaluation of the quality of life. With the aid of such types of patient poll, improvement in the quality of life and the state of health can be monitored over a period of time.

During the further course of the investigation, a delineation of the decision field must be undertaken. Environmental conditions can impede the realization of the goal and limit the room for decision. When it is noticed that certain measures cannot be carried out, they must be set aside. For the remaining realizable measures, concrete definitions of the goals of the alternatives to be investigated must be drawn up [6].

When investigating "the quality of life of patients with heart failure," for example, the following alternatives present themselves: (a) a comparison of old and new drugs and their effects on the NYHA stages; (b) a comparative investigation of old and new methods of treatment; (c) the cost and effectiveness of inpatient and outpatient treatment; (d) a comparison of treatments with and without psychological care; (e) the effects of various therapeutic concepts and rehabilitation processes on the quality of life; and (f) the cost and effectiveness of preventive measures.

Financial incapacity, personnel shortages (inadequate nursing personnel, too few psychologists, or specially trained doctors), as well as the inability to accept new patients in clinics, hospitals, and rehabilitation centers, are some of the environmental conditions which have a limiting influence. Furthermore, legal restrictions should be mentioned; for example, where legal regulations make it impossible for the health insurance plans to carry the costs of psychological care. Political influence on distribution can also stand in the way of the realization of goals. As heart failure is purported by some to be a disease of the lower social classes, decision-makers may, under political pressure from lobbies, neglect the treatment of heart failure in their policies.

A list of costs and the effectiveness of the treatment of heart failure patients is difficult without a concrete subject to investigate. Myrtek [24], among others, investigated the effects of the quality of life on the rehabilitation course of cardiovascular patients. He came to the conclusion that a more negative assessment of satisfaction with life had a negative effect on ability to work, which resulted in medically unnecessary services being claimed from the health care system. Although value judgments can certainly find their way into such investigations, the authors are nevertheless of the opinion that, on the basis of the cost-benefit relationship, an increased use of psychologists would "without doubt be positive" [24].

Set in this context, an attempt will now be made to list examples of cost and effectiveness of more intensive psychological care.

Costs

1. Higher direct costs owing to the employment of psychologists
2. Higher costs due to providing transport to place of treatment
   Minus costs
   a) saved from medical services
   b) saved indirect costs through a reduction in employment disability
   c) a saving in resources lost through premature death

Effectiveness

1. Increased ability to accept the illness
2. Change in the recognition of physical and mental limitations
3. Decrease in physical complaints
4. Decrease in depression and fear, in particular fear of death
5. Improved sexual potency
6. Happier marital and family life
7. Increased life expectancy
8. Increased ability to work
9. Prevention of loss of professional and social status

Thinking in terms of opportunity costs makes it necessary, when evaluating alternatives, to determine that part of the cost – the benefit – which can be saved through alternative means of therapy. This benefit has to be compared with the cost of treatment. In the above model, the changes have been contrasted in shortened form. As in the case of CBA, the cost in the CEA is determined on the basis of opportunity costs.

In order to assess effectiveness, points are allocated for the extent to which goals are achieved and, if possible, cardinal scales should be used [13]. For each alternative, the corresponding partial effectiveness is calculated by multiplying the partial weightings by the attainment of the goal. The addition of the partial effectiveness of an alternative produces the total effectiveness. However, a summary of this nature decreases clarity, and it is questionable whether such total effectiveness is free of value judgments (addition problem).

What is necessary but problematic is the incorporation of a time factor. The benefit of an improved quality of life for patients with heart failure may be distributed over several years. Discounting this is possible if the measurement of the partial effectiveness has been carried out on cardinal scales [13]. Another procedure is the forecasting of effectiveness over several years, in order to provide decision-makers with the temporal distribution of effectiveness. Forecasts of this nature are, however, linked to subjective value judgments. Sensitivity analyses should be carried out in order to reveal the significance of specific assumptions about the results [6]. Cost and effectiveness are, as a rule, determined under the premise of incomplete information. Input data and suppositions must again be verified for their plausibility.

With regard to the quality of life of patients with heart failure, one would have to show, for example, whether a tendency for the quality of life to improve can be expected in society as a whole in the foreseeable future, or whether the quality of life will, if anything, deteriorate. If necessary, one should again verify data on development of heart failure over the course of time without therapeutic intervention in relation to age structure, social class, and the like. If necessary, the effects of the changes in input data on the total result must be established.

A cost-effectiveness matrix can be drawn up in order to compare alternatives (see Fig. 2a). For each alternative, the sum of the costs is set against the partial or total effectiveness. It can be left to the decision-maker to combine the partial effectiveness values to produce the total effectiveness, so that she or he has a clear view of the degree to which the various individual goals have been attained. If the costs are compared with the total effectiveness, a critical value can be obtained as required, which states the costs per unit of effective-

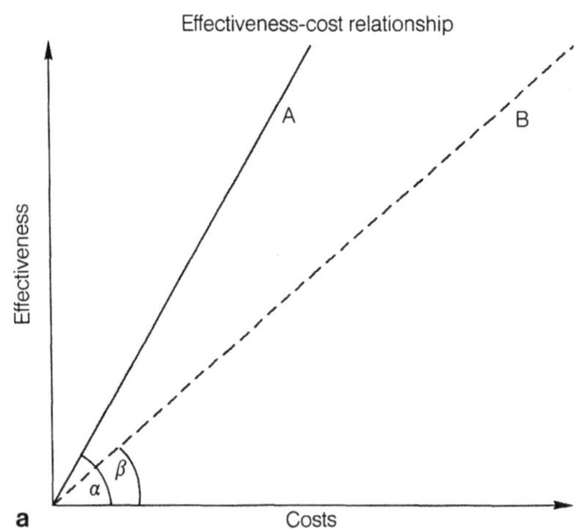

Fig. 2a.

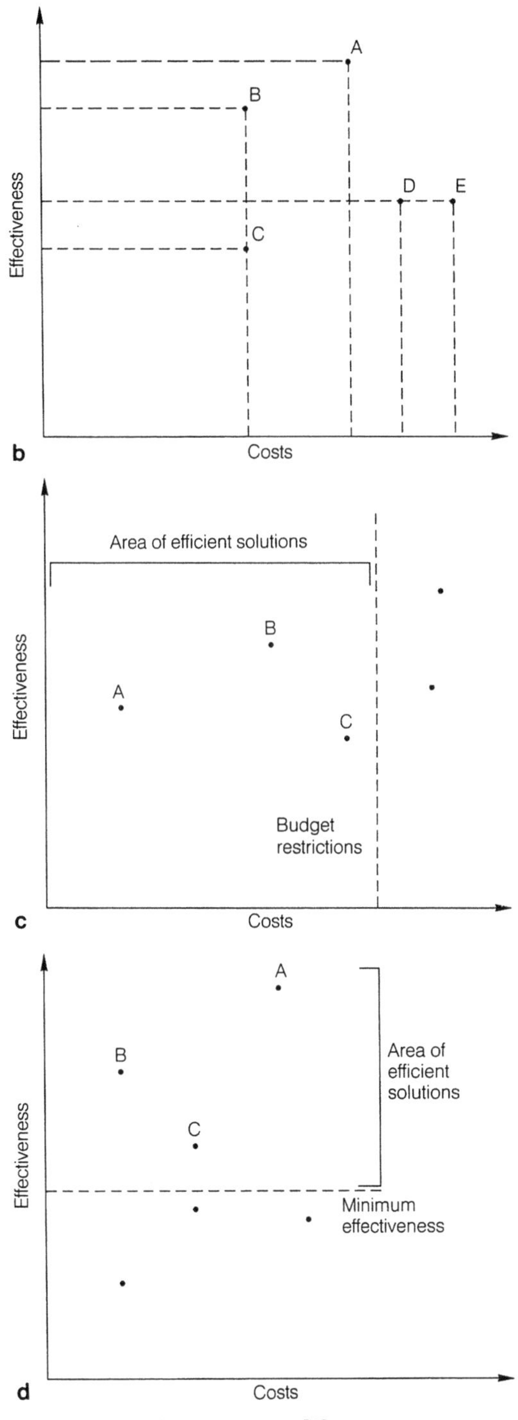

Fig. 2b–c. Possibilities for alternative assessment [6]

ness. By comparing this critical cost figure per unit of effectiveness in several projects, it is possible to classify these projects according to their cost-effectiveness.

The economic principle is to be used as the optimization criterion for an assessment of alternatives. When given funds are used, a maximum of effectiveness or a defined degree of attainment of goals should be achieved with a minimum of costs. Some examples of the possibilities for assessing alternatives have been selected and explained (Fig. 2b-d).

A comparison of alternatives is possible by means of effectiveness-cost or cost-effectiveness ratios. For each measure, the number of units of effectiveness the use of a monetary unit produces, or the costs required for one unit of effectiveness, are given [6]. When comparing pairs of alternatives, those alternatives must be selected in which equal costs are accompanied by higher effectiveness, equal effectiveness by lower costs, or higher effectiveness by lower costs. No decision has to be made in the case of higher effectiveness and higher costs compared with other measures. The introduction of a budget restriction or setting of a minimum effectiveness is conceivable. Those alternatives which, above the minimum effectiveness level, show the lowest costs or, within budget restrictions, show the highest effectiveness should be preferred. The CEA mentioned can help in making a decision, and in preparing a consensus between the diverging opinions to be found in a pluralistic society, but the final decision remains in the hands of the decision-maker. The necessary assessments and the value judgments which come into play should be disclosed in order to increase meaningfulness and to arrive at measurable results. The inclusion of greater expenditure during the years of life gained through treatment, the inclusion of indirect costs of prolonged life, and the resultant additional value created has led to disagreement and controversy [18].

Summary

As the question of quality of life becomes increasingly important in medicine, and due to the fact that the measurement and assessment of life and the quality of life is a highly emotive problem, the tasks of the future will be, among others, to increasingly consider the quality of life, to investigate improved measurement techniques and methods of assessment, and to analyze patient preferences by means of increased patient polling. Even if, for supposed ethical reasons, one would like to allocate all available resources to the health care system and to improving the quality of life, this unreasonable demand must be seen against the fact that the resources of society are limited and that the need for better quality of life for patients with heart failure competes with other needs. It has been shown that, in addition to taking account of the economic principle and the budget restrictions placed on all decision-makers, it is of paramount importance to think in terms of opportunity costs, because each currency unit spent on health care must be deducted from sums which could be put to alternative use inside or outside the health care system. Thus, economic thinking in the health care system is itself an ethical maxim. Despite all the euphoria regarding cost reduction, a meaningful allocation of funds should also take into account aspects of the quality of life, as false saving can lead to additional, avoidable economic costs. Both an increased use of economic methods and improvement in the data situation in the health care system require interdisciplinary cooperation, which should also include economists. Finally, it can be seen that economists have methods and processes at their disposal which not only provide greater clarity in the health care system, but also help to make emotionally charged discussions more rational.

References

1. Andersen HH, Schulenburg J-M Graf vd (1987) Kommentierte Bibliographie zur Gesundheitsökonomie. Sigma, Bonn
2. Badura B, Kaufhold G, Lehmann H, Pfaff H, Schott T, Waltz M (1987) Leben mit dem Herzinfarkt. Eine sozialepidemiologische Studie. Springer, Berlin Heidelberg New York
3. Biefang S, Pohlmeier H (1979) Gesundheitsindikatoren – ein Versuch, Gesundheit zu messen. Medizin Mensch Gesellschaft 4: 22–27
4. Biene P, Dennerlein R, Schneider M, Seeburger R (1987) Wohnortnahe Versorgung chronischer Polyarthritis, Konzeption der Kosten-Wirksamkeits-Analyse. BASYS, Augsburg
5. Brusis J, Biefang S (1980) Lebensqualität als Maßstab für Therapie-Evaluation. Medizin Mensch Gesellschaft 5: 10–16
6. Bundesminister für Arbeit und Sozialordnung (ed) (1984) Kostenwirksamkeitsanalysen im Gesundheitswesen. Forschungsbericht Nr 98, Bonn
7. Cooper B, Rice D (1978) Die volkswirtschaftlichen Kosten von Krankheiten. Ortskrankenkasse 23/24: 849–860
8. Dasgupta AK, Pearce DW (1972) Cost-benefit analysis, theory and practice. The Macmillan Press Ltd., London and Besingstoke
9. Dinkel R, Schulze-Röbbecke T (1982) Kosten-Effektivitäts-Analyse der Zytostatikatherapie von akuter Leukämie im Kindesalter. PROGNOS, Basel
10. Egger J (1982) Auswirkungen des Herzinfarktes auf die subjektive "Lebensqualität" der Patienten. Rehabilitation (Stuttg) 21 (2): 51–59
11. Forschungsgruppe Gesundheitsberichterstattung (1989) Aufbau einer Gesundheitsberichterstattung – Bestandsaufnahme und Konzeptvorschlag (Endbericht), vol 1. Kiel
12. Geigant F, Sobotka D, Westphal HM (1987) Lexikon der Volkswirtschaft, 5th edn. Verlag moderne Industrie, Landsberg
13. Hanusch H (1987) Nutzen-Kosten-Analyse. Vahlen, Munich
14. Henke K-D (1978) Ansätze für eine ökonomische Analyse des Gesundheitswesens. Das öffentliche Gesundheitswesen 40 (12): 789–794
15. Henke K-D (1983a) Ökonomische Analyse des Gesundheitswesens – wozu? Mimeo, Hannover, pp 1–19
16. Henke K-D (1983b) Gesundheitsplanung im Sinne makroökonomischer Ressourcenplanung. Das öffentliche Gesundheitswesen 45 (7): 349–361
17. Henke K-D (1986) Die direkten und indirekten Kosten von Krankheiten in der Bundesrepublik Deutschland im Jahre 1980. In: Henke K-D, Metze I (eds) Beiträge zur Gesundheitsökonomie. Finanzierung im Gesundheitswesen, vol 10. Bleicher Gerlingen, pp 209–274
18. Henke K-D (1988) Direkte und indirekte Kosten zur Beurteilung von Arzneimitteltherapien. Mimeo, Hannover, pp 1–25

19. Herder-Dorneich P (1986) Ordnungspolitik des Gesundheitswesens – eine analytische Grundlegung. In: Gäfgen G (ed) Ökonomie des Gesundheitswesens. Duncker and Humblot, Berlin, pp 55–70
20. Hort W (1984) Kreislauforgane. In: Eder M, Gedigk P (eds) Lehrbuch der allgemeinen Pathologie und der Pathologischen Anatomie, 31st edn. Springer, Berlin Heidelberg New York, pp 314–360
21. Inglehart R (1979) Lebensqualität: eine Generationenfrage. Psychologie heute 9: 24–30
22. Kind P, Rosser R, Williams A (1982) Valuation of quality of life. Some psychometric evidence. In: Jones-Lee MW (ed) The value of life and safety. North Holland Publishing Company, Amsterdam New York Oxford
23. Leu RE, Schaub T (1986) Ökonomie der Prävention. Nutzen-Kosten-Analyse einer gemeindeorientierten Interventionstudie zur Prävention von Herz-Kreislaufkrankheiten in der Schweiz. Mimeo, Hannover, pp 1–26
24. Myrtek M, Kreutel K, Wilk D, Welsch M, Herzog M (1987) Lebenszufriedenheit und Rehabilitationsverlauf – eine Untersuchung an Herz-Kreislauf-Patienten. Rehabilitation (Stuttg) 26 (1) 11–19
25. Pschyrembel W (1986) Klinisches Wörterbuch, 255th edn. De Gruyter, Berlin
26. Süssmuth R, Arnold M, Schulenburg J-M Graf v d (1987) Stellenwert der Prävention im Rahmen der Weiterentwicklung des Gesundheitswesens aus Sicht des Ökonomen. In: GVG (Ges. f. Versicherungswissenschaft und -gestaltung e. V.) (Hrsg) Stellenwert der Prävention im Rahmen der Weiterentwicklung des Gesundheitswesens. Vollmer, Bergisch-Gladbach, pp 30–51
27. Schulenburg J-M Graf v d, Klein S, Piojda U, Schöffski O (1989) The cost and benefits of the use of erythropoietin in the treatment of anaemia arising from chronic renal failure in the Federal Republic of Germany. Mimeo, Hannover, pp 1–35 (Veröffentlichung erfolgt in der Reihe des Centre for Health Economics, University of York)
28. Späth G (1988) Herzinsuffizienz. De Gruyter, Berlin
29. Statistisches Bundesamt (ed) (1987) Todesursachen. Gesundheitswesen, Fachserie 12, Reihe 4

The Questionnaire as a Tool for the Empirical Assessment of Quality of Life in Patients with Cardiac Failure

R. Rychlik, D. Urbahn, and P. Potthoff

Quality of Life in Heart Disease

The major cardiological syndromes that are of epidemiological importance, such as coronary heart disease, hypertension and heart failure, are chronic diseases which, as a rule, cannot be cured, but instead require treatment which is often for life. The aim of treatment is, above all, to arrest the progress of the disease and the complications associated with it, and thus make the patient's life pleasant and satisfying. In this context, it should be noted that prolongation of life, which may be the result of treatment, can be accompanied by a reduction in the quality of life. In such cases, prolonged life is, as a rule, very rarely acceptable to patients, their relatives and friends. On the other hand, patients expect that modern, technological medicine will be capable of relieving, to a large measure, a chronic disease. At the very least, extensive freedom from symptoms should be achieved.

Medication frequently also produces side effects which affect the previous state of well-being and thus lead to a reduction in the quality of life. The patient notices this, but is unable to establish a causal link with treatment and the clinical picture. It therefore seems to be a good idea to note these phenomena in the early stages and place them on record. This means that even at an early stage in drug development the quality of life should be taken into consideration as a standard measure. The subsequent evaluation within the context of drug licensing is, naturally, also bound up with this. Taking its effect on the mortality rate as the sole criterion for licensing a drug appears to be extremely problematical in this context.

Basic Principles for the Assessment of Quality of Life in Clinical Studies

The basic instrumentarium for the assessment of quality of life will not be discussed further at this point because it is summarised in several places in this book. It must, however, be pointed out that, in most cases, it is the so-called health-related quality of life that has been investigated hitherto in those clinical studies in which such an assessment has been included. The term "health-related quality of life" refers to the degree of well-being and contentedness associated with the routine and restrictions that are brought about by disease, accidents and treatment. The concept is all-embracing and covers all areas and activities, including work, leisure, family and home life. Not only the ability to fulfil social roles and perform tasks, but also the degree of satisfaction resulting therefrom, is taken into consideration. Other ideas that people consider valuable, such as a safe environment, adequate shelter, stable income, respect, love, and freedom are not part of this definition, although illness and treatment may have a thoroughly adverse effect upon them. The connecting links between diseases and their effects on the latter areas of life are, however, often too complex and multi-layered to incorporate reliably into empirical studies.

The quality of life can be considered both for individual patients and also for a group of patients. When the individual approach is used, it is merely a question of assessing the status of each individual patient, of balancing the possible therapeutic measures against each other in relation to that patient alone and of deciding upon the best possible treatment in that case. Only at a second stage are groups of patients examined before and after therapeutic measures by means of standardised criteria and the effects of the different types of treatment evaluated.

Differences in quality and length of life are important factors which are touched upon in the evaluation of the disease and the outcome of treatment, both for individual patients and for the patients considered as a group.

Results from studies of the quality of life can be used in a variety of ways for making clinical decisions:

- For monitoring and evaluating the condition of the patient.
- For choosing a type of treatment.
- For monitoring the effect of the chosen therapeutic measure.
- For developing a common view of the illness and the success of treatment by both doctor and patient.

Moriyama (quoted in [3]) identified six desirable features for evaluating health indices, and these may also, by analogy, be applicable to methods of measuring quality of life:

- The index should be meaningful and comprehensible.
- The index should be sensitive to changes in the phenomenon being measured.
- The assumptions upon which the index is based should be justifiable on theoretical grounds and should appear intuitively sensible.
- The index should be constructed from clearly defined elements.
- Each element should make an independent contribution to the changes in the phenomenon being measured.
- It should be possible to derive the index from information that is already available or that is easy to obtain.

In this context, it is now widely agreed that measurement of the quality of life, as a factor in clinical and epidemiological studies, is of use when the efficacy of drugs is being evaluated or when general therapeutic measures are being considered. Evaluation of the quality of life is of particular benefit in those cases in which standard clinical investigations have led to the diagnosis of a mild form of the disease only, but as the result of which the patient nevertheless feels wretched.

There is a great need to acquaint doctors and medical students with practical methods for measuring the quality of life. In this context, the most important thing is that the methods should be standardised, since their possible practical importance lies in the following areas:

- Improvement of decisions about treatment.
- Assessment of the risk-benefit ratio of a medical treatment.
- A basis for making decisions about health policy with reference to resource allocation and the cost-benefit analysis.

Planning of Studies: The Quality of Life of Patients with Cardiac Failure

The study presented here is in the planning stage at the time of completing this book. The aim is to assess the quality of life of patients with cardiac failure, and this will include as complete and specific an assessment as possible of:

- The disability resulting from the cardiac failure.
- Its subjective importance – i.e., the amount of impairment or disability that the individual feels.
- The expectation that patients with cardiac failure have from treatment.

On the other hand, such a study must include a control group which, in this case, should comprise members of the population of a similar age who do not regard themselves as having cardiac failure.

Definition of Terms

Before discussing the investigational plan, the way in which individual terms are used in the context of this study will be briefly described.

Quality of Life. Any description of the quality of life should take the following factors into account:

- Physical symptoms
- Psychological well-being
- Everyday activity
- Social activities

Patients with Cardiac Failure. It should be assumed that, in the majority of cases, patients with cardiac failure are elderly people with multiple pathology. If the proposed investigation were to confine itself to those who do not have multiple pathology, but who suffer from cardiac failure alone, an atypical group would be formed. It is, therefore, envisaged – despite all the problems that this creates in interpreting the results – that those with multiple pathology will not be excluded. It is also planned that only those with cardiac failure of classes II and III (NYHA) severity will be included in the investigation.

Control Group. The control group should consist of subjects of the same age who do not suffer from cardiac failure. Two possible ways of defining the control group may be considered:
- On the one hand, subjects who are as healthy as possible, i.e., those who are, at least, not suffering from chronic diseases.
- On the other hand, those who – apart from cardiac failure – have a "normal" spectrum of morbidity for their age.

Since those who do not suffer from chronic illness with advancing age constitute a rather atypical group, the control group should be chosen at random on the basis of a sample that is representative of the population to which no exclusion criteria apart from cardiac failure have been applied.

Study Plan

It is envisaged that target groups of patients with cardiac failure will be contacted via the doctors responsible for their treatment. Only by this approach will it be possible to link (objective) medical findings with (subjective) reports of impairment in the quality of life when the data are obtained. This means that a group of doctors in the community will first be recruited. The doctors will then select patients with cardiac failure from their practice, in accordance with criteria specified in advance, and will ask them to take part in the study.

Patient data on the subject of quality of life and cardiac failure can be obtained reliably and economically using standardised questionnaires. A written interrogation is also advisable in the light of the possible subsequent use of the survey method in the context of clinical trials. The doctors will record some additional information for each patient, and there will be discussion of their views and experiences in relation both to the quality of life and to the heart disease itself. Impairment of the quality of life in the patient group can then be evaluated with a high degree of certainty if the control group is selected on the basis of data available from the population sample. This is the case with studies designed to use a representative sample, which nowadays are increasingly carried out by telephone contact and recruitment with a subsequent printed questionnaire.

Sample

The way in which the plan presented so far can be transformed practically into a survey will now be demonstrated, using, by way of example, the study which we plan to carry out.

The basis for the sample of patients with cardiac failure is a sample of practising doctors. These will include locally resident medical specialists and general practitioners who carry out cardiological diagnosis using ECG and bicycle ergometry. According to the data available to us, this condition is fulfilled by about two-thirds of medical specialists and by one-third of general practitioners. In view of the subject area, an unrepresentative proportion of 75 medical specialists and 75 general practitioners seems to be appropriate. The doctors will be chosen by a quota technique, features of the quota being the area of specialisation and regular use of ECG/bicycle ergometry in the practice. In order to approach proportional representation of the regions, the distribution in the regions will be more or less the same as the distribution of community-based general practitioners and specialists in internal medicine as a whole.

Each doctor will be asked to recruit two patients with cardiac failure as participants and will be told the degree of severity and age group to select.

Significant advantages of this distribution of the sample (150 doctors with two patients each) are that it largely excludes clustering effects, and that it achieves a broad scatter of subjects – also from a regional point of view.

The sample of patients with cardiac failure should be distributed disproportionally with regard to age and degree of severity in order to have an adequate basis for comparative analysis in individual cells.

Participants in the Questionnaire

Doctors. An interview with the doctor is ideal for obtaining information on the following topics:

- Number and distribution of patients with cardiac failure in the practice.
- Treatment, requirements and expectations of treatment.
- The doctor's opinion on the quality of life of patients with cardiac failure.

The doctor will also record the severity of the cardiac failure, medication and also any additional illnesses for each patient selected.

Patients with Cardiac Failure. The patient with cardiac failure will receive a questionnaire to fill in, which covers the following subject areas:

- Physical Symptoms/General Complaints. Because it has been tested and standardised, the symptom list of von Zerssen [6] will be used.
- Psychological Well-Being. A well-tried method of assessment designed by von Zerssen is the "Scale of Well-Being" [7].
- Coping with Everyday Life. Here we will use the scale for the assessment of functional impairment in everyday life [5]. This is a German adaptation of Anita Stewart's scale, which was used in the RAND study. It has been used in Germany, for example, in the Munich blood pressure study and also in a study of the adverse effects of illness upon life in the resident population of Munich.
- Social Activities. By making use of questions from the "Life and Health in Germany" questionnaire used in the German Cardiovascular Prevention study [1], question designs can be used that have been tested in practice, and for which there are comparable data from large samples of the population.

- State of Health/Requirements. This group of questions should be given to both target groups for comparison, even if some of the answers (e.g., illnesses) are also obtained from the doctor in the case of patients with heart failure.
- Contentment. The information on the quality of life will be expanded by reports of satisfaction with health and with other areas of life.
- Additional Questions for Patients with Cardiac Failure. Important information is supplied by assessment of the degree of handicap experienced by patients with cardiac failure and their expectations from treatment. Open questions on this topic raise the possibility of identifying specific aspects that appear to be particularly important to the patient in the context of his/her own illness and situation in life.
- Sociodemographic Data. In order to assess sociodemographic features, a standardised, but abbreviated version should be selected.

Control Group. For methodological reasons, the control group should also be given a printed questionnaire, the contents of which are the same as of that given to those with cardiac failure – excluding questions relating to the expectation of treatment. Symptoms and diseases, as well as any medication taken, will be recorded for the control group as well.

Technical Procedure for a Survey

Patients with Cardiac Failure. It is intended that the survey will be carried out in the following manner:
- Recruitment of the doctor by the interviewer, completion of an interview: at this visit the interviewer gives the doctor the patient questionnaire and explains the selection criteria.
- The doctor gives the questionnaire to the next two patients who match the criteria quoted (and who are prepared to participate).
- The completed patient questionnaires will be delivered to the practice in a sealed envelope and the interviewer will make an appointment with the doctor to collect all the papers – including the additional information sheets recorded by the doctor for each patient.

Control Group. There are, basically, two approaches to the survey in the control group:
- Purely in writing, i.e., carried out by post.
- Delivery and collection of the written questionnaire completed by the interviewer.

In view of the relatively low rate of return of written questionnaires, this survey method should be considered only if there is telephone contact in addition (advance notice, follow-up). When the interviewer delivers and collects the questionnaire him-/herself, motivation is likely to be better and completion rate higher, but costs will be higher, too.

Evaluation

The data on the quality of life will be assessed in two stages, in accordance with the usual procedure for testing quality of data. The first stage of assessment will be to make a table for counting purposes, in which the individual results are entered, separated according to the different target groups or sample subgroups.

In a further step, the scales relating to elements in the quality of life are summarised to give scale or index values, and comparisons of the extent of impairment in those with cardiac failure and those without cardiac failure and between the two grades of severity of heart failure will be undertaken.

References

1. Güther B, Schwarz E (1986) Der Gesundheitssurvey – Design, Methoden, Feldergebnisse. Munich Infratest
2. Hoffmeister H, Hoeltz J, Schön D et al. (1988) Nationaler Untersuchungs-Survey und regionale Untersuchungs-Surveys der DHP, vol 1. DHP-Forum 1
3. Patrick DL, Erickson P (1988) Assessing health-related quality of life for clinical decision making. In: Walker SR, Rosser RM (eds) Quality of life: assessment and application. MTP Press Ltd. pp 9–49
4. Potthoff P (1982) Materialien zur Studie "Entwicklung von Indikatoren zur Messung subjektiver Gesundheit". GSF-Bericht 540/82, Munich
5. Potthoff P (1988) Krankheitsbedingte Lebensbeeinträchtigungen in der Münchner Wohnbevölkerung. Öff Gesundheitswes 50: 329–333
6. von Zerssen D (1976a) Die Beschwerdenliste. Beltz, Weinheim
7. von Zerssen D (1976b) Die Befindlichkeitsskala. Beltz, Weinheim

Quality of Life: A Relevant Endpoint

H. Troidl

Quality of Life – An Overriding Factor in the Ambience of the Consulting Room and in the Operating Theatre

In many cases – if not most – the patient presents him-/herself to the surgeon complaining of a marked impairment to his or her feeling of well-being. Examples of this are inguinal hernia, symptomatic gall stones, obstructive arterial disease, or joint injuries. Operations for inguinal hernia and on the gall bladder are the most common surgical procedures, and it would certainly not be an exaggeration to say that almost 80% of peripheral vascular operations are carried out because of the general debility felt by the patient and not on account of a life-threatening situation.

Since time immemorial it has been taken for granted and generally recognised that surgeons often consider that the patient's impaired feeling of well-being is an indication for surgical intervention and select treatment with this in mind. After treatment, the surgeon and the patient – in the "consulting room" – talk over the hoped-for improvement in or restoration of well-being as a criterion for the success of treatment.

It is therefore more than surprising that this sensible attitude is confined to the ambience of the consulting room and the operating theatre. On the scientific level, however (congresses, publications), discussion is almost exclusively concerned with objectives such as operative mortality, operative complications, survival curves, and the results of pathophysiological tests. "Improved hospital morbidity, mortality and survival after the Whipple procedure" [1] is the typical title of a new study from a famous department of surgery. No mention as to how these mortally ill patients may feel about the complications or whether they do or do not suffer!

Definition: Endpoint

A relevant endpoint is a variable, i.e., a measurable feature of a patient or a therapeutic procedure that allows the effects of an intervention (e.g., a surgical operation) to be recognized or described convincingly and in a manner acceptable to all concerned.

Thus, when a cholecystectomy is carried out for gallstones because of food intolerance, pain, etc. and the associated impairment in well-being, the relevant endpoint for assessing the therapeutic approach is patient well-being and, only secondarily, transaminases, white cells, complications of wound healing or operative mortality. This is all the more apparent in the treatment of oesophageal regurgitation, chronic pancreatitis, haemorrhoids, tennis elbow, torn menisci or the above-mentioned peripheral vascular disease.

In principle, the quality of life becomes a relevant endpoint when the surgical intervention itself has some effect on most of the dimensions that make up the quality of life [2]. One must also add that this effect is the rule rather than the exception [2, 3].

The Question Arises: Why This Discrepancy?

This serious situation is mostly due to the natural and absolutely essential interest of the surgeon in surgical techniques. Testing these techniques has therefore been and must continue to be one of his or her main occupations. In this context the relevant objectives are operative complications, operative mortality, and functional disturbances of a physiological and biochemical nature. In many areas, however, surgical technique is so well-developed that surgeons "have time" to be concerned about patient well-being as well, but the conventional endpoints for assessing effectiveness are either inadequate or inappropriate [2].

A further reason why well-being is so seldom used as an endpoint is the difficulty of defining the term "well-being" or even that of "quality of life". It is thanks in particular to sociologists, psychologists, philosophers, and physicians that this aspect is characterised time and again as quasi-metaphysical and is made needlessly complicated.

A much more important reason for this fact, however, is the way in which disease is interpreted and the resulting type of medical education. Chemistry, biochemistry, and physiology, and, above all, the "measuring" techniques, arising from the methods employed in these fields, are regarded as essential for the description and understanding of disease. "Looking ill", nausea and vomiting, pain or the "bloated stomach", the "bad knee", pains in the hips and, above all, their effects on well-being are considered to be so-called "soft" data because it is thought that they cannot be measured by methods accepted as scientific.

This is, moreover, a reason why pain or nausea, etc. are not to be found in graphs! Although, as any clinician can confirm daily and as has already been mentioned here several times, these so-called soft data often aid him or her in reaching decisions or act as criteria for assessment.

The main criteria for "hard data", namely, comprehensive measurement, are objectivity, consistency, every possible mensurability within dimensional systems and, above all, the most exact reproducibility possible (reliability). The leucocyte count, the titre-measurement of toxins in body fluids as well as the description of X-ray pictures, or the pathologists' observations of what they have seen under the microscope are accepted as hard data.

This necessary, successful and scientifically comprehensible way of looking at things, however, does lead to excessive reliance on hard data and to the neglect of the soft data. This development or, rather, this attitude, is described by Feinstein [4] as "the dehumanisation of clinical science". In this connection, it is clear that methods now exist which can render "soft" data "hard". If the criteria for hard data outlined above are applied strictly, much hard data falls rapidly into the category of soft data, e.g., that supplied by the pathologists and radiologists.

A further reason is a lack of interest and the overall ignorance of clinicians, in general, of this aspect of medicine. This becomes plain from an analysis of the literature which we made for this very reason. Using the Medline databank and the key words "surgery" and "surgery and quality of life", we analysed German and English surgical journals from the year 1980 up to and including October 1988. Over this period, 471 papers with the keywords "quality of life" and "surgery" appeared out of a total of 115,113 papers published under the keyword "surgery" (Fig. 1). Using this source analysis, only 98 papers appeared with the words "quality of life" in the title. In other words, during those 9 years, each journal published only one single paper with "quality of life" in the title.

Qualitative analysis, i.e. research work in which generally accepted instruments for measuring the quality of life were employed or which at least provided information about the instrument used was even more sobering (Fig. 2). This revealed an excessive influence of sociologists, psychologists, psychosomatic specialists, and epidemiologists. For my own part, I think that this is the reason why current ideas about the definition of and, above all, about the instrumentation for measuring quality of life are not yet fit for use in clinical medicine. "Co-operative theorists may, it is true, provide invaluable assistance, but the basic research, which the clinician requires above all, must be developed by himself ..." [5].

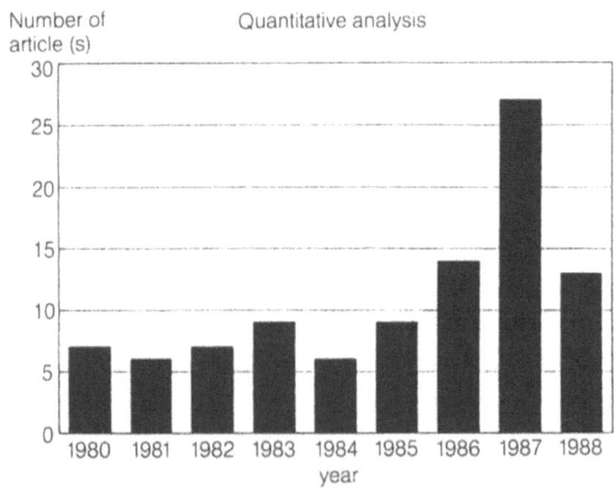

Fig. 1. Quantitative analysis of papers with the keyword "quality of life" in the title from journals containing surgical subject matter. Total number of papers: 98 articles in 75 journals from the year 1980 to 29.11.1989. (Source: MEDLINE databank)

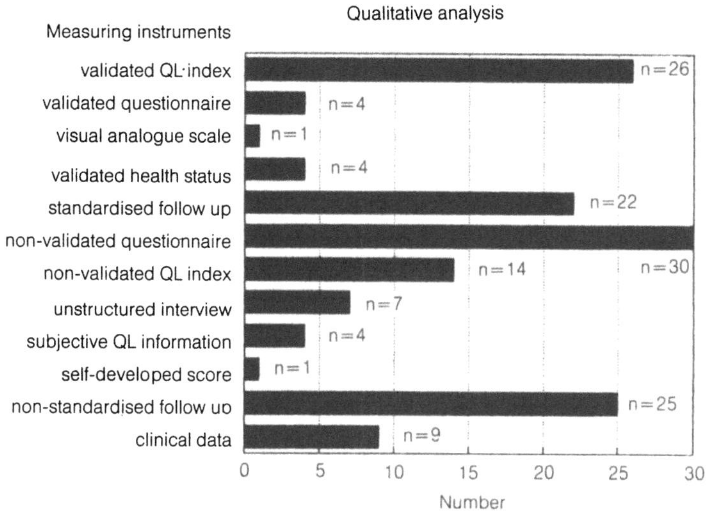

Fig. 2. Qualitative analysis of papers with the keyword "quality of life" in the title from journals with surgical subject matter. Measuring instruments chosen to assess quality of life and how often they were used; 72 articles in English or German in 51 journals from the year 1980 to 29 Nov 1989. (Source: MEDLINE databank)

Historical Aspects of the Endpoint "Quality of Life in Surgery"

However, before sociologists, psychosomatic specialists, psychologists, and even epidemiologists gave any thought to this dimension of medicine, certain clinicians – though few in number – and surgeons did change aspects of surgery by using well-being and their assessment of well-being to help them in their work. Typical examples of this are two controlled studies from Leeds dating from the 1964, which were published in the middle and at the end of the 1960s. In order to obtain treatment-relevant information about the possibly very unpleasant pain experienced following the various types of operation in the surgical treatment of haemorrhoids, Watts et al. [6] used five simple questions in a study to obtain information about pain after surgical treatment: "Pain is: a) almost none, b) less than average, c) average, d) more than average, e) severe" [6]. On the basis of the information obtained, Goligher (one of the co-authors) recommended and himself carried out the Milligan-Morgan method in cases of operable haemorrhoids [7].

Certainly, the most important example of this development, which is still effective in the field of surgery, was the introduction into gastroenterology of a so-called general assessment, an evaluation of general health designed by Visick [8]. Goligher transformed gastric surgery – and not only that – by using this instrument in his famous York-Leeds trial. The question was whether a chronic, symptomatic duodenal ulcer should be managed by resection, the combination of resection and section of nerves or only by nerve section, namely vagotomy. Since the conventional endpoints, i.e. operative mortality and operative complications, were almost identical, it was possible to recognise one or other method as superior only on the basis of well-being. This was so despite the fact that the "Visick" instrument [8] is predominantly symptom-orientated and had not been developed or tested in accordance with the current methodological knowledge [9, 2].

In addition to these methodological problems, i.e. development of an instrument and testing of that instrument in accordance with clinical research [10], a further major aspect of this question needs to be recognised and solved. This is the purely symptom-orientated way in which clinicians – including surgeons – look at things, which is also clearly recognisable in the Visick instrument. The step – the absolutely necessary step – which really affects the quality of life is the reaction of the patient to a symptom or the ability to cope with the symptom in different circumstances of life, e.g., in the elderly and in the young, and also in different cultural circles [2]. An example of this is a very recently published paper from Göteborg [11] which describes the effect of a so-called ileal pouch following total resection of the colon for colitis, using a "function score". This function score is a mixture of pathophysiological test findings and symptoms, but it lacks the crucial step which states how the individual patient copes with these changes. We had ourselves made exactly the same mistake in our study [12]. Our "scoring system" (Table 1) was and is predominantly symptom-orientated. Moreover, even this score has not, as is still usual in surgery, been developed and tested in accordance with the rules of clinical research.

However, a recognisable increase of interest and significance in this important aspect of medicine is noticeable, and this is also occurring in surgery. Publications in the fields of cardiac surgery [3], abdominal surgery [3] and, of course, in cancer surgery [3] are examples of this. Not least is the positive trend towards acceptance of the quality of life by surgeons indicated by the fact that the most recent German surgical conference had as its main theme "Quality of life and progress in surgery", while *Der Chirurg ("The Surgeon")*, which is probably the most important journal for surgeons in the German language, has recently dedicated a whole issue to the topic "quality of life" [13].

Well-being as a Relevant Endpoint Following Gastrectomy for Advanced Carcinoma of the Stomach – A Controlled Study

More than 10 years ago, it became clear to me in the course of systematic follow-up [14, 15] that, after successful removal of the stomach for advanced cancer, the general picture of patients' well-being was, with a few exceptions shortly after the operation, pitiable. On the day of their discharge, patients were still in good physical shape and most were even happy. At the first follow-up 6 weeks after surgery, their general mood remained good in most cases. The patients, however, were already aware of the lack of hunger and appetite. Almost all had lost more weight. Many already expressed their worry and anxiety. This was even more clear 3 months after the operation. Loss of appetite and hunger together with anxiety were the predominant characteristics and feelings of the patients [15, 12].

With regard to the permanent weight problem which had, after all, been familiar for a long time to clinical researchers [16, 17, 18], we also thought at first that changes in absorption were largely responsible. We adopted a "scientific" approach to this problem by investigating absorption, wishing naturally, to obtain information in this way and then to treat the patient successfully [15, 12]. These investigations, however, showed that changes in absorption could indeed be observed; that the duodenal passage did not show the importance that had been repeatedly attributed to it and that is still so attributed, and that adaptation very markedly minimises the whole problem. The possibility of a purely quantitative intake of nourishment following loss of the stomach as a reservoir appears to be extremely significant. This fact was also known to pioneers in this field [16, 18].

What really remained was the patients' fear of cancer, the short survival time in general and the recognition that the possibility of eating a lot of food

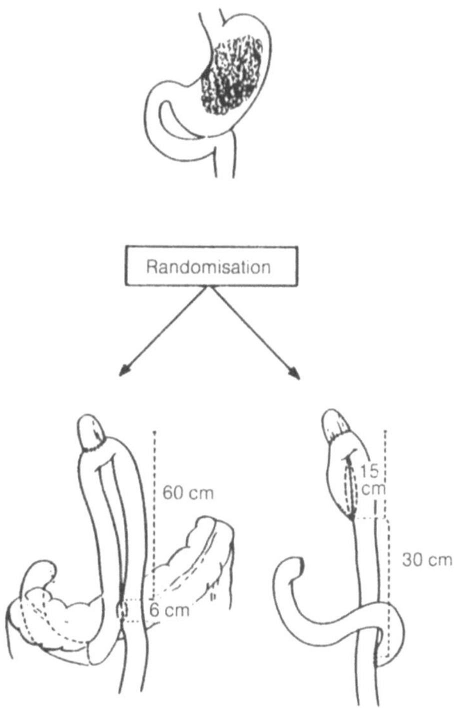

Fig. 3. Operative techniques in a randomised controlled study of the surgical treatment of gastric carcinoma. Principle: *"gastrectomie de nécessité"*, i.e., in advanced gastric cancer, total gastrectomy is only performed when the tumour cannot be controlled by a less extensive operation. *Left,* operative details of oesophagojejunostomy; *right,* pouch reconstruction after Hunt-Lawrence-Rodino

at each meal is very important for these patients [12]. This situation led us to formulate the following hypothesis: In advanced cancer of the stomach, gastric resection is palliative. Palliation, in the absence of any real possibility of saving the patient's life, means avoiding, if at all possible, any negative effects upon the course of the illness [19]. In view of this fact and the large measure of equivalence in the so-called conventional endpoints (operative mortality, immediate post-operative complications, results of pathophysiological tests and equal or limited survival times), the relevant endpoint here is well-being.

Thus a typical situation for surgeons, such as I have described [2], shows that well-being is the relevant endpoint when two available therapeutic approaches cannot be differentiated using conventional endpoints, or if the aim of therapy is pure palliation [2]. Quality of life or well-being as an endpoint should help surgeons to decide which method of reconstruction to choose. Among the possible methods of reconstruction [20, 21], we chose the so-called most simple, namely oesophagojejunostomy and a method that has proved itself a relatively good reservoir [16, 12] – the Hunt-Lawrence-Rodino pouch (Fig. 3).

Controlled Study: "Substitute Stomach" Versus Oesophagojejunostomy after Total Gastrectomy for Advanced Carcinoma of the Stomach

After conducting a pilot survey showing that a study of this type is, in principle, possible, and which also gave us the necessary training experience, this study was started in September 1976. The most important results of the pilot study showed that the vicious circle consisted of weight loss, lack of appetite, hunger and, above all, fear of the disease. The latter definitely affected psychological factors as well as the low intake of nutrients (see above). Immediate or post-operative complications were not significantly changed, and the survival rates were the same, i.e. equally poor. Duodenal passage, in particular, did not bring about any outstanding improvement in the post-operative situation [12].

Patients who had advanced gastric carcinoma, which had, above all, to be localised by endoscopy and demonstrated histologically, were admitted to the randomised study. Gastrectomy was only carried out if a less extensive procedure was inappropriate to the case, i.e. the principle *"gastrectomie de nécessité"* was adhered to (see Fig. 3). All measures were standardised; preoperative preparations, intraoperative procedures, particularly systematic removal of the lymph nodes, and the post-operative management as well as, of course, a systematic follow-up. All endpoints were defined exactly, as, for example, appetite, hunger, food intake, weight, operative mortality and survival times [12]. The randomisation was planned to include 60 patients, with 30 patients per arm [12]. Our intention of using "well-being" as a relevant endpoint was essential. Since, at that time, there was no usable, well-tested instrument for evaluating well-being, we had to solve this problem in a variety of ways. We chose variables which would describe the effects of gastrectomy to a clinician who had been involved in this type of problem for a number of years; thus, for example, opportunities for eating, including the amount per meal (food intake); then, feelings of hunger and appetite, and also information about fatigue, the ability to get refreshing sleep, etc. All these variables were put in the form of a standardised questionnaire which could be scored (Table 1).

Naturally, we had neither developed nor even tested this score in accordance with the, now familiar, rules of clinical research. Certainly, when we started work on this problem in 1974, this type of tested and usable instrument did not yet exist, as has already been emphasised more than once. Visick's index [8], which we altered after lengthy discussion with Professor Goligher, showed little capacity for sensitive assessment. A global index, such as that developed almost 10 years later by Spitzer et al. [22], remains of little use, now as then, in this type of problem [23].

In the light of my present understanding, however, I would like to make clear that our scoring system was also dominated excessively by a clinical way of thinking insofar as its excessive stress upon symptoms is concerned. In practice, this means that we were not then aware of the extent to which just those very variables, i.e. both the disease-specific and the sociodemographic variables, could interfere with the general well-being of the patient. In other words, how far does a patient suffer if he or she does not experience hunger or

Table 1. Scoring system for evaluation of general well-being (quality of life) of patients with gastric carcinoma before and after gastrectomy. Maximal scores: disease specific, 8; social and personal, 6

Variables	Score	
	Yes	No
Disease-specific		
Have you any appetite?		
Do you suffer from vomiting?		
How much can you eat:		
Normal-sized portions?		
More than half?		
Less than half?		
Do you have to belch?		
Weight		
Increasing		
No change		
Loss of 1–5 kg		
Loss of >5 kg		
Social and personal		
Are you tired?		
Can you sleep?		
Do you suffer pain (do you use analgesics)?		
Are you able to work (or carry on with a hobby)?		
Are you able to get out and about?		
Are you bedridden?		

if he or she is unable to sleep? Does this type of symptom represent a very marked impairment of well-being for some patients? Our score gave and gives us no real access to this particular aspect. The main results shown by this study which are of interest in this context, are as follows:

Food Intake

After both types of reconstruction, there is an extremely restricted food intake immediately after operation and for up to 3 months. This is found both in patients dying in the 1st year (Fig. 4) and also in patients who survived 1 year or more (Fig. 5). An important finding is that the ability to eat larger amounts of food improves after 3 months, and that this is particularly the case with the so-called Hunt-Lawrence-Rodino pouch.

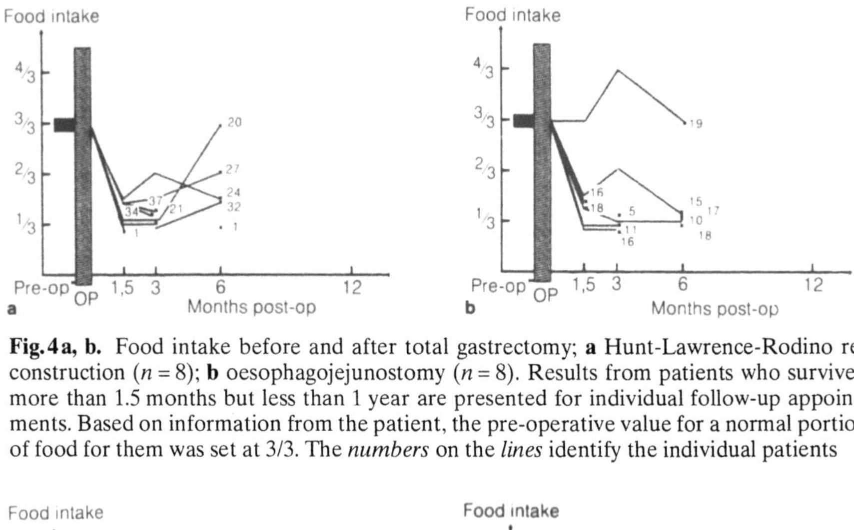

Fig. 4 a, b. Food intake before and after total gastrectomy; **a** Hunt-Lawrence-Rodino reconstruction ($n = 8$); **b** oesophagojejunostomy ($n = 8$). Results from patients who survived more than 1.5 months but less than 1 year are presented for individual follow-up appointments. Based on information from the patient, the pre-operative value for a normal portion of food for them was set at 3/3. The *numbers* on the *lines* identify the individual patients

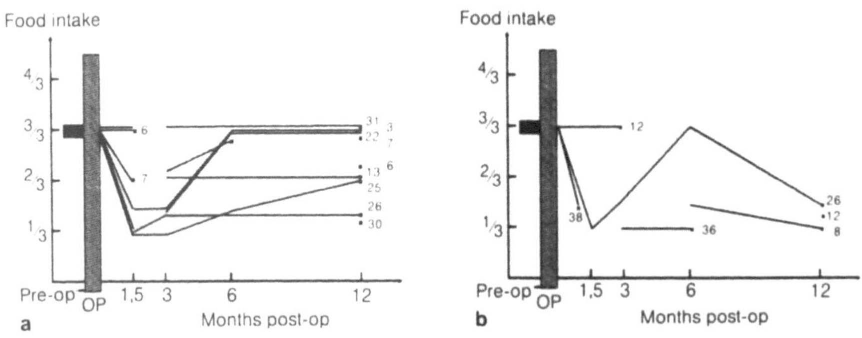

Fig. 5 a, b. Food intake before and after total gastrectomy; **a** Hunt-Lawrence Rodino reconstruction ($n = 10$); **b** oesophagojejunostomy ($n = 5$). Results from patients who survived more than 1 year are presented for individual follow-up appointments. Based on information from the patient, the pre-operative value for a normal portion of food for them was set at 3/3. The *numbers* on the *lines* identify the individual patients

Weight

Initially, even after discharge from hospital and despite intensive advice about nutrition, it is impossible to avoid a very obvious loss of weight. This improves only from the 3rd or 6th month after surgery [12].

Hunger and Appetite

With regard to the problem of hunger and appetite, a similar improvement could also be observed. Immediately after the operation, i.e. in the first 6 weeks or in the first 3 months, the sensation of hunger was markedly reduced, particularly in patients with oesophagojejunostomy. The longer the interval following surgery, the more this situation improves and this is clearly seen in the patients with a so-called substitute stomach (Fig. 6).

Well-being

The most interesting results, however, were achieved with our "inadequately" developed and "imperfectly" conceived scoring system. This is all the more clear when the patients are divided into those who died within the 1st year –

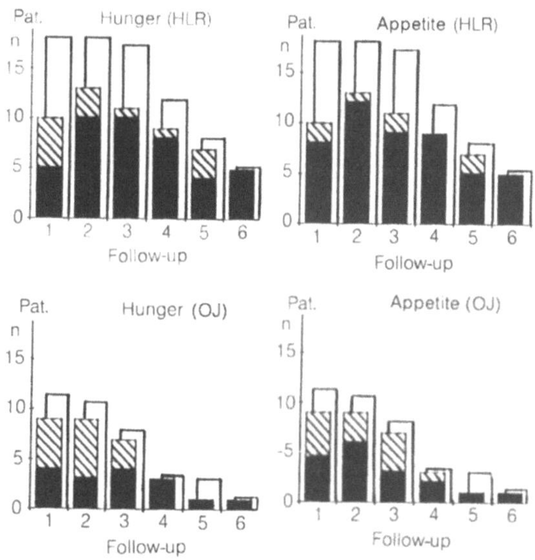

Fig. 6. Post-operative changes in hunger and appetite in patients after Hunt-Lawrence-Rodino reconstruction *(HLR)* and after oesophagojejunostomy *(OJ)*. Questions to the patients at the fixed follow-up appointments: "Do you experience hunger?" and "Do you have any appetite?" Answer: no, *hatched columns;* yes, *filled columns. Open columns,* total number of patients. Follow-up number: *1,* 6 weeks post-op.; *2,* 3 months post-op.; *3,* 6 months post-op.; *4,* 1 year post-op.; *5,* 2 years post-op.; *6,* 3 years post-op.

and this is the majority (almost 60%) – and patients who survived the operation for longer than 1 year.

All patients (Fig. 7a) who died in the 1st year showed in the "total score" an extreme impairment of well-being which, throughout the whole post-operative course, never came close to achieving the pre-operative status. This is all the more clear for the "disease specific variables" (Fig. 7b) with regard to food intake, appetite, hunger, swallowing difficulties, etc., described in the example of the Hunt-Lawrence-Rodino pouch. This result is interesting from two points of view and should be placed on record:

1. It shows that patients who appear to have an extremely poor prognosis do not benefit in the slightest from the operation, expressed in rough terms, if one takes this scoring system as a basis. It also shows that, when using this instrument alone, it is possible to obtain this type of information, particularly when the disease-specific variables are taken into account.
2. In contrast, in patients who survive *more* than 1 year (Fig. 8a–c), these same variables show a similarly interesting result. First, it can be seen that the adverse effect upon well-being reaches its nadir at about 6 weeks to 3 months

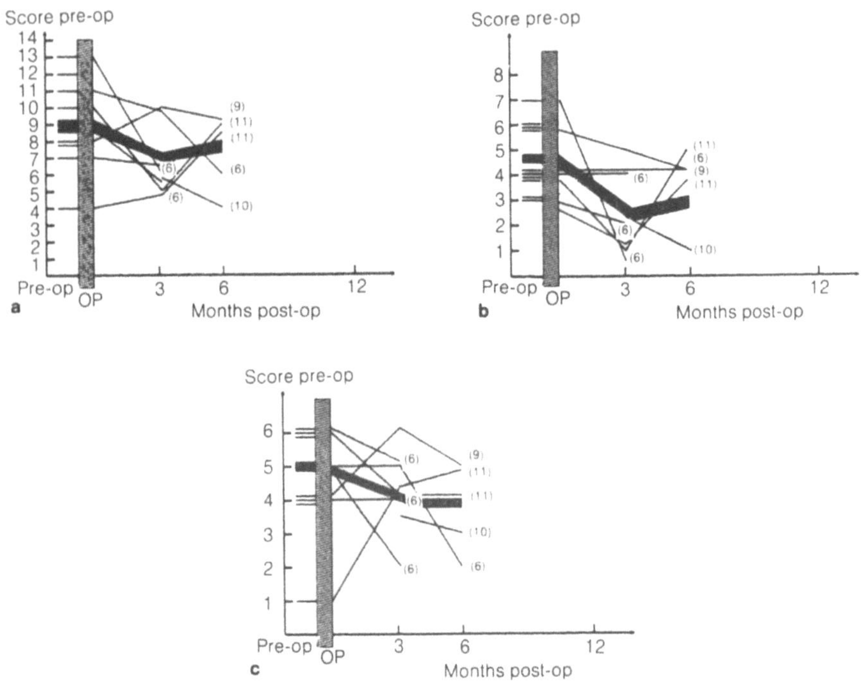

Fig. 7a–c. Quality of life after total gastrectomy and Hunt-Lawrence-Rodino reconstruction. Results from patients who survived more than 3 months but less than 1 year are presented as a pre-operative score ($n = 10$) and for the post-operative course ($n = 8$). Each *line* represents one patient, *broken lines* show missing results at the appropriate follow-up appointment. *Numbers in brackets,* survival times in months. **a** Total score, maximum 14 points; **b** disease-specific, maximum 8 points; **c** social and personal, maximum 6 points

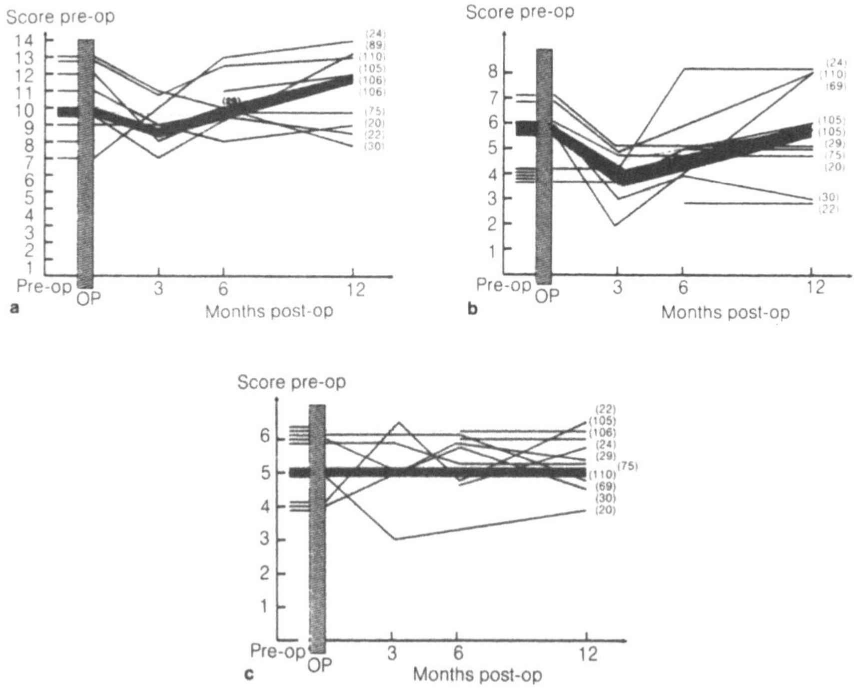

Fig. 8 a–c. Quality of life after total gastrectomy and Hunt-Lawrence-Rodino reconstruction. Results from patients ($n = 10$) who survived more than 1 year are presented as a pre-operative score and for the post-operative course. Each *line* represents one patient, *broken lines* show missing results at the appropriate follow-up appointment. *Numbers in brackets,* survival times in months. **a** Total score, maximum 14 points; **b** disease-specific, maximum 8 points; **c** social and personal, maximum 6 points

and that well-being improves thereafter so that one can say that patients having a favourable prognosis and surviving the so-called trough after 3 months have, to some extent, achieved good health after 6 months and even then continue to experience improvement. An important piece of clinical information!

Again, it emerges that our rather "harmless" scoring system makes possible information such as this to be obtained. It has also become clear that, even with this analysis, the disease-specific scale of variables here provides more convincing and clearer results. The "social and personal" variables, as we applied them, are completely ineffectual. If one compares the two reconstruction procedures using all the variables which we thought would give us useful information about well-being, it is clear that the substitute stomach should be favoured from this point of view.

In summary, this investigation perhaps allows us to formulate more than just a hypothetical statement, to the effect that the variables recognised by clinicians as important for well-being provide information of relevance to management.

Removal of the stomach for advanced gastric cancer means an extreme impairment of well-being for all patients. The nadir of this impairment occurs after 6 weeks to 3 months. If the patient has no opportunity of surviving for longer because of the advanced state of his or her disease, surgery does not benefit him or her as far as well-being is concerned. It seems worthwhile to develop instruments in the future which have been tested and developed in accordance with current views of clinical research in order to confirm these investigations.

References

1. Christ DW, Sitzman JV, Cameron JL (1987) Improved hospital morbidity, mortality and survival after the Whipple procedure. Ann Surg 206: 358
2. Troidl H (1989) Lebensqualität: ein relevantes Zielkriterium in der Chirurgie. Chirurg 60: 445–449
3. Troidl H, Kusche J, Vestweber K-H, Eypasch E, Bouillon B (1987) Quality of life: an important endpoint both in surgical practice and research. J Chronic Dis 40: 523–528
4. Feinstein A (1977) Clinical biostatistics XLI. Hard science, soft data and the challenges of choosing clinical variables in research. Clin Pharmacol Ther 22: 485
5. Feinstein A (1986) Clinimetric perspectives. In: The Portugal conference measuring quality of life and functional status in clinical and epidemiological research. J Chronic Dis 40: 635
6. Watts JM, Bennett RC, Duthie HL, Goligher JC (1964) Healing and pain after different forms of haemorrhoidectomy. Br J Surg 51: 808–817
7. Goligher JC, Pulvertaft CN, Watkinson G (1964) Controlled Trial of Vagotomy and Gastro-enterotomy, Vagotomy and Antrectomy, and Subtotal Gastrectomy in Elective Treatment of Duodenal Ulcer: Interim Report. Brit Med J 1: 455–460
8. Visick AH (1948) A Study of failures after gastrectomy; Ann R Coll Surg (Edinb) 3: 266–284
9. Wood-Dauphinée S, Troidl H (1986) Endpoints for clinical studies: Conventional and innovative variables. In: Troidl H, Spitzer WO, McPeek B, Mulder DS, McKneally MF (eds) Principles and practise of research. Springer, Berlin Heidelberg New York
10. Wood-Dauphinée SL, Troidl H (1989) Assessing quality of life in surgical studies. Theor Surg 4: 35
11. Öresland T, Fasth S, Nordgren S, Hulten L (1989) The clinical and functional outcome after restorative proctocolectomy. A prospective study in 100 patients. Int J Colorectal Dis 4: 50
12. Troidl H, Kusche J, Vestweber K-H, Eypasch E, Maul U (1987) Pouch versus esophagojejunostomie after total gastrectomy: a randomized clinical trial. World J Surg 11: 699–712
13. Chirurg (1989) 60: (7)
14. Troidl H (1978) Einrichtung einer systematischen Kontrolluntersuchungsklinik. Langenbecks Arch Chir 347: 467–480
15. Troidl H, Hamelmann H, Maul U, Thiede A, Jostarndt L (1979) Organisation der systematischen Kontrolluntersuchungsklinik bei Erkrankungen des Dickdarms und Analbereichs. In: Stock W (ed) Nachsorge beim kolorektalen Karzinom. Springer, Berlin Heidelberg New York, pp 33–44
16. Lawrence W Jr (1962) Reservoir construction after total gastrectomy. An instructive case. Ann Surg 155: 191
17. Scott HW, Law DH, Gobbel WG, Sayers JL (1968) Clinical and metabolic studies after total gastrectomy with a Hunt-Lawrence jejunal food pouch. Am J Surg 115: 148
18. Bradley EL, Isaacs J, Hersh T, Davidson E, Millikan W (1975) Nutritional consequences of total gastrectomy. Ann Surg 182: 415

19. Troidl H, Spangenberger W (1989) Metastasenchirurgie und Lebensqualität – die Befindlichkeit als entscheidendes Zielkriterium bei palliativer Chirurgie. In: Rothmund M (ed) Metastasenchirurgie. Pathologie, Bildgebende Diagnostik, Indikation, Operative Strategie. Thieme, Stuttgart
20. Herfarth C, Schlag P, Buhl K (1987) Surgical procedures for gastric substitution. World J Surg 11: 689–698
21. Largiader F, Säuberli H (1972) Die totale Gastrektomie, part I: Möglichkeiten der operativen Rekonstruktion nach Gastrektomie. Bruns Beitr Klin Chir 219: 601
22. Spitzer WO, Hall AJ, Chesterman JL , Shepherd R, Battista RN, Catchlove BR (1981) Measuring the quality of life of cancer patients. A concise QL-Index for use by physicians. J Chronic Dis 34: 585–597
23. Kusche J, Vestweber K-H, Troidl H (1987) Quality of life after total gastrectomy for stomach cancer. Results of three types of quality of life evaluative methods. Scand J Gastroenterol 22 [Suppl 133]: 96–101

The Risk Factor Model and Life-Styles Relevant to Health

J. von Troschke

Measures relating to preventive medicine are sanctioned by the state of knowledge about the aetiology of disease prevailing at the time. At the beginning, knowledge about edible and inedible plants was applied empirically. Later, observations were made which were converted into hygienic measures as a protection against infectious diseases. From time immemorial, an important role in the advice given by doctors with regard to the maintenance of health has been played by ethically based precepts relating to life-style and, associated with them, ideas about how to strengthen the power of resistance to disease. In keeping with the culturally and socially determined view of man prevailing at the time, individuals were either urged to adopt an attitude of responsibility for their own health, or government measures for normative regulation, designed to protect the health of the population, were implemented. With the growing knowledge of scientific medicine, the possibilities in the therapeutic field gained in importance, resulting in a reduced interest in prevention. The art of medicine expanded, above all, in the fields of diagnosis and treatment, the claims of which were justified by scientific principles. In comparison with this, preventive measures were far less convincing and were, moreover, hampered in that they were less positively effective.

Since World War II, with the development of the concept of the risk factor, modern epidemiology has fundamentally altered the preconceived notions of preventive medicine. Risk factors have been identified by mathematical methods and their effect upon pathological processes demonstrated statistically. A differential method was developed for identifying potentially pathogenic factors in retrospective case-control studies, and for testing the effect of a reduced spectrum of risk factors in systematically controlled intervention studies. The concept of the risk factor is in keeping with the scientific model of medicine in that it allows a systematic, objective testing of hypotheses aided by mathematical and statistical methods. With the investigation of large population groups, it was possible to work out relationships between a large number of factors (risk factors) and the incidence of disease. Particularly in relation to diseases of the cardiovascular system, which are especially common in industrially developed societies, it was possible not only to describe obvious risk factors, but also to demonstrate dose/effect relationships. The greater the number of cigarettes smoked, and the higher the total blood cholesterol, the weight and the blood pressure, the greater is the probability of cardiovascular disease developing. On the other hand, it was shown that a reduction in risk factors was associated with reduced rates of morbidity and mortality. This provided a clear rational basis for preventive measures: in primary prevention with the avoidance of the development of risk factors; in secondary prevention with the diagnosis and early treatment of risk factors; and, finally, in tertiary prevention, leading to rehabilitative reduction in risk factors with the aim of preventing disease recurrence. It was also possible to substantiate the epidemiologically based concept of the risk factor by retrospective clinical studies and pathological investigations. This provided a scientific basis for developing intervention strategies against the modern "diseases of civilisation" that were rational and goal directed, and for the effective representation of their relevance to health policy. Thus it is that, since World War II both research into methods of prevention and prevention at a practical level have been of increasing importance in health policy and have been able to obtain increasing financial support. All organisations engaged in prevention in our society have benefited from this in recent times.

A growing criticism of the concept of the risk factor, with the emphasis on particular problems, is, however, noticeable. On the one hand, research into risk factors has concentrated predominantly on the investigation of pathogenic processes, involving itself solely with diseases having high rates of morbidity and mortality, and, in so doing, it has largely neglected the investigation of phenomena responsible for health. In other words, scientific medicine has collected a large amount of information for understanding disease, but very much less for understanding health, i. e. for answering the question as to why people remain healthy – even under unfavourable conditions. This is a fundamental criticism and attacks medicine's understanding of itself. With regard to risk factors, one could take account of these objections by looking more energetically for protective factors, i. e. characteristics that are found with particular frequency and to a pronounced extent in people who do not become ill, but who remain healthy.

A further objection is based on the fact that known risk factors are never able to explain more than a part of the variability of pathogenic processes. Using the risk factors for cardiovascular disease that have been identified so far, only about 50% of the variability can be explained. The force of this objection is reduced when it is pointed out that more intensive research is necessary to investigate the relationships between risk and possible protective factors, and also to work out hitherto unidentified co-factors. The information that is currently available is not sufficient to create a model for explaining the effective mechanism of risk factors. Schaefer and Blohmke [1] produced a hypothesis of the aetiology of myocardial infarction based on the principle of the regulatory loop, and this still remains a valuable guide in the development of comprehensive hypotheses, despite a variety of criticisms. The rigid, one-dimensional view of risk factors was abolished by the systematic distinction between physical risks and health-related life-styles, and between genetic dispositions and variability in the structure of the person-

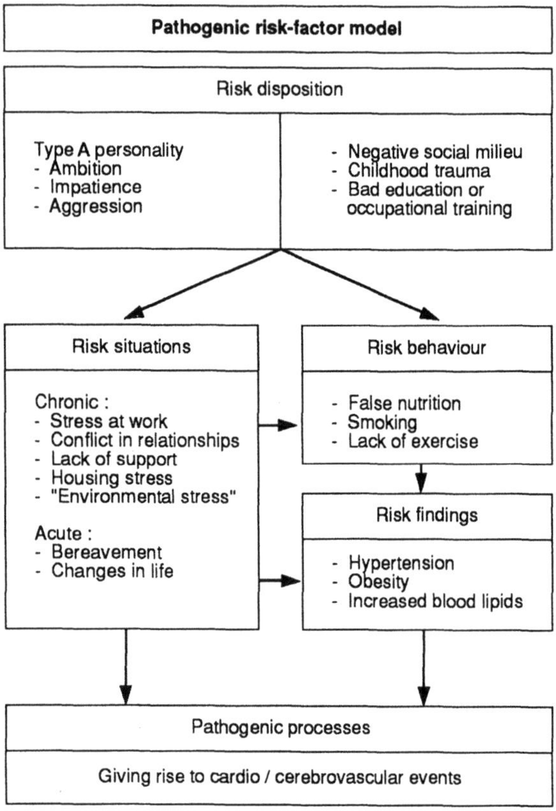

Fig. 1. Pathogenic risk-factor model

ality. Figure 1 illustrates a new approach, defining those risk-dispositions (genetic factors, social conditioning, variability in personality structure) which become active in risk situations (chronic or acute) and result in risk characteristics, eventually leading to risk states and finally to pathogenic processes ending in overt disease.

This model takes into account socioepidemiological data on pathogenic living conditions, actual findings arising from the social positions of different social groups. This has revealed a connection between physical risk states (such as raised blood pressure, obesity, increased blood lipid values), individual lifestyles that represent a health risk (such as false nutrition, smoking, and lack of exercise) and the general living conditions of social groups.

A fundamental problem of the risk-factor model lies in the statistical basis of the data concerning the relationships investigated. Information about risk factors and their expression, obtained from statistical correlations, is, strictly speaking, only valid for large population groups. Because of this, epidemiological intervention studies are aimed not at individuals but at large population groups, with the aim of achieving statistically significant reductions in the incidence of risk factors or even in the mean values for prevalence rates. This type of intervention strategy ran into difficulties when it was applied to individual people and tried to motivate them to change their behaviour. The statistical probabilities calculated for large population groups are of limited validity for individual subjects and thus have limited powers of persuasion. It is somewhat like casting a die. At each throw, one has the same chance of throwing one of the six alternatives. Even when one has thrown a six three times in succession, the probability of getting another six on the fourth throw is just as great as before. The statistical probability of finally throwing all numbers with equal frequency, if the die is not loaded, is achieved only with a very high number of throws. The statistical evidence for an increased probability of disease in those with risk factors can be related only to large population groups.

Decisions about human behaviour are only rarely made in the light of statistically calculated probabilities. More often they are made on the basis of the desirability of the target. Everyone who takes a risk does so because he or she reckons that it will not "get him or her", that he or she has a chance of being protected from the potential negative consequences.

This explains the fact that those in medically related jobs do not necessarily behave in an exemplary fashion in relation to the avoidance of risk factors although they are directly confronted every day with the negative consequences of this type of behaviour. Although the health risks of smoking have been clearly demonstrated, about 30% of doctors, 43% of chemists and 45% of nurses in the Federal Republic of Germany are smokers.

A dilemma of the risk-factor model lies in its conversion into persuasive tactics which will convince people of the importance of prevention. Since these often have little effect on individuals, provided they feel well, epidemiological intervention studies increasingly employ mass media methods which are directed at large population groups and aim to influence the social climate in such a way as to develop norms for risk avoidance.

A sociological approach to solving the problems of the risk-factor model is to be found in the "life-style concept" as propagated by the World Health Organisation in particular. The blanket terms "human life-style" or "way of life" describe the totality of patterns of behaviour incorporating not only types of behaviour that put health at risk but also those that promote health. This does away with the restrictive investigation of isolated risk factors and relates it to a behavioural context relevant to those people being treated. Systematic investigations have revealed that those types of behaviour putting health at risk are seen as purposeful within the situational context, and that the risks associated with them have therefore been weighed up. Insofar as the decision to smoke is concerned, we have developed a model of situation-related cost-benefit analysis which distinguishes ten different motives that have been made socially acceptable and reinforced by the community (Fig. 2).

Anyone wishing to light a cigarette has had the decision affected by various motives, as briefly characterised below.

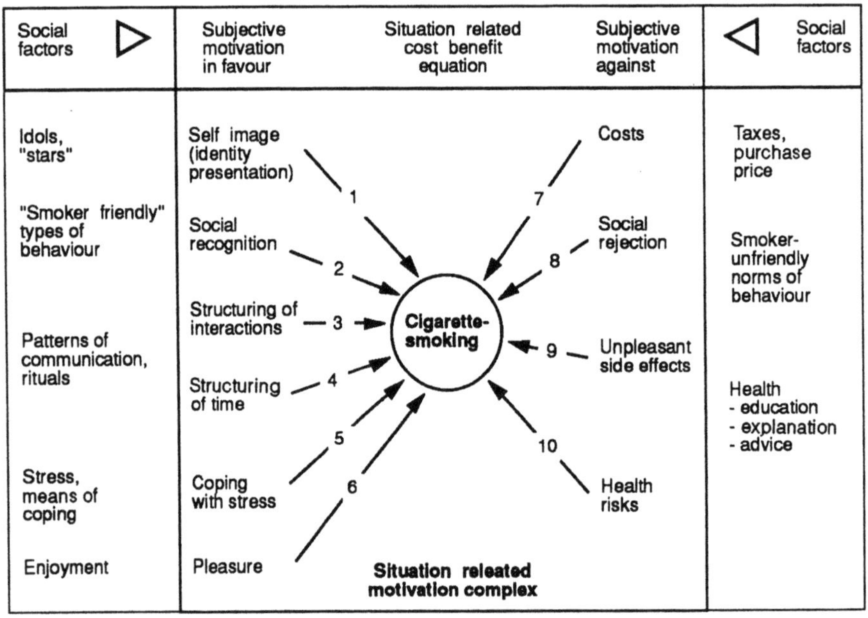

Fig. 2. From: Troschke [9]

Self Image (Identity Presentation). Smoking as a social activity offers individuals many ways of presenting themselves in public or in relation to others, particularly by means of the associated rituals [2–4].

Social Recognition. The individual is confirmed and recognised as a member of a group insofar as he or she conforms to group-specific expectations of behaviour and thus endorses the group and its norms. Within a group of smokers, members receive social recognition if they conform to the usual rituals of smoking.

Structuring of Interactions. As a social transaction, smoking is a method of communication, i.e. people are able to communicate verbally and non-verbally with each other by offering a cigarette, providing a light, etc. The offer of a cigarette is frequently used to overcome social barriers [5] or to establish contact.

Structuring of Time and Events. The function of "time structuring" [2] means that smoking can be used to fill in spare periods of time, e.g., when one is waiting for someone or something. Smokers can use the unit of time "one cigarette long" to structure events (event marking).

Overcoming Psychosocial Tension (Negative Affect Smoking). Smoking can help people to come to terms with their own feelings and moods. Since Tomkins [6], two groups have been distinguished in relation to affect management: "positive affect smoking", by which smoking produces or expresses pleasant feelings, and "negative affect smoking", by which smoking is used to cope with unpleasant feelings in times of psychosocial stress.

Enjoyment (Positive Affect Smoking). It is clear that, among other things, the relaxing effect of smoking causes positive feelings. Cigarettes are used as a means of enjoyment and are able to provide the smoker with a readily available pleasure with a rapid onset.

Cost. Smoking is associated with financial costs, but also with a certain expenditure of time (e.g., obtaining cigarettes).

Social Rejection. In social groups where smoking is viewed negatively, the smoker must expect to provoke a rejection reaction by lighting a cigarette. Because of the public stigma attached to smoking, social pressure upon the smoker has increased.

Unpleasant Side Effects. Smoking is not only pleasurable. It may also cause unpleasant feelings in the smoker him-/herself. Physical symptoms and disturbances of well-being often lead to temporary reduction or even to the giving up of smoking.

Health Risks. Last but not least, the smoker must accept for him-/herself that cigarette smoking has clearly been shown to be a medical risk factor and is capable of causing sub-chronic and chronic diseases.

In the great majority of cases, smoking – like other types of behaviour that damage health – is situation-related in that the subjects weigh up the advantages and disadvantages and then come to a largely unconscious decision as to the significance of the possible consequences for themselves.

Ways of life are relatively persistent patterns of behaviour which are accepted by a specific social group and reinforced, either positively or negatively, by successful or unsuccessful experiences in everyday life. Types of behaviour that are relevant to health are embedded in overall behavioural patterns which often do not serve one single purpose only, but have also become ritualised. Thus, eating habits are not concerned only with eating but also with social communication or with a form of pleasure-seeking because of the enjoyment associated with it. Sociological investigations in the field of ways of life seek to discover the meaning and importance of types of behaviour that are a risk to health within the life-style of specific social groups and how far they depend upon the living conditions determined by society. They investigate the extent to which the health factor affects behaviour, under what circumstances conscious risks are taken or have to be taken, what significance the threat of disease, mortality and death has for everyday conduct, how far acknowledged environmental dangers can affect behaviour, and how the interaction between socially determined circumstances and types of human behaviour can be systematically described and analysed.

In association with the strategy developed by the WHO for the European region, "Health for all towards the year 2000", the Federal Republic of Germany has also adopted the life-style as a starting point for the further development of the risk-factor concept in relation to effective and appropriate preventive measures. Thus, in a vote taken on 13th April 1989, the Federal Board of Health [7] stated:

> The target of all measures promoting health is not only the prolongation of life but, above all, the improvement both in health and in the quality of life. It is not just a question of prolonging life at all costs. It is far more that of living life to its full over the years. Strategies for promoting health should therefore not be restricted to the health sector but must fully include these areas as well [ways of life, working, environmental conditions]. Responsibility for a life-style that fosters health lies first of all with each individual himself. Supportive measures in the promotion of health may come from the family, the school, the world of work and professions and also from the municipal or regional community . . . Promotion of health is a comprehensive process whereby each person should have a greater degree of personal freedom to maintain his own state of health and should be enabled to reinforce this aim. Ideas about healthy life-styles should not therefore be directed so much towards individual targets but should relate to the whole span of life and should be comprehensive.

The programme of research and development in the health services, published by the Federal Republic in March 1988 [8], states specifically:

> The traditionally predominant bio-medical approach to research in health must therefore be extended and enlarged to include contributions from the social and behavioural sciences. . . . A fundamental task of research is to expand available knowledge about the effects on health of the factors comprising a way of life, to describe the extent of particular characteristics relating to health in the life-styles of various population groups . . . and to analyse their function in everyday life. (p. 24)

The life-style approach opens up a new dimension for policies on prevention, where equal account is taken both of medical experience of the risk-factor model and of sociological experience in socialisation and social reinforcement of types of human behaviour. New perspectives can enlarge not only the effectiveness of preventive measures but, above all, their adjustment to the social realities, i. e. the actual room for manoeuvre that exists within the framework of the conduct of human life under culturally and socially determined living conditions. Viewed in this way, future prevention need not be orientated exclusively either towards behaviour or towards outward circumstances but must take equal account of the interaction of both areas. For those engaged in the prevention of ill-health, this proves the need for combining both medical and sociological knowledge, so that the life-style approach can be developed further and can be made applicable in practice.

References

1. Schaefer H, Blohmke M (1978) Sozialmedizin. Thieme, Stuttgart New York
2. Best JA, Hakstian AR (1978) A situation-specific model for smoking behavior. Addict Behav 3: 79–92
3. Mc Kennell AC (1970) Smoking motivation factors. Br J Soc Clin Psychol 9: 8–22
4. Russell MAH, Peto J, Patel UA (1974) The classification of smoking by a factorial structure of motives. Jr Statist Soc 131: 313–346
5. Feinhandler SHJ (1981) Social function as component of marked value. Conference 'Analysis of consumer policy'. May 18–19, 1981, Wharton Applied Research Center, University of Pennsylvania
6. Tomkins SS (1968) A modified model of smoking behavior. In: Borgatta EF, Evans R (eds) Smoking, health and behavior. Chicago
7. Bundesgesundheitsrat (1989) Votum vom 13. April 1989 zur Verwirklichung der Einzelziele der WHO-Strategie 'Gesundheit 2000' in der Bundesrepublik Deutschland. In: Bundesvereinigung für Gesundheitserziehung eV (Hrsg) 40 Jahre Gesundheitserziehung in der Bundesrepublik Deutschland: Rückblick – Ausblick – Perspektiven
8. Bundesminister für Forschung und Technologie (Hrsg) (1988) Forschung und Entwicklung im Dienste der Gesundheit. Programm der Bundesregierung
9. Troschke Jv (1989) Motivierung im Gespräch. In: Bundesvereinigung für Gesundheitserziehung eV (Hrsg) Gesundheit im Gespräch. Zum Weltgesundheitsthema 1989. Bonn

II. Nitrate Therapy

The Role of Nitrates in Congestive Heart Failure

J. Abrams

The use of intravenous nitroglycerin and long-acting nitrates in acute and chronic heart failure (CHF) has been popular since the introduction of vasodilator therapy for CHF in the early 1970s. Although the angiotensin converting enzyme (ACE) inhibitors have become extremely popular as "unloading" agents in CHF, current evidence suggests that nitrates may be comparably effective to these agents. It is of interest that both classes of drugs exert their effects on the venous and arterial side of the circulation in heart failure, with a subsequent reduction in preload and afterload of the heart. This discussion relates only to the patient with CHF *and* depressed left ventricular contractile or systolic function (eg., ejection fraction less than 35%–40%).

The nitrates are beneficial in improving the disordered hemodynamics in the heart failure state. They consistently reduce pulmonary capillary wedge, pulmonary artery, and right atrial pressures. Although salutary nitrate effects on stroke volume and cardiac output are modest, cardiac pump performance is usually increased [1, 2]. In addition, the nitrates appear to enhance long-term exercise performance [3, 4]; of the vasodilator drugs, only the ACE inhibitors have been shown to have a similar positive effect on exercise tolerance in CHF [5]. Finally, oral nitrates have been shown to improve survival in patients with NYHA classes II–III CHF when used in conjunction with hydralazine [6].

Mechanisms of Action

Nitroglycerin and the long-acting nitrates are relaxants of vascular smooth muscle and induce vasodilatation of the veins and arteries, and, in high doses, the arterioles or resistance vessels (Fig. 1). Nitrate effects on the arterial circulation (afterload reduction) are critical to improvement of cardiac function in CHF. A pure venodilator would lead to a reduction in cardiac output, but the decrease in arterial impedance and systemic vascular resistance that occurs with nitrates results in an augmentation of stroke volume and cardiac output. After nitrate administration, the circulating blood volume is redistributed away from the lungs and heart towards the abdominal and splanchnic circulations as well as to the extremities. There is a decrease in return of blood to the right heart, resulting in a fall in preload of both right and left ventricles. Pulmonary artery systolic pressure falls, in part due to direct pulmonary artery vasodilatation. The decrease in right heart pressures appears to be a unique feature of this

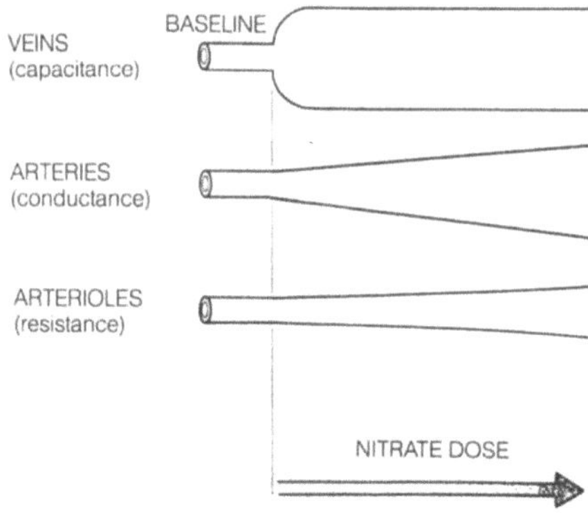

Fig. 1. Hemodynamic effects of nitrates on the major vascular beds. At very low doses the venous capacitance vessels are near maximally dilated. The systemic arteries begin to dilate at low doses; this is further accentuated with higher doses of nitrates. Systolic blood pressure begins to fall even in the absence of an effect on resistance vessels. Arteriolar vasodilatation occurs only with very high nitrate concentrations, which are probably not achieved with usual clinical doses of these drugs. (From [26])

group of vasodilators. Nitroglycerin compounds favorably effect the pressure-volume relationships of the left ventricle, perhaps in part because of altered pericardial restraining forces and/or a smaller right ventricular volume [7].

Acutely, nitrates do not improve exercise tolerance in heart failure subjects; this is similar to the ACE inhibitors. However, *long-term* administration of both classes of drug results in enhanced aerobic performance in patients with CHF [3, 4]. It may be that a sustained reduction in left ventricular filling pressure allows for a "training effect," such that patients are able to exercise better over time following chronic nitrate administration.

Clinical Studies

Many clinical investigations carried out in the 1970s and early 1980s documented the benefits of nitrates in patients with heart failure. In particular, oral isosorbide dinitrate (ISDN) and 2% nitroglycerin ointment have been shown to be effective as long-acting nitrate therapy [1]. Oral ISDN therapy results in long-term benefits on exercise tolerance (over a period of 12 weeks) in controlled trials [3, 4]. Sustained hemodynamic improvement at rest is manifest by a decrease in left ventricular filling pressure, a lower pulmonary artery pressure and resistance, with variable effects on arterial pressure, systemic vascular resistance, and cardiac output. Improvement in exercise hemodynamics during chronic nitrate therapy presumably reflects a dominant decrease in LV preload.

In the acute setting, intravenous nitroglycerin is extremely effective in improving the disordered hemodynamics in heart failure, producing a reduction in pulmonary wedge pressure and an improvement in cardiac output. Data with transdermal nitroglycerin patches in CHF is sparse [8–12]. Limited information documents the rapid appearance of vascular attenuation or tolerance, often within 24 h of patch application [9–12]. There is also a blunted hemodynamic response with long-term nitroglycerin patch therapy [8]. It appears that intermittent application of nitroglycerin patches allows for sustained vascular responsiveness [12]. Large doses of the patches may be necessary for a beneficial clinical effect [9–11] (see below).

Benefits of Nitroglycerin Therapy

Nitrates should be considered for use in addition to digitalis and diuretics in patients who continue to have signs and/or symptoms of CHF. As mentioned, several long-term studies document that administration of oral ISDN results in continued improvement in both hemodynamics and exercise performance [3, 4].

The United States Veterans Administration Cooperative Trial (V-Heft) demonstrated a substantial reduction in mortality in CHF when oral ISDN 40 mg (q.i.d.) was administered in conjunction with hydralazine 75 mg q.i.d., compared to a placebo control group as well as another cohort of patients treated with prazosin 5 mg q.i.d. [5]. There where mortality benefits of greater than 20% for up to 3 years in this study; the difference from placebo was a 38% reduction in deaths at 12 months. This was the first major trial with any drug documenting improvement in mortality in patients with CHF. Similar data in NYHA class IV CHF patients have been subsequently shown for enalapril [13].

Problems with Nitrate Therapy

Side Effects

The usual nitrate side effects of headache and hypotension appear to be less of a problem in heart failure. In general, patients with CHF tolerate nitrates quite well. Reflex tachycardia and hypotension are infrequent.

Nitrate resistance has been documented in several studies. This appears to be a primary failure of peripheral vasodilatation, such that hemodynamic parameters are unperturbed by nitrate administration [14–16]. Patients who are resistant to nitrates tend to be quite ill, often with severe biventricular CHF. Even very large doses may be ineffective.

Nitrate tolerance remains a major problem in the treatment of CHF as well as angina pectoris. At the present, all experts recommend an interval approach to nitrate therapy. This involves pulsed dosing employing a nitrate-free interval (one or more) thoughout a 24-h period. Thus, t.i.d. oral ISDN [16] or nitroglycerin ointment application, or nitroglycerin patch therapy with an 8–12-h off period [12], is recommended. Tolerance has been well documented in patients given large doses of transdermal nitroglycerin but may be avoidable with an intermittent or interval approach [12]. Acute administration of *N*-acetylcysteine has been shown to reverse hemodynamic tolerance to prolonged intravenous nitroglycerin infusion in CHF, but this therapy is experimental and impractical [7]. Carefully designed dosing strategies must be employed to avoid the vexing problem of nitrate tolerance. An interval or on-off performance approach is strongly recommended to provide protracted nitrate efficacy in CHF subjects.

Nitrate Dosage

In general, rather large amounts of nitrates are necessary for patients with CHF. Thus, 2.5–5 cm nitroglycerin ointment, 40–80 mg oral ISDN, and 20–40 mg per 24 h of transdermal nitroglycerin patches are recommended. Therapy should be started with smaller doses to make sure nitrates are well tolerated and that a fall in blood pressure does not occur.

Acute Myocardial Infarction

Intravenous nitroglycerin and other long-acting nitrates are effective drugs for controlling left ventricular failure associated with acute myocardial infarction; in addition, there is considerable animal and human data indicating that the *early* administration of intravenous nitroglycerin has important beneficial effects. This drug has previously been shown to reduce infarct size and decrease complications of acute myocardial infarction [18–22]. Pooled data from a number of small trials using intravenous nitroglycerin within the first 24–48 h of infarction have demonstrated a striking reduction in early mortality compared to placebo [23]. Important new data suggest that early administration of intravenous nitroglycerin also results in protection against the adverse remodeling that occurs (infarct expansion, wall thinning) in many patients following an acute myocardial infarction [24]. In addition to a meta-analysis showing that in-hospital mortality is reduced in acute myocardial infarction when intravenous nitroglycerin is started within the first few hours after the event [23], a recent large trial demonstrates that long-term mortality (1 year) has been found to be lower in anterior wall infarction initially treated for 48 h with an intravenous nitroglycerin drip [24].

Additional therapy with buccal nitroglycerin after an initial 48-h nitroglycerin infusion for 6 weeks has been shown to result in further improvement in LV size and function, as well as enhanced exercise performance [25].

Summary

Nitrates are extremely effective agents for adjunctive therapy of CHF. These drugs are safe, easy to use, and well tolerated. Nitrates have been shown to be truly cardioprotective, in addition to their beneficial effects on improving the disordered hemodynamics and reduced exercise tolerance in CHF. Exciting recent data also suggest that early administration of intravenous nitroglycerin following an acute myocardial infarction may have important short- and long-term benefits.

References

1. Abrams J (1979) Pharmacology of nitroglycerin and long-acting nitrates and their usefulness in the treatment of chronic congestive heart failure. In: Gould L, Reddy CVR (eds) Vasodilator therapy for cardiac disorders. Futura, Mount Kisco, pp 129–168
2. Herman B, Charlap S, Frishman WH (1984) Nitrates in congestive heart failure. Med Clin N Am 73 (2): 361–371
3. Leier CV, Huss P, Magorien RD, Unverferth DV (1983) Improved exercise capacity and differing arterial and venous tolerance during chronic isosorbide dinitrate therapy for congestive heart failure. Circulation 67: 817–822
4. Franciosa JA, Cohn JN (1980) Sustained hemodynamic effects without tolerance during long-term isosorbide dinitrate treatment of chronic left ventriculare failure. Am J Cardiol 45: 648–654
5. Abrams J (1988) A reappraisal of nitrate therapy. JAMA 259: 396–401
6. Cohn JN, et al (1986) Effect of vasodilator therapy on mortality in chronic congestive heart failure. Results of a Veterans Administration Cooperative Study. N Engl J Med 34: 1547–1552
7. Smith ER, Smiseth OA, Kingma I, Manyari D, Belenkie I, Tyberg JV (1984) Mechanisms of action of nitrates. Role of changes in venous capacitance and in the left ventricular diastolic pressure-volume relation. Am J Med 76(6 A): 14–21
8. Sharpe DN, Coxon R (1984) Nitroglycerin in a transdermal therapeutic system in chronic heart failure. J Cardiovasc Pharmacol 6: 76–82
9. Olivari MT, Carlyle PF, Levine B, Cohn JN (1983) Hemodyamic and hormonal response to transdermal nitroglycerin in normal subjects and in patients with congestive heart failure. J Am College Cardiol 2: 872–878
10. Jordan RA, Seith L, Henry DA, Wilen MM, Franciosa JA (1985) Dose requirements and hemodynamic effects of transdermal nitroglycerin compared with placebo in patients with congestive heart failure. Circulation 71: 980–986
11. Elkayam U, Roth A, Henriquez B, Weber L, Tonnemacher D, Rahimtoola SH (1985) Hemodynamic and hormonal effects of high-dose transdermal nitroglycerin in patients with chronic congestive heart failure. Am J Cardiol 56: 555–559
12. Sharpe N, Coxon R, Webster M, Luke R (1987) Hemodynamic effects of intermittent transdermal nitroglycerin in chronic congestive heart failure. Am J Card 59: 895–899
13. (1987) Effects of enalapril on mortality in severe congestive heart failure. Results of the Cooperative North Scandinavian Enalapril Survival Study (Consensus). N Engl J Med 316: 1429–1435
14. Armstrong PW, Moffat JA, Marks GS (1982) Arterial-venous nitroglycerin gradient during intravenous infusion in man. Circulation 66: 1273–1276
15. Magrini F, Niarchos AP (1980) Ineffectiveness of sublingual nitroglycerin in acute left ventricular failure in the presence of massive peripheral edema. Am J Cardiol 45: 841
16. Elkayam U, Jamison M, Roth A, Kulick D, Vasquez J, Rahimtoola SH (1989) Oral isosorbide dinitrate in chronic heart failure: tolerance development to four times daily versus three times daily regimen. 13: 1784
17. Packer M, Lee WH, Kessler P, Gottlieb S, Medina N, Yushak M (1987) Prevention and reversal of nitrat tolerance in patients with congestive heart failure. N Engl J Med 317: 799–804
18. Flaherty JT, Becker LC, Bulkley BH et al (1983) A randomized prospective trial of intravenous nitroglycerin in patients with acute myocardial infarction. Circulation 68: 576–588

19. Flaherty JT (1983) Comparison of intravenous nitroglycerin and sodium nitroprusside in acute myocardial infarction. Am J Med 74 (Suppl 6B): 53–60
20. Bussman WD, Passek D, Seidel W, Kaltenbach M (1981) Reduction of CK and CK-MB indexes of infarct size by intravenous nitroglycerin. Circulation 63: 615–622
21. Jugdutt BI, Becker LC, Hutchins C (1980) Effect of intravenous nitroglycerin on collateral blood flow and infarct size in the conscious dog. Circulation 63: 17–28
22. Jugdutt BI, Sussex BA, Warnica JW, Rossan RE (1983) Persistent reduction in left ventricular asynergy in patients with acute myocardial infarction by intravenous infusion of nitroglycerin. Circulation 68: 1264–1273
23. Yusuf S, MacMahon S, Collins R, Peto R (1988) Effect of intravenous nitrates on mortality in acute myocardial infarction: an overview of the randomised trials. Lancet I:1088–1092
24. Jugdutt BI, Warnica JW (1988) Intravenous nitroglycerin therapy to limit myocardial infarct size, expansion, and complications: Effect of timing, dosage, and infarct location. Circulation 78: 906–919
25. Jugdutt BI, Warnica JW (1989) Tolerance with low dose intravenous nitroglycerin therapy in acute myocardial infarction. Am J Cardiol 64: 531–537
26. Abrams J (1985) Hemodynamic effects of nitroglycerin and long-acting nitrates. Am Heart J 110: 216

Chemical Pathways Proposed for the In Vitro Metabolism of Nitrovasodilators and Their Conversion into Vasoactive Species

M. Feelisch and E. Noack

Introduction

Despite the extensive therapeutic use of nitrovasodilators and a fairly good knowledge of the hemodynamic alterations which lead to an improvement in myocardial performance, until recently surprisingly poor data were available concerning their mode of action at the molecular level. It now seems well documented that the pharmacodynamic action of these compounds is mediated by an activation of the soluble isoenzyme of guanylate cyclase (sGC, EC, 4.6.1.2). Relaxation studies with isolated vessel preparations revealed that there are at least two independent ways of sGC stimulation: an indirect endothelium-related process and a second one which results in sGC stimulation on bypassing the endothelium-dependent process. Substances like acetylcholine, ATP, or bradykinin on the one hand only behave in a vasodilatory manner in the presence of an intact endothelial layer while nitrovasodilators like nitrite (NO_2^-), glyceryl trinitrate (GTN), and sodium nitroprusside (SNP) on the other hand are relaxant agents in either case. That is why the existence of a humoral "endothelium-derived relaxing factor" (EDRF) was postulated, which after penetration into the smooth muscle cell is responsible for sGC activation by endothelium-dependent vasodilators. The consecutive raise in cytosolic cyclic guanosine monophosphate (cGMP) concentration induces a complex cascade of protein phosphorylations, finally resulting in smooth muscle relaxation [12, 34].

Although the chemical nature of the potential active metabolite is still the subject of controversy, it seems by now well established that nitrovasodilators have to be metabolized prior to revealing their vasodilatory activity [3, 29]. Brien et al. [4] recently demonstrated that GTN biotransformation and elevation of tissue cGMP precedes the onset of GTN-induced relaxation. Whether the conversion of nitrovasodilators to vasoactive compounds is achieved by enzymatic and/or nonenzymatic steps is as yet unknown.

Beside the proposed interaction with a specific organic nitrate receptor [35], there are three main hypotheses dealing with the mode of action of organic nitrates: (1) In an early developed concept, NO_2^- intracellularly formed upon enzymatic degradation was regarded as the ultimate active metabolite [17]. This concept, however, is contradicted by the fact that NO_2^- shows a vasodilatory effect at even higher concentrations than organic nitrates themselves. It is, therefore, highly unlikely that the comparably low amounts of

NO_2^- arising from intracellular cleavage of organic nitrates should have any effect on sGC activity, the more so as basal NO_2^- concentration within the smooth muscle is thought to be in the μmolar range. (2) Ignarro and co-workers [23] developed an elegant hypothesis by which released NO_2^- (in equilibrium with HNO_2) reacts with thiol groups to form S-nitrosothiols (RSNOs). Because of hemodynamic behavior similar to that of organic nitrates, nitrosothiols for a long time were regarded as the common active principle of sGC stimulation by all nitrovasodilators. (3) In contrast, Murad and colleagues [24] favored the nitric oxide free radical (NO) as terminal activator of sGC. By which molecular process(es) NO should, however, be liberated from the heterogeneous group of nitrovasodilators remains obscure (Fig. 1).

The aim of the current study was to cast some light on the mechanism of nitrovasodilator-induced sGC activation and, from a biochemical point of view, to test the reliability of these hypotheses.

In broken cell preparations organic nitrates were found to require the presence of the SH-containing amino acid cysteine to induce enzyme stimulation while other nitrovasodilators like SNP, NO_2^-, or SIN-1, the active metabolite of molsidomine, thiol-independently accelerate sGC activity. Previous studies in our laboratory have furthermore shown that incubation with thiols other than cysteine results in the decomposition of GTN in each case while a concomitant activation is only seen with a few compounds [10], indicating that a thiol-induced degradation of organic nitrates does not necessarily give rise to sGC activation. With these results in mind it seemed promising to look for a metabolite which is produced only in the presence of one of these "active" thiols. We suggested that under certain conditions enzyme activation is preceded by a non enzymatic metabolic step generating the active species responsible for sGC stimulation. We, therefore, decided to monitor the occurrence of NO_2^-, NO_3^-, RSNO, and NO directly and to investigate whether the pattern of metabolites arising upon incubation of GTN with several thiol compounds differs fundamentally from that of GTN with cysteine. By the choice of identical conditions for enzyme activation and measurement of metabolite

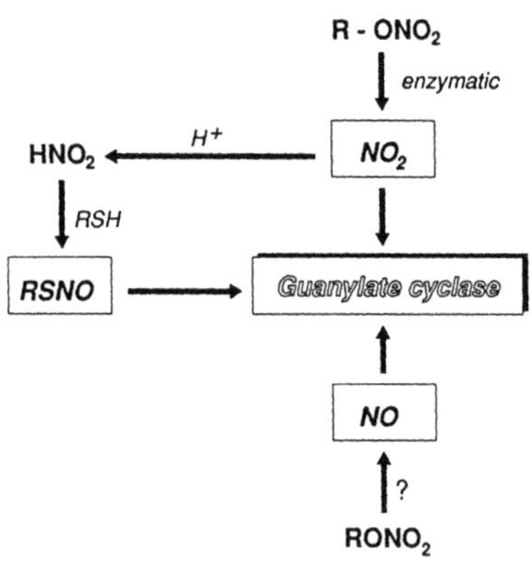

Fig. 1. Scheme of proposed modes of action of organic nitrates

generation it was possible to directly compare both processes. Furthermore, we were interested to clarify whether or not only one common metabolite is responsible for the sGC-activating properties of all nitrovasodilators.

Methods

Guanylate Cyclase

A crude extract of soluble sGC from rat liver was prepared according to Steurer and Schütz [39] and Ignarro et al. [22]. Activity of sGC was determined according to the method of Kimura et al. [27]. Concomitant glutathione *S*-transferase (EC, 2.5.1.18) activity was estimated by comparison with crystalline glutathione-*S*-transferase from equine liver (Sigma) using reduced glutathione and GTN as substrates, amounting to 3.6 mU/mg protein ($n = 4$). cGMP formed was determined by means of radioimmunoassay (RIA) with a commercial RIA kit for cGMP (Amersham/Buchler). Assay conditions were as follows: 50 m*M* phosphate buffer (pH 7.40), 4 m*M* $MgCl_2$, 1 m*M* GTP, 30 μg protein, incubation time 15 min, 37° ± 0.05°C.

Nitrite/Nitrate

Monitoring of NO_2^- and NO_3^- generation during decomposition of nitrovasodilators was simultaneously performed using an HPLC technique described in detail elsewhere [8, 30]. Incubations were carried out at 37° ± 0.05°C in 50 m*M* phosphate buffer, pH 7.70. All data are expressed as initial kinetics calculated from the slope of the linear part of the time-dependent concentration record.

Nitric Oxide

For the measurement of NO release we used a difference-spectrophotometric technique based on the rapid NO-induced conversion of oxyhemoglobin (HbO_2) to methemoglobin (MetHb) [8, 16]. According to the simple 1:1 stoichiometry of the underlying reaction ($HbO_2 + NO \rightarrow MetHb + NO_3^-$) [6] a continuous time-dependent recording of the methemoglobin concentration mirrors the rate of NO formation. This method is highly sensitive, is specific for NO, and does not suffer from interference by dissolved oxygen or released NO_2^-. The selectivity and accuracy of this indirect method have been verified by a chemiluminescence technique [7].

Nitrosothiols

The potential formation of *S*-nitrosothiols was monitored by using an HPLC technique recently described by Fung et al. [13] with small modifications. Briefly, measurements were performed on a Supersher RP-18 column (5 μm,

12,5 × 0,4 cm, Merck, Darmstadt), maintained constantly at 20°C, which was eluted with a mixture of 12% methanol and 88% 50 m*M* phosphate buffer pH 2.45 at a flow rate of 1 ml/min. Detection wavelength was set at 335 nm. Peak identity was verified by injection of freshly prepared aqueous solutions of crystalline *S*-nitrosothiols. Calibration was performed with aqueous solutions containing 50 μ*M* thiol and 500 μ*M* NO_2^-, preincubated in citrate/HCl buffer, pH 2.0, at 20°C for 10 min. The completeness of *S*-nitrosothiol formation was checked photometrically.

Stability constants of *S*-nitrosothiols were determined at 37°C in 50 m*M* phosphate buffer, pH 7.40. The respective rates of decomposition were calculated from the slope of the photometrically recorded decrease in absorbance at 335 nm. Absolute values for the determination of molar extinction coefficients were corrected for thiol and NO_2^- absorbance versus buffer.

Preparation of Aqueous NO Solutions

Double distilled water was deaerated by 10 min sonification, bubbled with argon for further 10 min, and then filled into an all-glass apparatus, which had previously been purged with argon. Commercially available argon gas (Linde) was freed from oxygen traces by scrubbing with an alkaline pyrogallol solution. Argon was passed in a gentle stream through the whole system for at least 60 min. Gas flow was then quickly switched from argon to NO by means of a three-way stopcock. Nitric oxide (>99.9%, AGA, Bottrop, FRG) had previously been scrubbed by passage over solid KOH and bubbling through a 10% aqueous KOH solution to remove trace amounts of higher N oxides. Within 30–40 min of equilibration the water was saturated with NO. The concentration of this aqueous NO solution was determined by means of HPLC as NO_2^-. The final concentration of the saturated NO solution was 1.75 ± 0.08 m*M* (25°C, $n = 7$).

Materials

Erythrityl tetranitrate was purchased from Chemische Fabrik Tempelhof Preuß & Temmler, Berlin, FRG, and stored in absolute ethanol. All other organic nitrates used in this study were the kind gift of Pharma Schwarz, Monheim, FRG, except for the glyceryl mononitrates, which were from Mack, Illertissen, FRG, and the isomeric teopranitols, which were from Schwabe, Karlsruhe, RFG. Amyl nitrite was obtained from Fluka, Switzerland. All sydnonimines and furoxans were the generous gift of Cassella, Frankfurt, FRG. SNP was obtained from Merck, Darmstadt, FRG. All nitrovasodilators were used as aqueous solutions, freshly prepared just before use. GTN was diluted from a commercial formulation (4.4 m*M* GTN in 5% aqueous glucose solution, Perlinganit, Pharma Schwarz, Monheim, FRG). Thiol compounds were obtained from Boehringer, Fluka, Merck, Merck-Schuchard, Serva, and Sigma, each of the highest grade available and further purified to HPLC purity if neccessary. 3-Mercaptopicolinic acid was the kind gift of

Dr. DiTullio, SK&F Res. Labs., Philadelphia, United States. All other reagents used were of analytical or research grade. Nitrosothiols were synthesized by the acid-catalyzed *S*-nitrosation of the respective thiols with NO_2^-. *N*-Acetyl-D, L-penicillamine thionitrite was prepared according to Field et al. [11] and identified by its IR and UV/VIS spectrum. The synthesis and separation of crystalline nitrosothiols of L-cysteine, *N*-acetyl-L-cysteine, L-glutathione, L-homocysteine, and L-cysteamine was achieved by modification of published procedures [5, 15, 31] with yields ranging from 30% to 70% ($n = 15$). Identity of prepared nitrosothiols was verified by elementary analysis (within 30 min after preparation), UV/VIS, mass spectroscopy (m^+–30), and HPLC.

Data Analysis

Data presented in figures, tables, and the text are means ± SEM. Differences were assessed using Student's two-tailed *t*-test for paired variables. Differences are regarded as statistically significant when $p < 0.05$. EC_{50} values of the concentration/response curves at the sGC were calculated using the logit transformation according to Hafner et al. [14]

Results

Nitrite Formation

With all organic nitrates a more or less pronounced NO_2^- formation was seen in the presence of cysteine. The rate of NO_2^- production was linearly dependent on the concentration of both thiol and organic nitrate (0.01–100 m*M* cysteine, $r = 0.996$, $n = 14$; 0.05–2 m*M* GTN, $r = 0.993$, $n = 13$). By determining the rates of NO_2^- formation at varying pHs, we obtained evidence that NO_2^- formation is due to a nucleophilic attack of the thiolate anion (RS-) on the nitrate ester (RSH does not react with organic nitrates prior to deprotonation, data not shown). The rate of NO_3^- formation is comparably lower and is thiol-independently brought about by alkaline hydrolysis of the ester. Its absolute rate depends on the hydrolytic stability of the nitrate ester (about 1/50 of that of NO_2^- formation with 0.2 m*M* GTN/1 m*M* cysteine at pH 7.4 and 37°C).

A kinetic comparison of the rates of cysteine-induced NO_2^- formation for different organic nitrates with the concentration/response curves of the respective compounds at the isolated sGC revealed a close correlation ($r = 0.997, p < 0.001, n = 11$; Fig. 2). At first glance one may be tempted to consider this as a clue for a causal linkage between both processes. Furthermore, the decomposition rate of GTN in the presence of several thiol compounds closely corresponded to the NO_2^- formation seen under the same conditions (results not shown). But by investigating more than 60 thiol compounds as potential substitutes for cysteine, with some of these compounds we observed a considerably higher degree of NO_2^- production than that seen with cys-

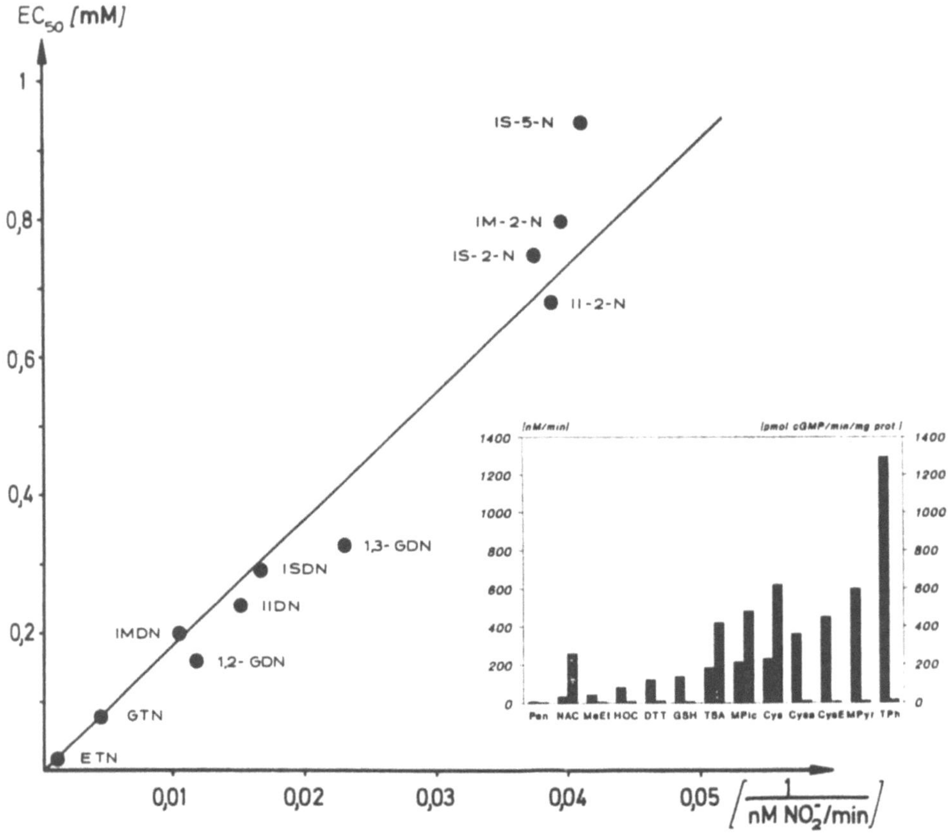

Fig. 2. Comparison of the cysteine-induced NO_2^- formation rate with the sGC-activating potency of different organic nitrates. *NO_2^-*, organic nitrates 0.2 m*M*, cysteine 1 m*M; sGC,* cysteine 5 m*M. ETN,* erythrityl tetranitrate; *GDN,* glyceryl dinitrate; *GNT,* glyceryl trinitrate; *IIDN,* isoidide dinitrate; *IMDN,* isomannide dinitrate; *ISDN,* isosorbide dinitrate; *II-2-N,* isoidide 2-mononitrate; *IM-2-N,* isomannide 2-mononitrate; *IS-2-N,* isosorbide 2-mononitrate; *IS-5-N,* isosorbide 5-mononitrate. *Inset,* Comparison of NO_2^- formation rate with sGC activation by GTN in the presence of different thiols. NO_2^-, GTN 0.2 m*M,* thiols 1 m*M; sGC,* GTN 0.5 m*M,* thiols 5 m*M. Cys L-cysteine,* CysE *L*-cysteinemethylester; *Cysa L-cysteamine; DTT,* 1,4-dithiothreitol; *GSH, L*-glutathione; *HOC,* L-homocysteine; *MeEt,* mercaptoethanol; *MPic,* 3-mercaptopicolinic acid; *MPyr,* mercaptopyruvic acid; *NAC, N*-acetyl-L-cysteine; Pen, L-penicillamine; *TPh,* thiophenol; *TSA,* thiosalicylic acid

teine, although most of them have been proven to be ineffective as GTN-costimulators at the sGC (inset, Fig. 2).

RSNO Formation

To examine the RSNO theory we tested the activity of several *S*-nitrosothiols at the sGC. By molar comparison the observed enzyme stimulation was higher only in the case of those compounds which have a lower stability under the

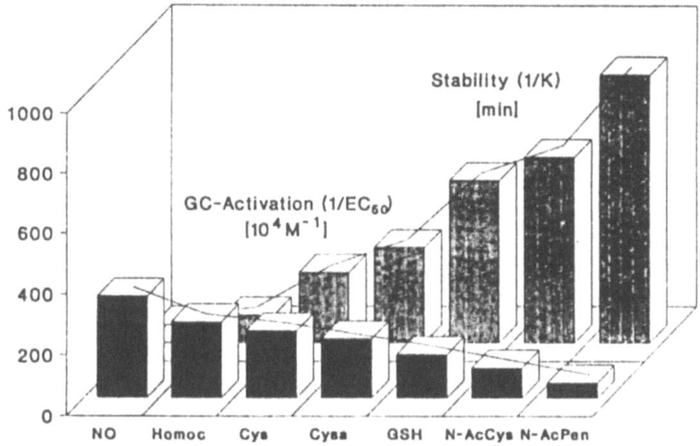

Fig.3. Relation between stability and sGC-stimulating potency of *S*-nitrosothiols. *Cys,* L-cysteine-SNO; *Cysa,* L-cysteamine-SNO; *GSN,* L-glutathione-SNO; *Homoc,* L-homocysteine-SNO; *N-AcCys,* *N*-acetyl-L-cysteine-SNO; *N-AcPen,* *N*-acetyl-DL-penicillamine-SNO; *NO,* nitric oxide

conditions used. These data led us to conclude that not the intact RSNO compound but rather a metabolite produced by its decomposition is responsible for enzyme stimulation. A direct comparison with authentic NO confirmed that this metabolite is most probably NO itself, as this radical is the strongest activator of sGC so far tested by us (Fig.3).

In contrast to enzyme activation, GTN surprisingly showed low RSNO formation with all the thiol compounds tested (GTN 2 m*M*, thiols 20 m*M*, $p < 0.01$, $n = 5$, data not shown). At pH 7.40 nitrosothiol formation from GTN in the presence of cysteine was still so modest that it can neither account for the NO formation observed nor for enzyme activation. Although we also noticed low RSNO formation with all other thiols, no corresponding sGC stimulation could be observed. A quantitative comparison demonstrated that RSNO formation is due to an equilibrium reaction between NO_2^- produced and the present thiol. Thus, intermediary formed *S*-nitrosothiols seem, as well as NO_2^-, equally unlikely to account for the sGC-stimulation potency of organic nitrates.

Nitric Oxide Formation

Beside the described NO_2^- generation, we measured the formation of NO in all incubation runs of an organic nitrate with cysteine. As already seen with NO_2^-, the rate of this NO formation correlates well with the stimulatory potency of the respective nitrates at the isolated sGC ($r = 0.994$, $p < 0.001$, $n = 14$; Fig.4). Enzyme stimulation was higher (the corresponding EC_{50} values accordingly lower) the more NO was liberated per unit of time. The highest NO liberation was seen with GTN, followed by the group of dinitrates and mononitrates. In contrast to NO_2^- and nitrosothiol formation, a generation of

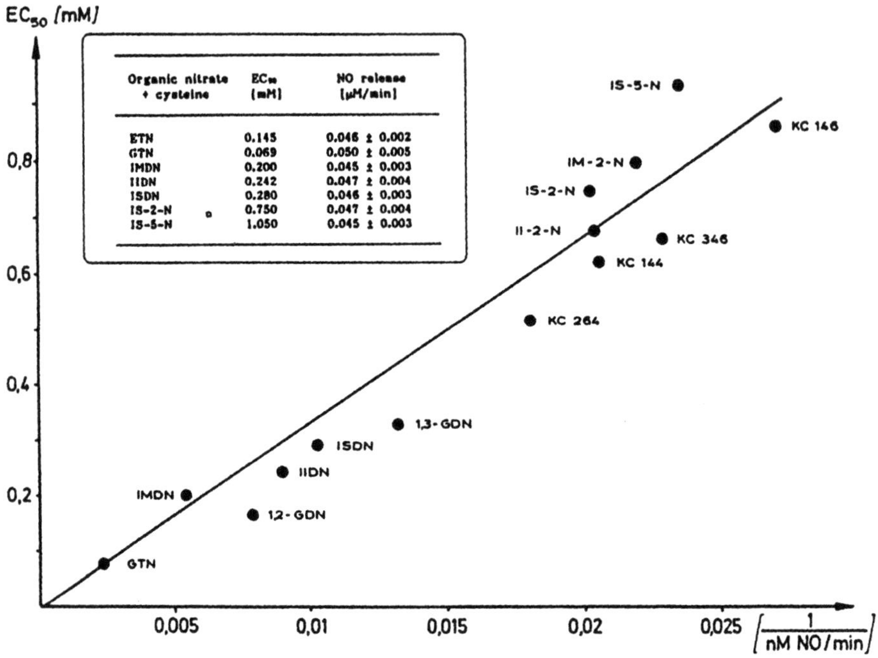

Organic nitrate + cysteine	EC_{50} (mM)	NO release [μM/min]
ETN	0.145	0.046 ± 0.002
GTN	0.069	0.050 ± 0.005
IMDN	0.200	0.045 ± 0.003
IIDN	0.242	0.047 ± 0.004
ISDN	0.280	0.046 ± 0.003
IS-2-N	0.750	0.047 ± 0.004
IS-5-N	1.050	0.045 ± 0.003

Fig. 4. Correlation between cysteine-mediated NO release from different organic nitrates with sGC stimulation. *NO,* organic nitrates 1 m*M*, L-cysteine, 5 m*M*, 37°C, pH 7.70, $n = 7$; *sGC,* L-cysteine 5 m*M*, 37°C, pH 7.40, $n = 3$. *KC-xxx,* isomeric teopranitols, all other abbreviations, see legend to Fig. 2. *Inset,* Uniform NO release from different organic nitrates at their EC_{50} concentrations

NO was only measurable under the influence of such thiol compounds which coincidently mediate enzyme activation. The dependence of initial NO liberation rates on the cysteine concentration displays a sigmoidal characteristic comparable to that found at the sGC[8], while the rate of NO_2^- formation was found to be linearly dependent on the cysteine concentration. When organic nitrates were tested at those concentrations which induce half-maximal enzyme stimulation (EC_{50}), a nearly identical rate of NO liberation was recorded for each compound (inset, Fig. 4).

Additional experiments furthermore confirmed the significance of NO liberation for enzyme stimulation: NO formation accelerates at higher temperature and more alkaline conditions, being in close agreement with a concomitant shift of the concentration/response curves at the sGC to the left; a semilogarithmic plot of measured NO formation rates versus amounts of cGMP produced revealed a linear correlation (data not shown). The addition of methylene blue and related dyes reduced both NO formation and enzyme activation in a concentration-dependent manner but did not significantly alter the initial kinetics of NO_2^- formation ($n = 3$). The addition of 10–100 μ*M* oxyhemoglobin (but not methemoglobin) to our sGC assay completely abolished enzyme stimulation by all organic nitrates tested ($n = 5$) being in accordance with its scavenging effect on generated NO.

Mechanism of Interaction Between Organic Nitrates and Thiols

Concerning the possible reaction mechanism between organic nitrates and cysteine it was remarkable that for the cleavage of organic nitrates to NO_2^- a SH-containing compound of either structure seems to be sufficient while for additional NO formation (and concomitant enzyme activation) obviously a supplemental factor is required, because only very few of more than 60 tested thiols have proven to be active in both respects. These are cysteine and its *N*-acetyl analogue, thiosalicylic acid and 3-mercaptopicolinic acid. Structural differences are thought to account for the reason that these few thiols not only decompose organic nitrates to NO_2^- but simultaneously to NO as well. As an essential feature for this metabolism, indeed a common chemical structure could be determined: a mercapto group separated by two carbon atoms from a carbonyl function, oriented in plane. Besides these sterical requirements additional electronical parameters seem to be limiting; for an isomeric compound of the "active" 3-mercaptopicolinic acid, 2-mercaptonicotinic acid was found to be ineffective, although meeting the structural demands for NO generation from organic nitrates ($n = 4$).

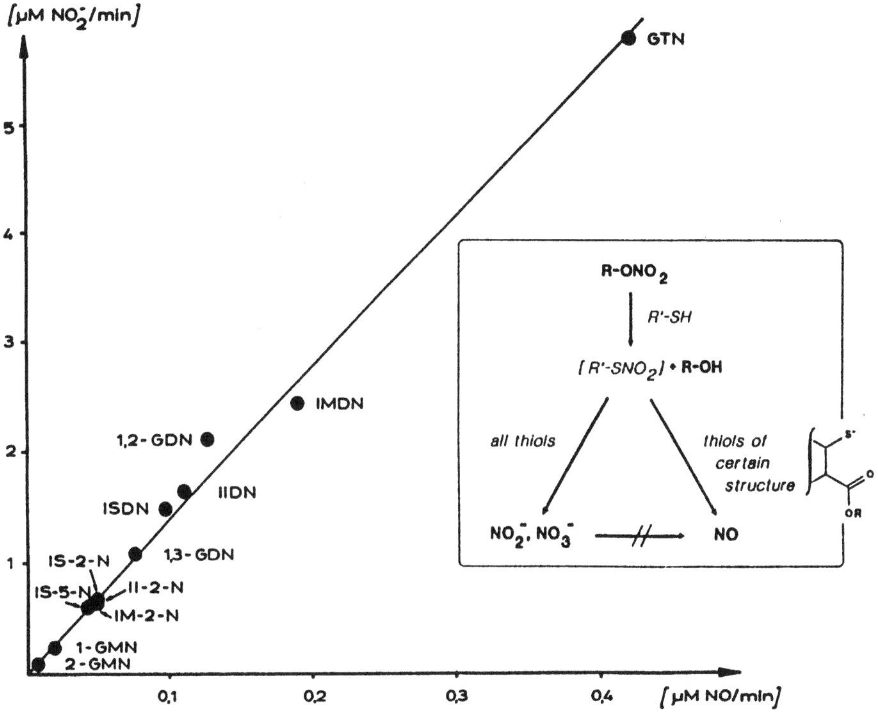

Fig. 5. Relation between cysteine-induced NO and NO_2^- formation rates from GTN. Conditions at both assays: organic nitrates 1 m*M*, cysteine 5 m*M*, 37°C, $n = 3$ (NO_2^-), $n = 7$ (NO). *GMN*, glyceryl mononitrate, all other abbreviations as in legend to Fig. 2. *Inset*, Proposed mechanism of the reaction between thiols and organic nitrates. *R-OH*, alcohol; *R-ONO*$_2$, organic nitrate; *R'-SH*, thiol; *R'-SNO*$_2$, thionitrate

There was an obvious relation between the cysteine-induced NO_2^- and NO formation rates from different organic nitrates (Fig. 5). With cysteine as thiol compound, both rates revealed a constant ratio of 14:1 regardless of the structure of the organic nitrate. Any reduction of generated NO_2^- to NO was experimentally excluded, for neither definite amounts of NO_2^- (1, 10, 100 ν*M*) were reduced by incubation with 1 m*M* cysteine (pH 7.4, 37°C, 1 h, $n = 3$), nor any NO formation was measurable under these conditions ($n = 5$). We, therefore, assume that in the course of the reaction between organic nitrate and the SH-containing compound a common thiol-specific intermediate arises which decomposes with the release of NO_2^- and – under special circumstances – also of NO. A thionitrate ($RSNO_2$), i.e., a thioester arising from transesterification from the organic nitrate to the respective thiol, appears to be the most likely candidate for such an intermediate. A carbonyl function in a distinct spatial orientation toward the mercapto group (as in cysteine or thiosalicylic acid) may favor an intramolecular rearrangement of this intermediate with consecutive NO split-off.

Nitric Oxide Release from Other Nitrovasodilators

Comparative measurements with other nitrovasodilators concerning their thiol-induced or spontaneous product formation uniformly confirmed the dependence of enzyme stimulation on the rate of NO release. The pathway of NO formation, however, is different for each class of compound.

Amyl nitrite, as a representative of the organic nitrites, releases NO only in the presence of thiols although it appears to be less discriminating in respect of the chemical structure of the SH compound, as NO liberation was detected with all tested thiols. Interestingly, on incubation of amyl nitrite with several SH-containing compounds a strong nonlinear formation of *S*-nitrosothiols was observed. The measurable rates of NO liberation thus arise as the product of *S*-nitrosothiol concentration present at a certain time and its respective decomposition rate under the prevailing conditions.

SNP spontaneously releases NO in a nonlinear manner seemingly independent of pH over a wide range. At high concentrations (> 1 m*M*) saturation conditions are achieved [8], which may be brought about by reaching a certain dissociation limit for NO from the pentacyano complex.

Sydnonimines like SIN-1 are highly unstable in aqueous solution and pH-dependently hydrolyze to the ring-opened "A" forms which in turn decompose with the release of NO. All tested compounds ($n = 29$) spontaneously release NO. This NO formation seems to be initiated by molecular oxygen (data not shown). Those sydnonimines with a high tendency toward NO release are again the most effective sGC stimulators. The simultaneous presence of cysteine enhanced neither NO formation nor enzyme activation (cysteine 5 m*M*, $n = 3$). Besides NO, NO_2^- and NO_3^- are generated as well, the rate of which however does not indicate any relation to sGC stimulation.

As a rather new vasodilatory group, furoxans were also tested for potential NO liberation. None of the compounds investigated liberates NO spontaneously, but all show strong NO formation upon addition of thiols. The rate of NO formation closely correlates with enzyme activation in each case. Again, those compounds showing the highest rate of NO release display the strongest enzyme activation [9]. As already observed with amyl nitrite, upon incubation with SH compounds furoxans show a strong nitrosothiol formation, the initial rates of which run in parallel with the measured rates for NO formation.

Discussion

In this study we present clear evidence that neither NO_2^- nor nitrosothiols directly activate sGC but do so by decomposition to NO. Collectively, our data strongly support the concept that NO is the ultimate enzyme-stimulating metabolite of all nitrovasodilators.

In the case of organic nitrates we observed marked NO_2^- formation with all tested thiols, although noticeable sGC activation was observed only in the presence of a few compounds. With cysteine as thiol compound we even demonstrated an apparently close correlation between NO_2^- generation and sGC-stimulating potency with different organic nitrates. This clearly shows that a nice correlation between two parameters does not necessarily demonstrate a cause-effect relationship. Our data are consistent with the results by Romanin and Kukovetz [37], who demonstrated that comparable amounts of NO_2^- are formed with GTN and either cysteine or cysteinemethylester, whereas a sGC activation was only observed in the case of cysteine. Hence, because of the specific need for cysteine in the activation of sGC by organic nitrates, this compound must have another function for sGC activation in addition to that of NO_2^- production. Hence, the denitration rate is obviously an unsuitable correlate for the judgment of new developed organic nitrates, because NO_2^- formation is not coupled to activation of the target enzyme sGC.

In addition to NO_2^- generation we measured the formation of NO when organic nitrates were incubated with such thiols which are effective costimulators at the sGC. The rate of NO liberation appears to be closely related with enzyme stimulation. The dependence on cysteine concentration, the uniform rate of NO release at enzymatically equipotent concentrations, and the inhibitory effects of added methylene blue and oxyhemoglobin furthermore confirm the role of NO as terminal activator of sGC by nitrovasodilators.

The pH dependence of both NO formation and sGC stimulation by organic nitrates makes it highly unlikely that nitrosothiols are the precursors of NO formation for this class of nitrovasodilator. In contrast to organic nitrates, nitrosothiols are presumably the precursors of the thiol-induced NO formation from organic nitrites and furoxans. The inverse correlation between stability and sGC-stimulating potency of several nitrosothiols however clearly demonstrates that nitrosothiols do not directly activate sGC. It is even more likely that NO liberated by decomposition of these instable compounds is responsible for enzyme activation. This idea is further supported by the finding that authentic NO is the strongest activator of sGC. In contrast to a published procedure for the synthesis of nitrosothiols [21], a separate experiment with NO and thiols confirmed that NO does not react with thiols directly ($n = 4$; data not shown).

As a résumé of the experimental data dealing with the mechanism of NO liberation from different classes of nitrovasodilators we propose the scheme in Fig. 6.

All nitrovasodilators have to penetrate through the membrane of the smooth muscle cell first in order to become metabolized intracellularly. NO already released in the plasmatic compartment presumably does not account for the vasodilatory effect of these drugs, because of its extremely short half-life in the presence of oxygen (5.6 s in air-saturated aqueous buffer [26]). It is therefore rather unlikely that NO intraluminally generated from nitrovasodilators crosses the endothelial diffusion barrier to reach the underlying smooth muscle in concentrations able to induce a marked relaxation. Organic nitrates will be degraded to the alcohols and NO_2^-/NO_3^- in both an enyzmatic and a non enzymatic step. The enzymatic method is mainly catalyzed by the glutathione-*S*-transferase [18] and a so-called "nitrate-forming enzyme" [40], the non enzymatic one by free thiols present in the cytosol [10]. In both cases a thionitrate may be the common intermediate. NO_2^- and NO_3^- interchange

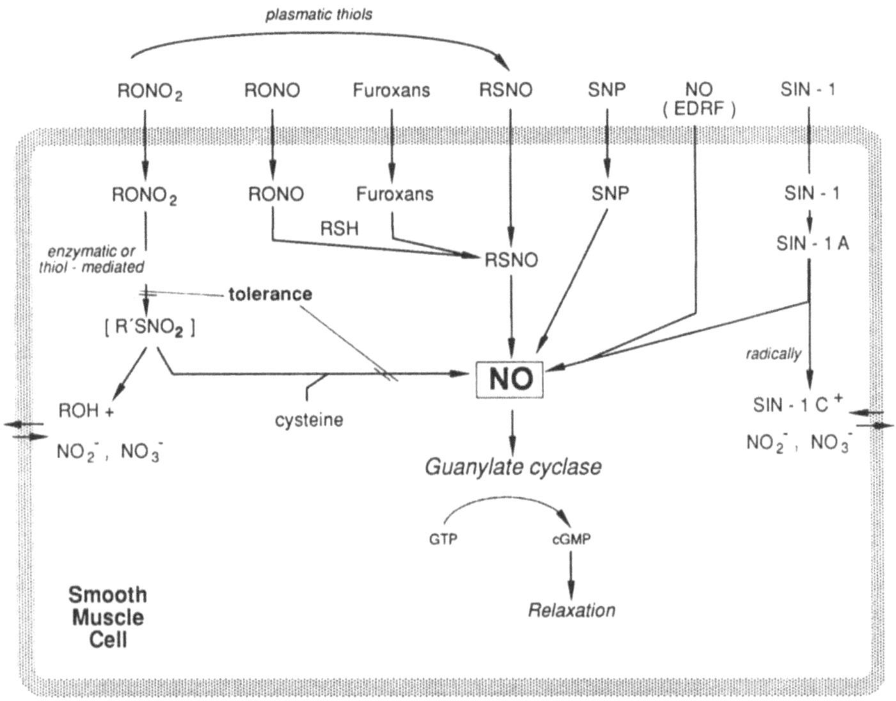

Fig. 6. Proposed scheme of sGC activation by nitrovasodilators including their different pathways of NO formation. *EDRF,* endothelium-derived relaxing factor; *Furoxans,* 3,4-disubstituted furazan-2-oxides; *NO,* nitric oxide; *RONO,* organic nitrite; *$RONO_2$*, organic nitrate; *RSH,* thiol; *RSNO,* nitrosothiol; *SIN-1,* active metabolite of molsidomine; *SNP,* sodium nitroprusside

with extracellular compartments [36] without affecting enzyme activity. From all thiols present within the smooth muscle cell only cysteine or its *N*-acetyl analogue (either free or at the active site of any enzyme) are able to induce additional NO liberation. Organic nitrites and furoxans react intracellularly with all available thiols (presumably predominantly with glutathione) to form *S*-nitrosothiols which – as thermodynamically unstable compounds – decompose to NO by radical cleavage. *S*-Nitrosothiols possibly formed by a plasmatic process from organic nitrates as recently proposed by Fung et al. [13], decompose by the same way to the disulfide and NO after entering the cell. SNP releases NO spontaneously, and sydnonimines by a presumably oxygen-dependent radical process following base-catalyzed hydrolysis to the ring-opened A-forms. Simultaneously arising NO_2^- and NO_3^- are clearly not responsible for the observed sGC stimulation.

The apparent similarities between the denitration profile in vitro and the biotransformation in vivo lead to the conclusion that the non enzymatic metabolism of organic nitrates and other nitrovasodilators plays an essential role in the pharmacodynamic action of these compounds in vivo. For organic nitrates there obviously exist two mutually independent pathways of decomposition, one leading to NO formation and by this to enzyme activation and the other leading to NO_2^- formation. Although the latter predominated, it only represents an inactivation pathway. One may assume that within the smooth muscle cell a combination of enzymatic and nonenzymatic processes is acting, the output of which depends on both concentration of free thiols and enzymatic activity. It is furthermore conceivable that the reactions observed in vitro are a projection of enzymatic metabolism in vivo. If so, such processes may be of importance not only for cGMP-regulated systems but also for many other biological processes, which are controlled by enzymes containing a thiol/haem group at their regulatory or catalytic site.

A better understanding of the structural determinants that govern the reaction between organic nitrates and thiols is suggested to be a helpful tool for the design of drugs which are more suited for a therapeutic application. Our results give new support according to the chemical mechanism of NO and NO_2^- formation from organic nitrates, postulating the transient formation of a thionitrate ($RSNO_2$). Such a chemical species has previously been assumed to represent an enzymatic intermediate of glutathione-*S*-transferase-mediated organic nitrate decomposition [25] and as intermediate which gives rise to NO_2^- in the reaction of organic nitrates with thiols [41, 43]. We suggest that such a thionitrate represents the common intermediate for both NO_2^-/NO_3^- and NO formation from organic nitrates. This would easily explain the close relation between NO_2^- and NO formation rates. A certain sterical prerequisite is thought to play an essential role for a given thiol to shift the decomposition profile of this intermediate from NO_2^- to NO. In addition, we here report 3-mercaptopicolinic acid to be an as yet unknown "active" thiol compound in regard to NO release from organic nitrates and costimulating activity at the sGC.

The demonstrated failure of cross-tolerance between organic nitrates like GTN and thiol-independent nitrovasodilators like SNP or SIN-1 [28, 33], together with the recent finding of Slack et al. [38] that the NO-induced relaxation of rabbit aortic rings was unaffected by organic nitrate tolerance, indicates that a decrease in the biotransformation of organic nitrates may be more likely by determining rather than desensitization of sGC. Thus, apart from possible alterations of the target enzyme sGC itself [2, 42], tolerance is thought to be related to an intracellular event like depletion of cysteine stores and/or diminished activity of an as yet unidentified enzyme which converts organic nitrates to NO. We, therefore, suggest that tolerance of smooth muscle to organic nitrate-induced vasodilation involves decreased formation of NO. It would be interesting to investigate whether or not the induction of nitrate tolerance in vascular tissue is accompanied by a diminished biotransformation to NO.

In this context it is interesting to note that in recent studies by Ahlner et al. [1] with bovine mesenteric arteries and by Hütter et al. [19] with a working rat heart preparation, a biphasic behaviour in the concentration/response curve for GTN has been described. As already discussed by Bennett et al. [2] there may exist two sites of biotransformation: a high-affinity component of enzymatic nature and a low-affinity component which is represented by nonenzymatic, cysteine-dependent NO formation. The enzymatic method of bioactivation in intact cells is presumably much more effective than the nonenzymatic one but suffers from an easy desensitization on exposure to high concentrations of organic nitrates. Upon high-dose exposure the enzymatic capacity will be completely abolished and only the nonenzymatic pathway of NO formation remains, which corresponds to a strong rightward shift of the concentration/response curve for GTN or other nitrates but not for nitrovasodilators which spontaneously liberate NO or for NO itself.

Much effort has been spent in the past 2 years identifying the chemical nature of the "endothelium-derived relaxing factor." Some groups recently provided evidence that EDRF is actually identical to NO [20, 32]. That means, the endothelium-dependent vasodilators are acting in the same way as the nitrovasodilators, namely by the induction of NO formation. In the light of this discovery it becomes clear that the synthetic nitrovasodilators imitate a physiological process of blood vessel regulation. From this point of view they thus represent different prodrugs of EDRF and have thereby left the stage of empirical therapy to revive as physiologically and rationally established drugs. The application of nitrovasodilators may, therefore, be regarded as causal therapy, inasmuch as these compounds are able to compensate the decreased production and/or effectiveness of an endogenous factor under conditions of endothelial dysfunction.

Summary

Recently it has been certified that the heterogeneous group of the so-called nitrovasodilators is acting through a common mechanism, i. e., the generation of nitric oxide (NO). These compounds thus represent prodrugs which mimic a physiological process of EDRF release responsible for endothelial controlled vasodilation. The analysis of the chemical pathways by which NO is generated will enhance our understanding of stimulation of soluble guanylate cyclase (sGC) and consecutive vasodilation. We, therefore, examined the in vitro metabolism of organic nitrates and nitrites, nitrosothiols, furoxans, sydnonimines, and nitroprusside using difference spectrophotometry, chemiluminescence, and HPLC techniques in direct comparison to enzyme activation. As cysteine is known to act as an essential cofactor for sGC stimulation in vitro, measurement was performed in the presence and absence of several related thiol compounds. We found quite different ways of "bioactivation" depending on the chemical nature of both nitrovasodilator and SH component. Organic nitrates, nitrites and furoxans require the preceding interaction with a mercapto group to liberate NO radicals while sydnonimines, nitrosothiols, and nitroprusside spontaneously release NO. The rate of NO generation always closely correlates with enzyme activation. The reaction of organic nitrates with cysteine seems to be highly specific, because SH compounds with slightly different sterical and electronic characteristics only decomposed organic nitrates into nitrite ions (NO_2^-) without producing NO or affecting sGC activity. A scheme is presented which summarizes the corresponding molecular events finally leading to NO liberation including hypothetical and detected intermediates. We suggest that these new data will be also helpful for the design of new drugs by which the respective therapeutic demands should be better accomplished and which may furthermore help to overcome nitrate tolerance.

Acknowledgement. We would like to thank Mrs. I. Beier for skilful technical assistance. This work was in part supported by the Deutsche Forschungsgemeinschaft (SFB 242, coronary heart disease, Düsseldorf) and by a scholarship from SKD, Göttingen, FRG.

References

1. Ahlner J, Axelsson KL, Ljusegren M, Grundström N, Andersson RG (1987) Demonstration of a high-affinity component of glyceryl trinitrate-induced vasodilatation in the bovine mesenteric artery. J Cyclic Nuleotide Protein Phosphor Res 11: 445–456
2. Bennett BM, Schröder H, Hayward LD, Waldman SA, Murad F (1988) Effect of in vitro organic nitrate tolerance on relaxation, cyclic GMP accumulation and guanylate cyclase activation by glyceryl trinitrate and the enantiomers of isoidide dinitrate. Circ Res 63: 693–701
3. Brien JF, McLaughlin BE, Breedon TH, Bennett BM, Nakatsu K, Marks GS (1986) Biotransformation of glyceryl trinitrate occurs concurrently with relaxation of rabbit aorta. J Pharmacol Exp Ther 237 (2): 608–614
4. Brien JF, McLaughlin BE, Kobus SM, Kawamoto JH, Nakatsu K, Marks GS (1988) Mechanism of glyceryl trinitrate-induced vasodilation. I. Relationship between drug biotransformation, tissue cyclic GMP elevation and relaxation of rabbit aorta. J Pharmacol Exp Ther 244 (1): 322–327
5. Cantoni C, Bianchi MA, Beretta G (1975) Stabilita di nitrosoderivati (nitrosotioli, nitrosofenoli e nitrosoemoglobina) a pH alkalino (in Italian). Ind Aliment 14 (7–8): 79–81
6. Doyle MP, Hoekstra JW (1981) Oxidation of nitrogen oxides by bound dioxygen in hemoproteins. J Inorg Biochem 14: 351–358

7. Feelisch M, Noack E (1987) Nitric oxide formation from nitrovasodilators occurs independently of hemoglobin or non-heme iron. Eur J Pharmacol 142: 465–469
8. Feelisch M, Noack E (1987) Correlation between nitric oxide formation during degradation of organic nitrates and activation of guanylate cyclase. Eur J Pharmacol 139: 19–30
9. Feelisch M, Noack E (1989) Thiol-induced generation of nitric oxide (NO) accounts for the vasodilatory action of furoxans (abstract). Naunyn Schmiedebergs Arch Pharmacol 339 [Suppl]: R 67
10. Feelisch M, Noack E, Schröder H (1988) Explanation of the discrepancy between the degree of organic nitrate decomposition, nitrite formation and guanylate cyclase stimulation. Eur Heart J 9 [Suppl A]: 57–62
11. Field L, Dilts RV, Ravichandran R, Lenhert PG, Carnahan GE (1978) An unusually stable thionitrite from *N*-acetyl-*D, L*-penicillamine. X-ray crystal structure of 2-(acetylamino)-2-carboxy-1,1-dimethylethyl thionitrite. J Chem Soc Chem Commun 1157: 249–252
12. Fiscus RR, Rapoport RM, Murad F (1984) Endothelium-dependent and nitrovasodilator-induced activation of cyclic GMP-dependent protein kinase in rat aorta. J Cyclic Nucleotide Protein Phospor Res 9 (6): 415–425
13. Fung HL, Chong S, Kowaluk E, Hough K, Kakemi M (1988) Mechanisms for the pharmacologic interaction of organic nitrates with thiols. Existence of an extracellular pathway for the reversal of nitrate vascular tolerance by *N*-acetylcysteine. J Pharmacol Exp Ther 245 (2): 524–530
14. Hafner D, Heinen E, Noack E (1977) Mathematical analysis of concentration-response relationship. Method for the evaluation of the EC_{50} and the number of binding sites per receptor molecule using the logit transformation. Arzneimittelforschung 27: 1871–1873
15. Hart TW (1985) Some observations concerning the *S*-nitroso and *S*-phenylsulfonyl derivatives of *L*-cysteine and glutathione. Tetrahydron Lett 26 (16): 2013–2016
16. Haussmann HJ, Werringloer J (1987) Mechanism and control of the denitrosation of *N*-nitrosodimethylamine. In: Bartsch H, O'Neill IK, Schulte-Herman R (eds) Relevance of *N*-nitroso compounds to human cancer: exposure and mechanisms. IARC Sci Publ 84: 109–112
17. Hay M (1983) The chemical nature and physiological action of nitroglycerin. Practitioner 30: 422–433
18. Heppel LA, Hilmoe RJ (1950) Metabolism of inorganic nitrite and nitrate esters: II. The enzymatic reduction of nitroglycerin and erythritol tetranitrate by glutathione. J Biol Chem 183: 129–138
19. Hütter J, Schmidt M, Rittler J (1988) Effects of sulfhydryl-containing compounds on nitroglycerin-induced coronary dilation in isolated working rat hearts. Eur J Pharmacol 156: 215–222
20. Ignarro LJ (1989) Endothelium-derived nitric oxide: actions and properties. FASEB J 3: 31–36
21. Ignarro LJ, Gruetter CA (1980) Requirement of thiols for activation of coronary arterial guanylate cyclase by glyceryl trinitrate and sodium nitrite. Possible involvement of *S*-nitrosothiols. Biochim Biophys Acta 631: 221–231
22. Ignarro LJ, Kadowitz PJ, Baricos WH (1981) Evidence that regulation of hepatic guanylate cyclase activity involves interactions between catalytic site SH-groups and both substrate and activator. Arch Biochem Biophys 208 (1): 75–86
23. Ignarro LJ, Lippton H, Edwards JC, Baricos WH, Hyman AL, Kadowitz PJ, Gruetter CA (1981) Mechanism of vascular smooth muscle relaxation by organic nitrates, nitrites, nitroprusside and nitric oxide: evidence for the involvement of *S*-nitrosothiols as active intermediates. J Pharmacol Exp Ther 218: 739–749
24. Katsuki S, Arnold W, Mittal C, Murad F (1977) Stimulation of guanylate cyclase by sodium nitroprusside, nitroglycerin and nitric oxide in various tissue preparations and comparison to the effects of sodium azide and hydroxylamine. J Cyclic Nucleotide Res 3: 23–35

25. Keen JH, Habig WH, Jakoby WB (1976) Mechanism for the several activities of glutathione-*S*-transferases. J Biol Chem 251: 6183–6188
26. Kelm M, Feelisch M, Spahr R, Piper HM, Noack E, Schrader J (1988) Quantitative and kinetic characterization of nitric oxide and EDRF released from cultured endothelial cells. Biochem Biophys Res Commun 154 (1): 236–244
27. Kimura H, Mittal CK, Murad F (1975) Activation of guanylate cyclase from rat liver and other tissues by sodium azide. J Biol Chem 250 (20): 8016–8022
28. Kukovetz WR, Holzmann S (1986) Mode of action of nitrates with regard to vasodilation and tolerance. Z Kardiol 75 [Suppl 3]: 8–11
29. Kuropteva ZV, Pastushenko ON (1985) Change in paramagnetic blood and liver complexes in animals under the influence of nitroglycerin (in Russian). Dokl Akad Nauk SSSR 281 (1): 189–192
30. Leuenberger U, Gauch R, Rieder K, Baumgartner E (1980) Determination of nitrate and bromide in foodstuffs by high-performance liquid chromatography. J Chromatogr 202: 461–468
31. Mirna A, Hofmann K (1969) Über den Verbleib von Nitrit in Fleischwaren. I. Umsetzung von Nitrit mit Sulfhydrylverbindungen (in German). Fleischwirtschaft 10: 1361–1366
32. Moncada S, Palmer RMJ, Higgs EA (1988) The discovery of nitric oxide as the endogenous nitrovasodilator. Hypertension 12: 365–372
33. Mülsch A, Busse R, Bassenge E (1988) Desensitization of guanylate cyclase in nitrate tolerance does not impair endothelium-dependent responses. Eur J Pharmacol 158: 191–198
34. Murad F (1986) Cyclic guanosine monophosphate as a mediator of vasodilation. J Clin Invest 78: 1–5
35. Needleman P, Johnson EM (1975) The pharmacological and biochemical interaction of organic nitrates with sulfhydryls: possible correlations with the mechanism of tolerance development, vasodilation and mitochondrial and enzyme reactions. In: Needlemann P (ed) Organic nitrates. Springer, Berlin Heidelberg New York, pp 97–114 (Handbook of experimental pharmacology, vol 40)
36. Parks NJ, Krohn KA, Mathis CA, Chasko JH, Geiger KR, Gregor ME, Peek NF (1981) Nitrogen-13-labeled nitrite and nitrate: distribution and metabolism after intratracheal administration. Science 212 (4490): 58–60
37. Romanin C, Kukovetz WR (1988) Guanylate cyclase activation by organic nitrates is not mediated via nitrite. J Mol Cell Cardiol 20: 389–396
38. Slack CJ, McLaughlin BE, Nakatsu K, Marks GS, Brien JF (1988) Nitric oxide-induced vasodilation of organic nitrate- tolerant rabbit aorta. Can J Physiol Pharmacol 66: 1344–1346
39. Steurer G, Schütz W (1984) Guanylate cyclase stimulation by nitro-compounds is dependent on free Ca^{2+} Experientia 40: 970–971
40. Tsuruta H, Hasegawa H (1970) Studies on nitroglycol poisoning – on some properties of an enzyme which decomposes nitroglycol into inorganic nitrate. Ind Health 1970; 8: 99–118.
41. Tsuruta H, Hasegawa H (1970) Studies on nitroglycol poisoning – decomposition mechanism of nitroglycol by nitrite forming enzyme. Ind Health 8: 119–140
42. Waldman SA, Rapoport RM, Ginsburg R, Murad F (1986) Desensitization to nitroglycerin in vascular smooth muscle from rat and human. Biochem Pharmacol 35 (20): 3525–3531
43. Yeates RA, Laufen H, Leitold M (1985) The reaction between organic nitrates and sulfhydryl compounds – possible model system for the activation of organic nitrates. Mol Pharmacol 28: 555–559

Effect of Intravenous Isosorbide Dinitrate Versus Nitroglycerin in Patients with Acute Myocardial Infarction and Elevated Pulmonary Artery Wedge Pressure*

G. B. Cintron and S. P. Glasser

During the last 10 years, the use of intravenous nitroglycerin has become widespread in the therapy of patients with acute myocardial infarction [1–4]. The usual clinical indications for the use of this agent in patients with acute myocardial infarction are either early postinfarction angina or left ventricular failure. The goals of therapy are to prevent further episodes of myocardial ischemia and to improve left ventricular function by decreasing ventricular load. The prolonged use of cutaneous or intravenous nitroglycerin is limited by the rapid development of tolerance [5–8]. We have compared the acute and sustained (24-h) hemodynamic effect of intravenous isosorbide dinitrate with that of intravenous nitroglycerin in patients with acute myocardial infarction and elevated pulmonary artery wedge pressure.

Patients and Methods

Patients presenting to the coronary care unit with the clinical picture of acute myocardial infarction were evaluated and the diagnosis was documented by enzymatic and electrocardiographic criteria. Those patients who within 96 h of admission had a pulmonary artery wedge pressure $\geqslant 18$ mm Hg (two consecutive readings, 10 min apart) and agreed to participate in the study were randomized to receive either isosorbide dinitrate or nitroglycerin. The randomization procedure was weighted with two patients receiving isosorbide dinitrate for each patient treated with nitroglycerin. The assigned medication was administered in progressive doses and during titration and infusion the patients underwent continuous hemodynamic and electrocardiographic monitoring. The starting doses were nitroglycerin 4 μg/min or isosorbide dinitrate 33 μg/min. The goal of initial titration was to decrease baseline pulmonary artery wedge pressure by 25% or to obtain a pulmonary artery wedge pressure of 12–18 mm Hg.

* These studies were performed with support from a grant by Ives Laboratories Inc., Philadelphia, PA, with support by the Veterans Administration and the participating institutions. Reprinted with permission from Cintron et al. (1988) [20].

Results

One hundred and eleven patients were evaluated. Of these, 74 received isosorbide dinitrate and 37 received nitroglycerin. Both groups of patients had similar clinical characteristics and no significant treatment group differences were found. A similar number of patients in both groups reached stabilization. Sixty-one (82%) of the isosorbide dinitrate-treated patients and 34 (92%) of the nitroglycerin-treated patients achieved the goal pulmonary artery wedge pressure. The mean dosage of isosorbide dinitrate ($\pm$ standard error of the mean) at the time of stabilization was $203 \pm 17 \mu g/min$ and the mean dosage of nitroglycerin was 55 ± 7 µg/min. The mean titration time (from baseline to stabilization) was longer for the isosorbide dinitrate group than for the nitroglycerin group (1.4 ± 0.1 vs 1.1 ± 0.1 h, $p = 0.03$). In the stabilized patients, the initial titration with both agents produced significant decreases ($p < 0.05$) in pulmonary artery wedge pressure (35% for isosorbide dinitrate, 36% for nitroglycerin) and systolic blood pressure (6% for isosorbide dinitrate, 4% for nitroglycerin). Cardiac output increased with both medications ($p = 0.05$); however, the 17% increase observed with nitroglycerin was significantly higher ($p = 0.02$) than the 9% increase observed with isosorbide dinitrate. There was no change from baseline in heart rate with either medication. The time course of changes in dose, pulmonary artery wedge pressure, systolic blood pressure and cardiac output for the patients who completed the 24-h infusion are shown in Fig. 1.

In the isosorbide dinitrate group, the dosage required to maintain the target pulmonary artery wedge pressure remained essentially unchanged from the initial stabilization dose. In contrast, in the nitroglycerin group the dosage required to maintain the target pulmonary artery wedge pressure gradually and progressively increased to more than two times the initial stabilization dose. This increase in the nitroglycerin dosage was statistically significant ($p < 0.05$) from 18 h on.

Compared with baseline values, mean pulmonary artery wedge pressure values were significantly lower ($p < 0.05$) throughout treatment with both agents. Although these decreases were consistently greater in the isosorbide dinitrate group than in the nitroglycerin group from 3 to 24 h, there were no significant differences between the groups. Relative to the values at the time of stabilization, mean pulmonary artery wedge pressure in the nitroglycerin group was significantly higher ($p < 0.05$) at the 6- and 18-h intervals. In contrast, the pulmonary artery wedge pressure for the isosorbide dinitrate group remained stable during the infusion period.

Mean cardiac output for both treatment groups remained significantly higher ($p < 0.05$) than baseline values. Compared with the stabilization values, mean cardiac output was significantly higher ($p < 0.05$) in the isosorbide dinitrate group at 24 h and lower in the nitroglycerin group at 18 h.

The mean number of upward dose adjustments per patient was significantly higher ($p < 0.01$) for the nitroglycerin group (7.6 ± 1.5) than for the isosorbide dinitrate group (2.0 ± 0.4). The proportion of patients requiring upward titration of isosorbide dinitrate was 31% compared with 59% on nitroglycerin

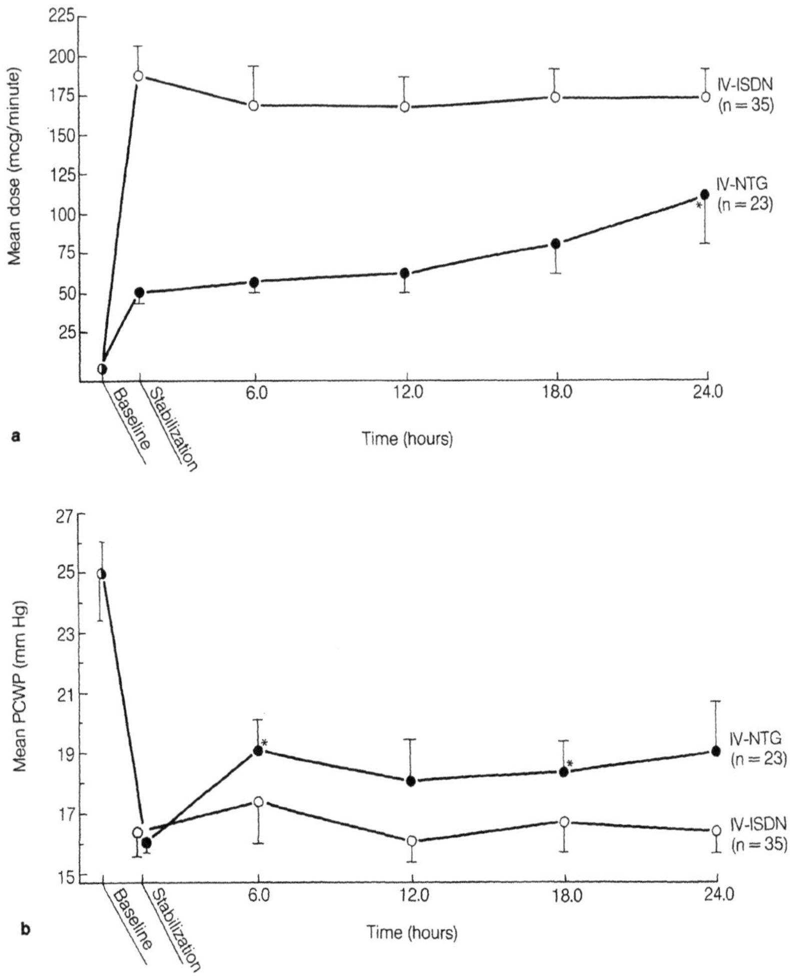

Fig. 1. a, b.

($p = 0.02$). The mean number of downward dose adjustments per patient was similar in both groups (2.9 vs 2.3).

Patients who achieved the target pulmonary artery wedge pressure in both groups remained at that level throughout the 24-h infusion. Of the patients who failed to complete the infusion, one in the isosorbide dinitrate group did so because of mild hypotension and one in the nitroglycerin group did so because of headache, nausea, and dizziness. The isosorbide dinitrate infusion in another patient was stopped because he was treated with furosemide, a drug prohibited by the protocol. All other patients in both groups who failed to complete the infusion did so for reasons unrelated to treatment; the primary reason was that their condition improved sufficiently for them be removed from therapy.

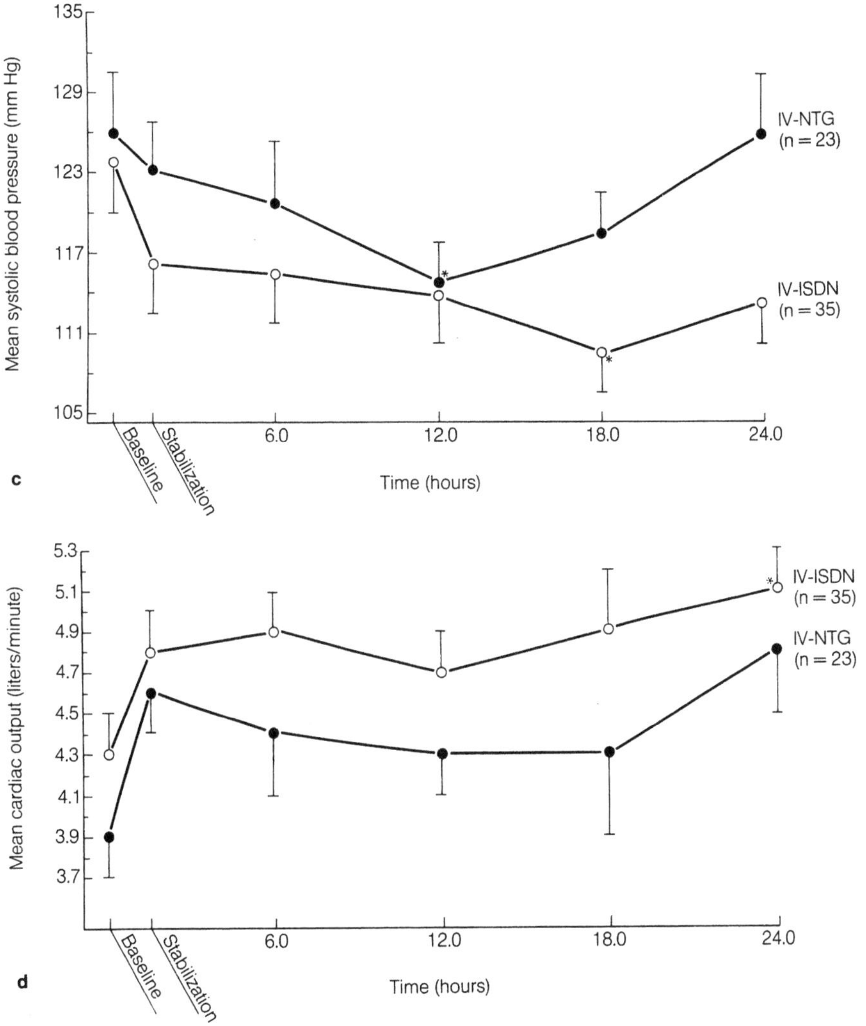

Fig. 1 a–d. a Mean changes in dose. **b** pulmonary capillary (artery) wedge pressure *(PCWP)*. **c** systolic blood pressure, and **d** cardiac output, over a 24-h period, for 35 patients receiving intravenous isosorbide dinitrate *(IV-ISDN; open circles);* and *23* receiving intravenous nitroglycerin *(IV-NTG; closed circles). Baseline* and *stabilization* values are indicated. Significant difference from stabilization ($p \leq 0.05$) indicated by an asterisk

Discussion

The results of this study demonstrate that the short-term administration of nitroglycerin or isosorbide dinitrate produced similar beneficial hemodynamic effects in patients with acute myocardial infarction complicated by left ventricular failure. This is a further confirmation of previously published studies [9–11]. The mode of action of both medications appears to be venous and arterial dilation resulting in a decrease in left ventricular filling pressure and systemic vascular resistance, with an increase in cardiac output and a slight lowering of systemic blood pressure.

A notable difference between isosorbide dinitrate and nitroglycerin was the apparent absence of the development of tachyphylaxis with the use of isosorbide dinitrate. Partial tachyphylaxis developed rapidly in the patients receiving nitroglycerin, but not in the isosorbide dinitrate-treated patients. Progressive increases in the dose of nitroglycerin were necessary to maintain desirable hemodynamic effects, i. e., decreased pulmonary artery wedge pressure and increased cardiac output. Even with these dosage increases, mean pulmonary artery wedge pressure increased after stabilization only in the nitroglycerin treated patients. The number of upward titrations and dose adjustments was also much higher in the patients who received nitroglycerin than in those who received isosorbide dinitrate. The development of tachyphylaxis has been reported previously in patients given nitroglycerin or cutaneous nitroglycerin patches [5–8, 12, 13]. The possibility that the loss of efficacy with nitroglycerin was related to differences in drug absorption or decreased circulating drug levels is unlikely, because both agents were delivered directly into the systemic circulation. In addition, a recent report by Zimrin et al. has documented a progressive decrease in the antianginal effect of nitroglycerin, despite constant plasma drug levels. The possibility that the nitroglycerin-treated patients had more serious hemodynamic problems requiring increasing doses of medication is also unlikely, considering that the initial clinical characteristics were similar for both groups. The mechanism for the loss of efficacy or the development of tachyphylaxis with nitroglycerin is not clear, but is probably related to depletion of sulfhydryl groups necessary for the conversion of nitrate into active vasodilating compound [14]. The response to nitrates, however, may be re-established after a brief washout period or by the infusion of *N*-acetylcysteine, a sulfhydryl donor [15, 16].

Two studies have documented the short-term beneficial effect of isosorbide dinitrate [9, 10]. Rabinowitz et al. [9] studied the effect of isosorbide dinitrate in 22 patients with chronic pump failure and 18 patients with acute pump failure due to acute myocardial infarction and the response of the latter patients was almost identical to that of our isosorbide dinitrate-treated patients. Although the isosorbide dinitrate infusion in the study of Rabinowitz et al. was maintained for a 24-h period, no mention is made of continued hemodynamic efficacy. Leier et al. [10] compared isosorbide dinitrate with nitroglycerin and sodium nitroprusside; however, the maintenance infusion period was only 1.5 h. There are no previously published studies comparing the long-term hemodynamic effects of intravenous infusions of isosorbide dinitrate versus nitroglycerin. When used orally in patients with angina or heart failure, isosorbide dinitrate has been demonstrated to sustain its hemodynamic and antianginal efficacy over a relatively long period of time [17, 18]. Other investigators have shown that continued and frequent use of oral isosorbide dinitrate result in the attenuation of its hemodynamic effect after a 48-h period

[15, 19]. Thus, it is possible that prolonged infusion of isosorbide dinitrate would also result in partial hemodynamic tolerance.

Summary

The results of this study indicate that, in the patient with acute myocardial infarction complicated by high pulmonary artery wedge pressure who requires intravenous nitrates, isosorbide dinitrate may offer the benefit of a stable hemodynamic effect.

References

1. Hill NS, Antman EM, Green LH, Alpert JS (1981) Intravenous nitroglycerin. A review of pharmacology, indications, therapeutic effects and complications. Chest 79: 69–77
2. Kaplan K, Davison R, Parker M, Przybylek J, Teagarden JR, Lesch M (1983) Intravenous nitroglycerin for the treatment of angina at rest unresponsive to standard nitrate therapy. Am J Cardiol 51: 694–698
3. Roberts R (1983) Intravenous nitroglycerin in acute myocardial infarction. Am J Med 74: 45–52
4. Flaherty JT, Becker LC, Bulkley BH, Weiss JL, Gerstenblith G, Kallman CH, Silverman KJ, Wei JY, Pitt B, Weisfeldt ML (1983) A randomized prospective trial of intravenous nitroglycerin in patients with acute myocardial infarction. Circulation 69: 576–588
5. Jordan RA, Seth L, Casebold P, Hayes MJ, Wilen MM, Franciosa J (1986) Rapidly developing tolerance to transdermal nitroglycerin in congestive heart failure. Ann Intern Med 104: 295–298
6. Thadani U, Hamilton SF, Olson E, Anderson J, Voyles W, Prasad R, Teague SM (1986) Transdermal nitroglycerin patches in angina pectoris. Dose titration, duration of effect, and rapid tolerance. Ann Intern Med 105: 485–492
7. Zimrin D, Reichek N, Bogin K, Cameron S, Douglas P, Fung HL (1985) Antianginal effects of IV nitroglycerin. Circulation 72 [Suppl III]: III–460 (abstr)
8. Packer M, Medina N, Yushak M (1984) Rapid attenuation of hemodynamic efficacy of continuous transcutaneous nitroglycerin in severe heart failure: lack of cross tolerance of isosorbide dinitrate. Circulation 70 [Suppl II]: II–113 (abstr)
9. Rabinowitz B, Tamari I, Elazar E, Neufeld HN (1982) Intravenous isosorbide dinitrate in patients with refractory pump failure and acute myocardial infarction. Circulation 65: 771–778
10. Leier CV, Bambach D, Thompson MJ, Cattaneo SM, Goldberg RJ, Unverferth DV (1981) Central and regional hemodynamic effects of intravenous isosorbide dinitrate, nitroglycerin and nitroprusside in patients with congestive heart failure. Am J Cardiol 48: 1115–1123
11. Nelson GI, Silke B, Ahuja RC, Hussain M (1983) Haemodynamic advantages of isosorbide dinitrate over furosemide in acute heart failure following myocardial infarction. Lancet I: 730–733
12. Parker JO, Fung HL (1984) Transdermal nitroglycerin in angina pectoris. Am J Cardiol 54: 471–476
13. Reicheck LN, Priest C, Zimrin D, Chandler T, Sutton MS (1984) Antianginal effects of nitroglycerin patches. Am J Cardiol 54: 1–7
14. Leier CV (1985) Nitrate tolerance. Am Heart J 110: 204–232
15. Parker JO, Fung HL, Ruggirello D, Stone JA (1983) Tolerance to isosorbide dinitrate: rate of development and reversal. Circulation 68: 1074–1080

16. Packer M, Lee WH, Kessler P, Medina N, Yushak M (1986) Induction of nitrate tolerance in human heart failure by continuous intravenous infusion of nitroglycerin and reversal of tolerance by N-acetylcysteine, a sulfhydryl donor. J Am Coll Cardiol 7: 27 A (abstr)
17. Lee G, Mason DT, De Maria AN (1978) Effects of long-term oral administration of isosorbide dinitrate on the antianginal response to nitroglycerin. Absence of nitrate cross-tolerance and self-tolerance shown by exercise testing. Am J Cardiol 41: 82–87
18. Franciosa JA, Cohn JN (1980) Sustained hemodynamic effects without tolerance during long-term isosorbide dinitrate treatment of chronic left ventricular failure. Am J Cardiol 45: 648–654
19. Parker JO, Vankoughnett KA, Fung HL (1984) Transdermal isosorbide dinitrate in angina pectoris: effect of acute and sustained therapy. Am J Cardiol 54: 8–13
20. Cintron GB, Glasser SP, Weston BA, Linares E, Conti R (1988) Effect of intravenous isosorbide dinitrate versus nitroglycerin on elevated pulmonary arterial wedge pressure during acute myocardial infarction. Am J Cardiol 61: 21–25

Effects of Isosorbide-5-Mononitrate on Haemodynamics and Exercise Tolerance in Patients with Heart Failure: Comparisons with Digoxin and Placebo

I. Dews, C. Marks, D. Woodings, J. Stephens, and M. VandenBurg

Introduction

The pre-eminence of diuretics as first-line treatment in heart failure remains virtually unchallenged but there is less certainty concerning the drugs available for second-line treatment. The longest established of these is the inotrope digoxin, but its efficacy in heart failure with sinus rhythm has recently been called into question [1, 2]. Oral isosorbide dinitrate has also been found to be effective when combined with diuretics in the treatment of heart failure. However, large doses are often required to achieve beneficial acute effects [3, 4].

Isosorbide-5-mononitrate is an active metabolite of isosorbide dinitrate and may be more potent than the parent compound [5]. In this paper we describe a programme of three studies designed to determine the acute and chronic haemodynamic effects of this drug in patients with heart failure and to investigate its potential for extending exercise tolerance in this condition.

Methods

Overall Design

The programme was made up of three studies:

1. An acute haemodynamic study of isosorbide-5-mononitrate alone, with Swan-Ganz catheterisation.
2. A 4-week comparison of isosorbide-5-mononitrate with digoxin, assessed by repeat Swan-Ganz catheterisation and by bicycle ergometry.
3. A 12-week double-blind placebo-controlled trial of the effects of isosorbide-5-mononitrate on exercise tolerance using treadmill exercise tests.

The patients assessed in each of the studies had congestive cardiac failure of non-valvular aetiology. All were in New York Heart Association functional classes II to IV while receiving treatment with diuretics. The entry criteria for the three studies are detailed below.

Statistical analysis of each study was by means of Student's paired *t* test for within-group comparisons and the unpaired *t* test for between-group comparisons.

This research programme was performed in accordance with the Declaration of Helsinki and all three parts of the programme were approved by the local ethical review committee. Written informed consent was obtained from all patients before inclusion in a study.

Acute Haemodynamic Study

Sixteen patients, aged 52–74 years, with heart failure secondary to ischaemic heart disease or congestive cardiomyopathy were studied. Ten were in functional class II, four in class III and the remaining two in class IV. All had been receiving stable doses of diuretics as their only treatment for heart failure.

Haemodynamic assessment was by right heart catheterisation using the Swan-Ganz technique. On the day of the study all medications were withdrawn at least 12 h before commencement. Measurements were taken from mid chest level.

Baseline measurements of heart rate, systemic blood pressure (cuff), pulmonary capillary wedge pressure, mean right atrial pressure and cardiac output (thermodilution) were made in triplicate. Providing that there was close agreement between pulmonary capillary wedge pressure and pulmonary arterial end-diastolic pressure, the latter was taken to represent left ventricular filling pressure. Having confirmed haemodynamic stability, increasing doses of isosorbide-5-mononitrate were given orally to lower pulmonary arterial diastolic pressure to below 20 mm Hg, provided that systemic arterial pressure was maintained. An initial dose of 20 mg was given, followed 90 min later by haemodynamic evaluation. If necessary, further doses of 30 and 50 mg were given with haemodynamic responses measured after 90 min.

Mean pressures were calculated from systolic pressure (S) and diastolic pressure (D) using the formula: mean pressure = (S–D)/3 + D.

Four-Week Comparison with Digoxin

This open study followed on directly from the acute haemodynamic study and enrolled a subset of the same patients. Exercise tolerance was determined by bicycle ergometry starting at 0 W and increasing by 10 W after each min until exhaustion.

Patients were randomised to receive either the dose of isosorbide-5-mononitrate that had been found to be optimal during the haemodynamic study three times daily or digoxin up to 0.25 mg daily. This treatment continued for 4 weeks during which time the patients were followed clinically, and plasma digoxin levels were monitored to ensure correct dosage. Those patients who completed 4 weeks' treatment with either isosorbide-5-mononitrate or digoxin underwent repeat bicycle ergometry and catheterisation. In the digoxin group measurements were made 3 h after the morning dose of digoxin while in the isosorbide-5-mononitrate group measurements were made at least 12 h after the previous dose. In both cases the evaluations were repeated 90 min after administration of the maintenance dose.

Twelve-Week Double-Blind Placebo-Controlled Trial

This part of the programme was based on a different group of patients recruited using similar criteria: chronic heart failure of functional classes II to IV, non-valvular aetiology, stable diuretic treatment, and ability to complete between 2 and 16 min of the modified Naughton treadmill protocol. All had chest X-ray confirmation of their heart failure. It was originally intended to study sufficient patients to allow for subgroup analysis between different doses of isosorbide-5-mononitrate. However, the trial was terminated early because of the difficulty in maintaining patients with severe heart failure on long-term placebo therapy in the face of mounting evidence of the value of second-line therapy [6]. A total of 16 patients had been recruited at the time of termination.

Provided that two baseline exercise tests were within 20% of each other in duration, patients were randomised to receive, in addition to their diuretics, either placebo or isosorbide-5-mononitrate at a dose of 20 mg three times daily or 40 mg twice daily. This treatment was continued for 12 weeks and exercise testing was repeated at the end of the treatment period. Patient response to treatment was further assessed by a scoring system for the signs and symptoms of heart failure and by the use of a visual analogue scale questionnaire to evaluate subjective well-being.

Results

All numerical results are given as mean ± standard deviation.

Acute Haemodynamics

Six patients received only 20 mg of isosorbide-5-mononitrate while six received a further dose of 30 mg (cumulative dose of 50 mg), and four received a further dose of 50 mg (cumulative dose of 100 mg) at 90 min intervals. Of the 12 patients who did not receive the maximum dose, pulmonary artery diastolic pressure fell below 20 mm Hg in ten and systemic diastolic pressure fell below 50 mm Hg in two patients. In two patients there was no haemodynamic response despite receiving a cumulative dose of 100 mg.

Table 1. Acute changes in mean values (± standard deviation) for haemodynamic variables in sixteen patients in response to maximum tolerated dose of isosorbide-5-mononitrate (IS-5-MN)

Variable	Control	IS-5-MN
Heart rate (beats/min)	92 ± 14	91 ± 15
Mean systemic arterial pressure (mm Hg)	88 ± 10	76 ± 10**
Pulmonary arterial systolic pressure (mm Hg)	54 ± 19	38 ± 18**
Pulmonary arterial end-diastolic pressure (mm Hg)	27 ± 8	19 ± 9**
Mean right atrial pressure (mm Hg)	4.9 ± 3.7	3.0 ± 3.8*
Cardiac output (l/min)	4.1 (0.8)	3.8 (0.7)
Stroke volume (ml)	45 ± 13	42 ± 10

* $p < 0.05$; ** $p < 0.01$

The changes in mean haemodynamic paramters are shown in Table 1. In summary, the acute administration of isosorbide-5-mononitrate reduced systemic, pulmonary arterial and right atrial pressures significantly without significant changes in heart rate, cardiac output or stroke volume.

Chronic Haemodynamic Effects of Isosorbide-5-Mononitrate

Six patients who received long-term treatment with isosorbide-5-mononitrate underwent repeat right heart catheterisation at 4 weeks. As in the first haemodynamic study, the oral administration of isosorbide-5-mononitrate produced significant falls in the mean values for pulmonary arterial systolic and diastolic pressure and right atrial pressure with no significant change in cardiac output, heart rate or stroke volume (Fig. 1). The fall in mean systemic arterial pressure did not reach significance.

Chronic Haemodynamic Effects of Digoxin

Five patients who received long-term digoxin underwent repeat right heart catheterisation. Mean values for haemodynamic variables after long-term treatment with digoxin were unchanged from pretreatment values except that cardiac output increased from 3.6 ± 0.7 to 4.4 ± 0.7 l/min ($p < 0.01$) and stroke volume increased from 40 ± 8 to 54 ± 9 ml/min (p 0.01) (Fig. 2).

Comparison of Chronic Haemodynamic Effects

The principal long-term haemodynamic effect of isosorbide-5-mononitrate was reduced in left ventricular filling pressure, whereas digoxin increased cardiac output without a significant effect on left ventricular filling pressure.

Four-Week Exercise Study

Treatment with isosorbide-5-mononitrate over 4 weeks increased mean exercise duration from 9.0 ± 1.9 min to 9.7 ± 2.3 min. This increase was not statistically significant.

After 4 weeks' treatment with digoxin, exercise duration decreased from 10.6 ± 1.3 min. to 9.5 ± 1.4 min. This change was again not significant. However, the difference in the change in exercise duration between the two treatment groups was significant ($p < 0.05$).

Fig. 1. Acute and long-term haemodynamic effects of isosorbide-5-mononitrate *(IS-5-MN)*. Mean values and standard deviation in six patients for pulmonary arterial *(PA)*, mean systemic arterial *(MAP)* and right atrial *(RAP)* pressures (mm Hg), cardiac output (*CO;* l/min), heart rate (*HR;* beats/min) and stroke volume (*SV;* ml/min). *Solid lines,* acute changes; *broken lines,* changes after long-term treatment; *open circles,* pulmonary arterial diastolic pressure; *NS,* not significant

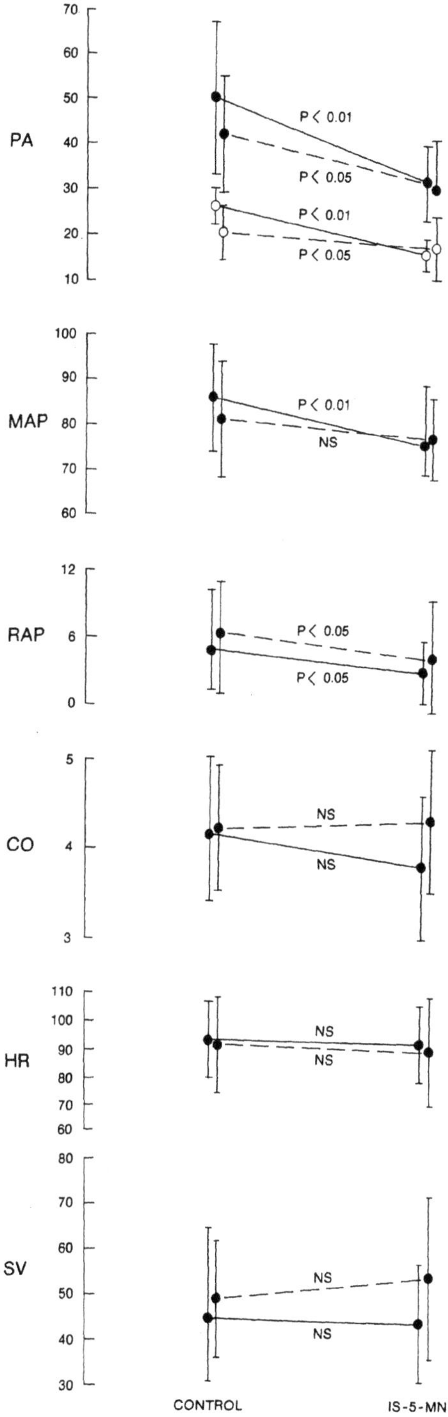
PA
P < 0.01
P < 0.05
P < 0.01
P < 0.05
MAP
P < 0.01
NS
RAP
P < 0.05
P < 0.05
CO
NS
NS
HR
NS
NS
SV
NS
NS
CONTROL
IS-5-MN

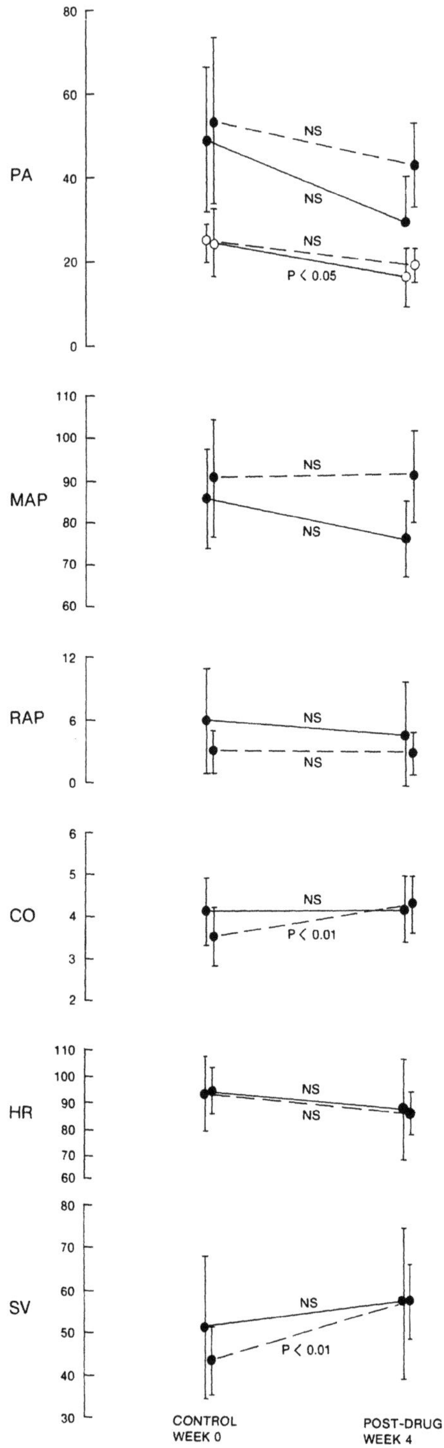

PA
80
60
40
20
0
NS
NS
NS
P < 0.05
MAP
110
100
90
80
70
60
NS
NS
RAP
12
6
0
NS
NS
CO
6
5
4
3
2
NS
P < 0.01
HR
110
100
90
80
70
60
NS
NS
SV
80
70
60
50
40
30
NS
P < 0.01
CONTROL
WEEK 0
POST-DRUG
WEEK 4

Twelve-Week Exercise Study

Eight patients completed 12 weeks on isosorbide-5-mononitrate while two completed the full period on placebo. Of the eight patients on active treatment, four received 20 mg three times daily and four received 40 mg twice daily.

After 12 weeks' treatment with isosorbide-5-mononitrate, mean exercise time was increased significantly from 8.9 ± 4.3 min to 11.8 ± 4.5 min ($p < 0.05$). There was no mean change in the placebo group, but the smallness of this group precluded any statistical analysis of the difference between the groups.

The scoring system for the signs and symptoms of heart failure was based on individual scores for weakness, fatigue, orthopnoea, daytime dyspnoea, paroxysmal nocturnal dyspnoea, jugular venous pressure, gallop rhythm, rales, oedema, and ascites which were totalled to yield an aggregate score between 0 and 28. After 12 weeks' treatment with isosorbide-5-mononitrate, the mean score of the actively treated group had improved by 32% from 8.50 ± 6.16 to 5.75 ± 6.50 ($p = 0.068$).

The small numbers resulting from the premature termination precluded useful analysis of the visual analogue scale data concerning patients' subjective perception of treatment effects.

Discussion

A clear picture of the principal effects of isosorbide-5-mononitrate emerges from this combination of studies. The acute haemodynamic study shows the compound to result in a marked lowering of left ventricular filling pressure, presumably on account of its pulmonary vasodilating action. This effect is similar to that seen with isosorbide dinitrate [3, 4]. It is noteworthy that stroke volume is maintained in the face of this fall in filling pressure, implying that myocardial function has been shifted to a more efficient part of the Starling curve.

The repeat haemodynamic studies after 4 weeks' treatment yield interesting data for both isosorbide-5-mononitrate and digoxin. Although there have been reports of tolerance developing to the antianginal effects of nitrates [7], the beneficial haemodynamic effects of isosorbide-5-mononitrate in heart failure patients were still readily observed after chronic treatment. Again this is parallel to the findings with isosorbide dinitrate [5, 8].

The increase in cardiac output without change in filling pressure seen with digoxin is in agreement with the well-known positive inotropic effect of the drug

Fig. 2. Long-term haemodynamic effects of isosorbide-5-mononitrate (six patients): comparison with digoxin (five patients). Mean values and standard deviation for pulmonary arterial *(PA)*, mean systemic arterial *(MAP)* and right atrial *(RAP)* pressures (mmHg), cardiac output (*CO;* 1/min), heart rate (*HR;* beats/min) and stroke volume (*SV;* ml/min). *Solid lines,* isosorbide-5-mononitrate; *broken lines,* digoxin; *open circles*, pulmonary arterial diastolic pressure; *NS,* not significant

[9, 10] and with a withdrawal study by Arnold et al. [11] which showed that long-term digoxin was associated with improved left ventricular dynamics. However, it is at variance with published acute data which suggest that digoxin has little or no effect on cardiac output or left ventricular filling pressure [10, 12].

The correlation between the haemodynamic effects of a drug and the improvement in exercise tolerance which is a primary goal of treatment is not always as clear as might be expected [13]. However, both bicycle ergometry after 4 weeks' treatment and treadmill testing after 12 weeks suggest that isosorbide-5-mononitrate improves exercise tolerance in patients with heart failure in parallel with its persisting beneficial haemodynamic effects. Nitrates have a clearly demonstrable effect in reducing peripheral vascular tone [14] and it has been suggested that the improvement in exercise tolerance seen with these compounds is brought about by an increase in muscle blood flow and oxygen utilisation, rather than a central effect on cardiac performance. The marked improvement in overall heart failure score ($p = 0.068$) after 12 weeks' treatment does, however, seem likely to represent an improvement in cardiac function.

Summary

Isosorbide-5-mononitrate has valuable haemodynamic effects in the failing heart which are maintained for at least 4 weeks and are associated with improved exercise tolerance and possibly with a general improvement in the signs and symptoms of heart failure.

References

1. Gold H, Greiner T, Kwit NT, Otto HL, Warshaw L (1959) Effect of digitalis and diuretics on exertional dyspnoea: study in ambulant patients with chronic heart disease. JAMA 169: 229–236
2. Johnston GD, McDevitt DG (1979) Is maintenance digoxin necessary in patients with sinus rhythm? Lancet I: 567–570
3. Franciosa JA, Mikulic E, Cohn JN, Jose E, Fabie A (1974) Haemodynamic effect of orally-administered isosorbide dinitrate in patients with congestive heart failure. Circulation 50: 1020–1024
4. Stephens JD, Camm AJ, Spurrell RAJ (1978) Improvement in exercise haemodynamics by isosorbide dinitrate in patients with severe congestive cardiac failure secondary to ischaemic heart disease. Br Heart J 40: 832–837
5. Stephens J, Marks C, Woodings D, VandenBurg MJ (1987) Acute and chronic use of isosorbide-5-mononitrate in patients with heart failure. Cardiology 74 [Suppl 1]: 69–71
6. Consensus Study Group (1987) Effects of enalapril on mortality in severe heart failure. N Engl J Med 316: 1429–1435
7. Thadani U, Manyari D, Parker JO, Fung H (1980) Tolerance to the circulatory effects of oral isosorbide dinitrate: rate of development and cross tolerance to glyceryl trinitrate. Circulation 61: 526–535
8. Leier CV, Huss P, Magorien RD, Unverferth DV (1983) Improved exercise capacity and differing arterial and venous tolerance during chronic isosorbide dinitrate therapy for congestive heart failure. Circulation 67: 817–822

9. Davidson C, Gibson D (1973) Clinical significance of positive inotropic action of digoxin in patients with left ventricular disease. Br Heart J 35: 970–976
10. Hayward R, Greenwood H, Stephens J, Hamer J (1983) Relationship between myocardial uptake and actions in heart failure of methyldigoxin. Br J Clin Pharmacol 15: 41–48
11. Arnold SB, Byrd RC, Meister W, Melmon K, Cheitlin MD, Bristow JD, Parmley WW, Chatterjee K (1980) Long-term digitalis therapy improves left ventricular function in heart failure. N Engl J Med 303: 1443–1448
12. Cohn K, Selzer A, Kersh ES, Karpman LS, Goldshlager N (1975) Variability of hemodynamic response to acute digitalization in chronic cardiac failure due to cardiomyopathy and coronary artery disease. Am J Cardiol 35: 461–468
13. Franciosa JA, Dunkman WB, Leddy CL (1984) Haemodynamic effects of vasodilators and long-term response in heart failure. J Am Coll Cardiol 3: 1521–1530
14. Mason DT, Braunwald E (1965) Effects of nitroglycerin and amyl nitrite on arteriolar and venous tone in the human forearm. Circulation 32: 755–766

Improved Left Ventricular Geometry and Function by Prolonged Nitroglycerin Therapy After Acute Myocardial Infarction*

B. I. Jugdutt, B. L. Michorowski, and W. T. Tymchak

Introduction

Early therapy with intravenous nitroglycerin during acute myocardial infarction reduces infarct expansion experimentally and clinically [1, 2]. However, the healing process after myocardial infarction takes place over 6 weeks in the dog model, with early expansion and late thinning [3, 4], and even longer in humans [4]. The aim of this study was therefore to test the hypothesis that prolonged reduction of preload and afterload during the healing phase after a first anterior transmural acute myocardial infarction (ATAMI) might further improve left ventricular (LV) geometry and function [5, 6].

Methods

The protocol is shown in Fig. 1: 43 consecutive patients admitted to the coronary care unit with a *first* ATAMI were evaluated. Informed consent was obtained if inclusion and exclusion criteria were met. The patients were given low-dose intravenous nitroglycerin infusion therapy for up to 48 h after onset of typical chest pain of infarction. At 48 h, *randomization* was done to buccal nitroglycerin (group 1) or placebo (group 2) for 6 weeks. During the first 48 h, clinical data, electrocardiograms, ST-segment maps, hemodynamics, two-dimensional echocardiograms, serial creatine kinase, and CK-MB data were obtained. After 48 h, clinical data, electrocardiograms, and echocardiograms were collected at 2, 4, 6, 8, and 10 days "and again" at 6, 12, and 24 weeks. Further inclusion criteria were age < 75 years, heart rate > 60 beats/minute, systolic blood pressure > 110 mmHg and < 200 mmHg, sinus rhythm, no past history or electrocardiographic evidence of infarction. Exclusion criteria included heart block, cardiogenic shock, rheumatic or valvular or congenital heart disease, other contraindications to nitroglycerin therapy, concomitant use of calcium channel blockers or beta-blockers or drugs that might alter parameters being measured after nitroglycerin.

The dose of intravenous nitroglycerin infusion was titrated to decrease mean blood pressure by 10% but not below 80 mmHg. The dose of buccal ni-

* Supported in part by a grant from the Canadian Heart Foundation, Ottawa, Ontario.

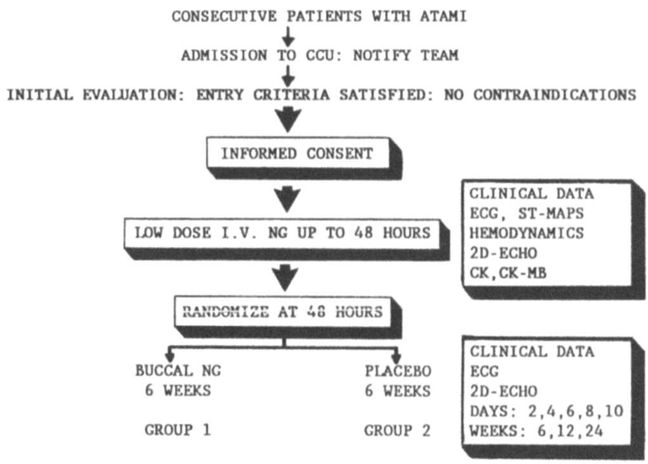

Fig. 1. The protocol. *CK,* creatine kinase; *CCU,* coronary care unit; *ECG,* electrocardiogram; *2D-ECHO,* two-dimensional echocardiogram; *NG,* nitroglycerin

troglycerin (Nitrogard-SR, Syntex) was 1–3 mg three times a day, in three divided doses given at 5-h intervals at 8:00, 13:00, and 18:00 h with an 8-h washout period to avoid vascular tolerance. The initial dose was titrated so that mean blood pressure did not fall by more than 10%. The placebo group received 2 mg nonactive buccal tablets (Syntex).

As described previously [2] and shown in Fig. 2, serial echocardiograms were used to measure: (a) the endocardial surface area of LV asynergy, defined as akinesis plus dyskinesis; (b) LV ejection fraction; and (c) LV geometry (expansion index, thickness ratio, peak regional shape distortion). Expansion index was defined as the ratio of asynergy containing endocardial segment length to normal segment length. The mid-point of the papillary muscles provided landmarks for segment lengths. Thickness ratio was measured as the ratio of thicknesses of mid-asynergic and normal zones.

All data were coded and final analysis done in blinded fashion at the end of the study. Serial data were analysed using analysis of variance (ANOVA) and multiple measures ANOVA. Values are expressed as mean ± standard deviation. Statistical significance was assumed for $p < 0.05$.

Results

Baseline Data

All 43 patients entered in the study with a first ATAMI received intravenous nitroglycerin over the first 48 h. Thereafter, they were randomly allocated to buccal nitroglycerin (n = 23, group 1) and buccal placebo (n = 20; group 2). The baseline data for the two groups were similar regarding sex (percentage

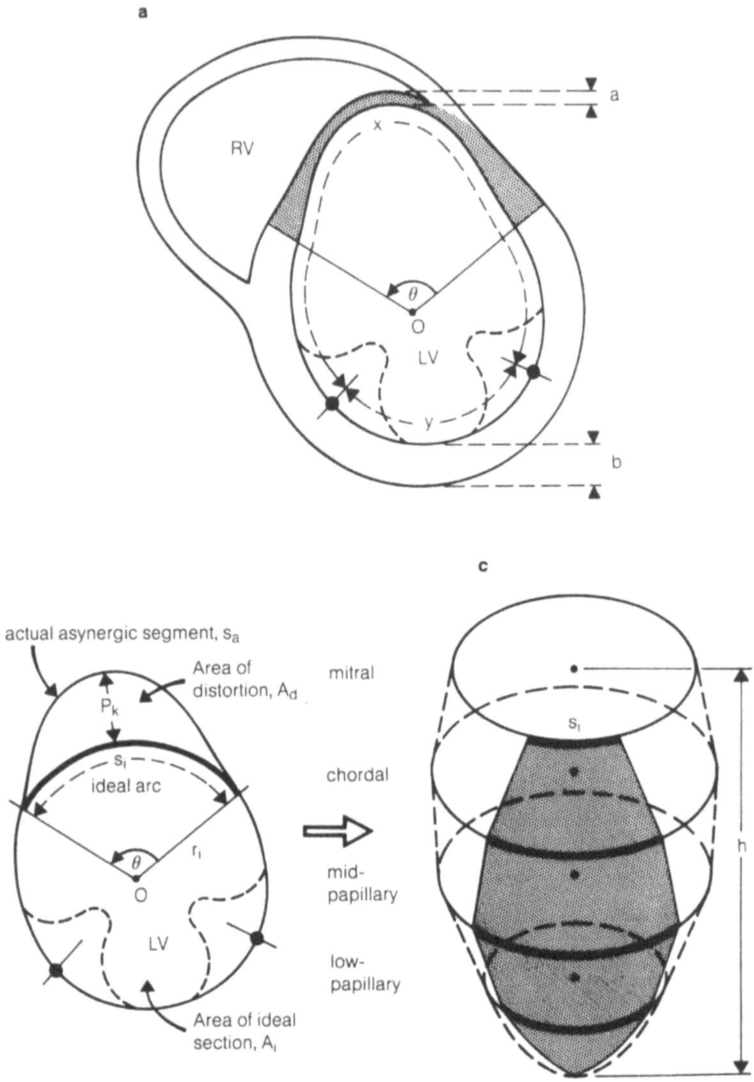

Fig. 2. a Method of calculating expansion index *(x/y)* and thinning or thickness ratio *(a/b)*. A mid-left ventricular *(LV)* section at end-diastole from a patient with anterior infarction; asynergy and regional shape distortion are shown. LV segment lengths were measured from the midpoints of papillary muscle landmarks. *RV*, right ventricle; θ, argular extent of asynergy; *x*, endocardial length of asynergy-containing segment; *y*, endocardiol length of segment without asynergy; *a*, average thickness of asynergic zone; *b*, average thickness of nonasynergic zone. **b** Method of quantifying regional shape distortion on the endocardial outline of a mid-papillary left ventricular section at end-diastole. The center *O* of the ideal circle and radius were computed using points from the nonasynergic segment. The angular extent of asynergy *(θ)* was measured and the ideal arc of asynergy (S_{i}) computed using the nonasynergic segment and center *O*, thus replacing the actual distorted asynergic segment (S_{a}). The areas of the regional shape distortion (A_{d}) and the ideal section (A_{i}) were measured. Peak distortion (P_{k}) was computed from the angular distribution of the average

male: 78% vs. 85%), age (61 vs. 57 years), body surface area (1.9 vs. 1.9 m^2), interval from chest pain to admission (13 vs. 13 h), interval from chest pain to nitroglycerin infusion (15 vs. 15 h), average nitroglycerin dose (71 vs. 77 µg/min), interval from admission to the first echocardiogram (24 vs. 23 h), and admission Killip class score (1.3 vs. 1.4). At 2 days, there was no difference between the two groups in blood pressure (systolic: 111 vs. 115 mm Hg, diastolic: 74 vs. 74 mm Hg, mean: 87 vs. 88 mm Hg), heart rate (84 vs. 87 beats/minute) and heart rate × mean blood pressure (7.6 vs. 8.0 mm Hg × beats/minute × 10^3), and Killip class score (1.3 vs. 1.4). Also, there was no difference between the two groups in peak creatine kinase (2583 vs. 2174 IU/liter) and creatine kinase infarct size (52 vs. 46 gEq).

Clinical follow-up

At 24 weeks, the functional class score (New York Heart Association) was less in the nitroglycerin than placebo group (1.3 ± 0.4 vs. 1.8 ± 0.6, $p < 0.005$). There were also fewer deaths at 24 weeks in the nitroglycerin group, but the difference was not statistically significant (0/23 vs. 3/20). There was no statistically significant difference in resting hemodynamics for the two groups at 2 days, 10 days, 6 weeks, 12 weeks, or 24 weeks.

Effect on LV Function

Asynergy decreased similarly between admission and 48 h with intravenous nitroglycerin in the two groups. Thus, total regional LV asynergy in the two groups was similar initially and after randomization at 2 days. However, total LV asynergy was persistently lower in the buccal nitroglycerin group compared to placebo at 6 weeks (14% vs. 21%, $p < 0.05$) and after buccal nitroglycerin was discontinued, at 12 and 24 weeks.

LV ejection fraction, computed by the modified Simpson's rule in the two groups was similar initially and at 2 days but was higher in the buccal nitroglycerin group compared to placebo at 6, 12, and 24 weeks; at 24 weeks, the mean values were 51% vs. 40% ($p \leq 0.05$, by multiple measures ANOVA).

Effect on LV Geometry

Figure 3 shows the increase in the expansion index after randomization in the placebo group but it did not change in the buccal nitroglycerin group. The divergence was significant at 6 weeks and persisted at 12 and 24 weeks

distances, within 5° windows, between pairs of corresponding points at intersections of radials with actual and ideal segments, respectively. **c** Total left ventricular asynergy *(stippled area)* was computed as percentage surface area of the endocardial shell constructed from the four short-axis sections and height *(h)* derived from the average long-axis length of the apical four- and two-chamber views

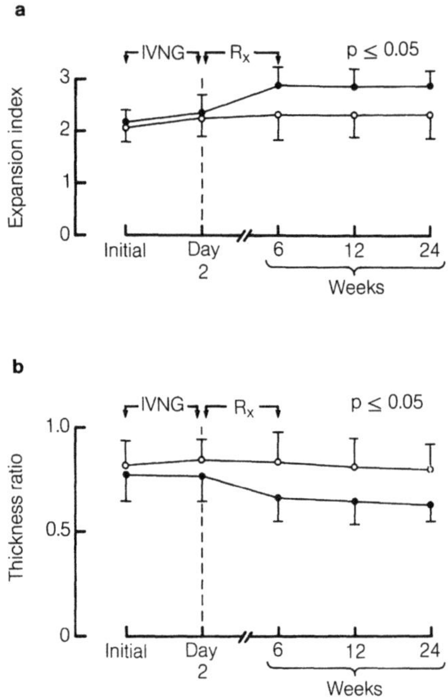

Fig. 3 a, b. Effect on infarct expansion (**a**) and thinning (**b**). *IVNG,* intravenous nitroglycerin; *solid circles,* placebo; *open circles,* buccal nitroglycerin

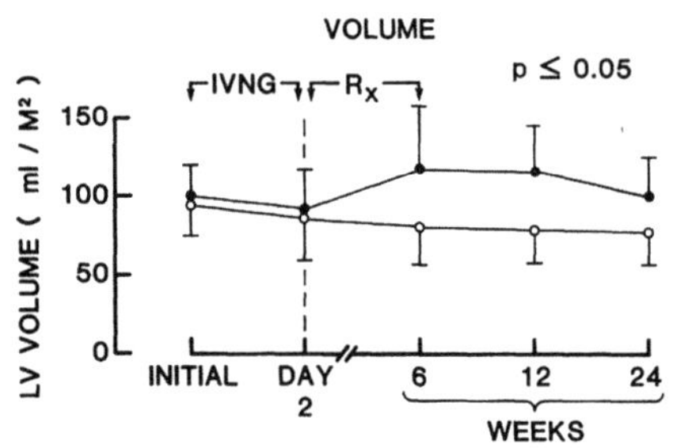

Fig. 4. Effect on LV volume. *IVNG,* intravenous nitroglycerin; *solid circles,* placebo; *open circles,* buccal nitroglycerin

($p \leq 0.05$). Similarly, the thickness ratio was persistently unchanged in the buccal nitroglycerin group but decreased in the placebo group, the divergence being significant at 6, 12, and 24 weeks ($p \leq 0.05$).

Figure 4 shows the effect on LV volume normalized to body surface area. There was no significant change after randomization in the buccal nitroglycerin group but a persistent increase in the placebo group. This divergence was statistically significant ($p \leq 0.05$) at 6, 12, and 24 weeks. A similar effect of nitroglycerin was seen on normalized LV diameters at the different short-axis levels and on the peak regional shape distortion.

Summary

The results of this study indicate that prolonged nitroglycerin therapy after a first ATAMI decreased LV volume, decreased infarct expansion, decreased infarct thinning, decreased LV asynergy, and increased LV ejection fraction. Prolonged nitroglycerin therapy during early and late phase of healing after ATAMI preserves LV topography and function. These beneficial effects on early and late remodeling with prolonged nitroglycerin therapy might prevent LV aneurysm formation. The most likely mechanisms include prolonged reduction of preload, afterload, LV volume, and LV wall stress.

References

1. Jugdutt BI (1985) Delayed effects of early infarct-limiting therapies on healing after myocardial infarction. Circulation 72: 907–914
2. Jugdutt BI, Warnica JW (1988) Intravenous nitroglycerin therapy to limit myocardial infarct size, expansion and complications: Effect of timing, dosage, and infarct location. Circulation 78: 906–919
3. Jugdutt BI, Amy RW (1988) Healing after myocardial infarction in the dog: Changes in infarct hydroxyproline and topography. J Am Coll Cardiol 7: 91–102
4. Jugdutt BI (1987) Left ventricular rupture threshold during the healing phase after myocardial infarction in the dog. Can J Physiol Pharmacol 65: 307–316
5. Michorowski BL, Tymchak WT, Jugdutt BI (1987) Improved left ventricular function and topography by prolonged nitroglycerin therapy after acute myocardial infarction. (abstract) Circulation 76 (Suppl IV): IV–128
6. Jugdutt BI, Michorowski BL, Tymchak WJ (1989) Improved left ventricular function and topography by prolonged nitroglycerin therapy after acute myocardial infarction. Z Kardiol 78 (Suppl 2): 127–129

Intravenous Nitroglycerin in Acute Myocardial Infarction*

B. I. Jugdutt

Introduction

It is generally recognized by clinicians that nitroglycerin (NG) should be used with caution in acute myocardial infarction (AMI) because of the danger of hypotension and reflex tachycardia. In fact, Prodger [1] in 1932, and later Friedberg [2] in 1966, warned against its use in AMI for those reasons. In the mid 1970s, when intravenous NG was being used in early AMI to improve left ventricular function and salvage ischemic myocardium [3–5], an attempt was made to titrate the dose to a discrete hemodynamic end point and avoid hypotension.

The importance of a modest decrease in blood pressure was demonstrated in a conscious dog model developed to objectively assess the effect of therapeutic interventions such as NG on infarct size [6]. Infarct size in that model was measured relative to the size of the risk region so as to correct for variable size of the coronary vascular bed [6]. The dose of NG was titrated to decrease the mean blood pressure by 10%, an end point that could be easily recognized as evidence of pharmacologic effect in the clinical setting [6]. This modest decrease in afterload by low-dose intravenous NG over 6 h after coronary occlusion was associated with a marked decrease in preload, marked increase in collateral blood flow, and marked decrease in pathologic infarct size [6]. It was later shown in the same model that higher doses of NG, to decrease mean blood pressure (MBP) by 45% to a level of 58 mm Hg, offset the increase in collateral flow and increased infarct size [7]. It appears that benefit at lower doses of NG was associated with marked venodilation and coronary collateral dilation but mild arterial dilation [6]. In contrast, higher doses of NG seemed to produce global vasodilation and compromised autoregulation, so that collateral flow became exquisitely sensitive to perfusion pressure [7].

Although tachycardia was not significant in these studies [6, 7], it would be expected to compromise ischemic myocardium by reducing diastolic filling time and increasing myocardial oxygen demand. The results of these experiments underscore the need for caution when using higher doses of NG in patients with AMI, especially when multivessel disease is likely to be present.

---

* Supported in part by grants from the Canadian Heart Foundation, Ottawa, Ontario.

Table 1. Beneficial effects in randomized trials of low-dose intravenous nitroglycerin in acute myocardial infarction

Effect	Reference
1. Decreased LV filling pressure	8–15
2. Improved global LV performance	9–11, 13, 15
3. Decreased LV volume	11, 15
4. Decreased ST-segment elevation (an index of ischemic injury)	9, 11, 15
5. Decreased creatine kinase index of infarct size	9, 11, 13, 15
6. Decreased LV akinesis and dyskinesis (an index of regional mechanical dysfunction) on two-dimensional echocardiograms	11, 15
7. Improved perfusion or defect scores on thallium-201 scintigrams	10
8. Decreased infarct-related complications, including infarct extension, infarct expansion, and mortality	10, 15

LV, left ventricular

Clinical Trials of Low-Dose Intravenous Nitroglycerin

To date, eight randomized, placebo-controlled studies of low-dose intravenous NG in AMI have been reported [8–15] and they all indicate benefit for one or more measured clinical and biologic parameters (Table 1). The conclusion from these studies is that low-dose intravenous NG is safe in AMI and low dose should be used for optimum benefit [15].

What Is Low-Dose Therapy?

It is clear from these clinical studies [8–15] that the dose of intravenous NG must be individualized and titrated to a discrete hemodynamic end point. The various end points used are summarized in Table 2. Most clinical researchers used a 10% decrease in blood pressure as cutoff. In our institution, a 10% decrease in MBP but not below 80 mm Hg was first used [11], as in the conscious dog studies [6, 7]. In our larger study [15], the regimen was modified to a decrease in MBP by 10% for normotensive patients and by 30% for hypertensive patients (defined as a blood pressure > 140/90 mm Hg) but still not below 80 mm Hg. These end points were reliably monitored by a standard sphygmomanometer cuff in 154 patients given NG and 156 patients given placebo [15].

It is also evident from these studies [8–15] that there is no safe fixed dose of intravenous NG in AMI. There are several reasons for this:

1. Wide variation in patient responsiveness.
2. Wide heterogeneity of patients with AMI especially with respect to hemodynamic, biochemical, metabolic, and neurohumoral profiles.
3. Unstable hemodynamics during infarct evolution.
4. Presence of concomitant pharmacologic agents [15].

Table 2. Dose and duration in randomized trials of intravenous nitroglycerin in acute myocardial infarction

Reference	Interval from onset of symptoms to therapy	Number of patients	Duration of therapy	Dose of therapy
Chiche et al. [8]	< 24 h	35 controls 35 NG	5–7 days (24 h for first 12 patients)	Begin 15 μg/min; increased to reduce SBP by 20 mm Hg
Bussman et al. [9]	< 8 h (mean 4.5) and > 8 h (mean 12.8)	29 controls 29 NG	48 h	Begin 12.5 μg/min; increased to 100 μg/min to reduce PCWP by 10%–30% and MBP no more than 10%.
Flaherty et al. [10]	< 12 h	42 controls 56 NG	48 h intravenous; 72 h ointment (1 inch)	Begin 5 μg/min; increased to reduce MBP by 10%; keep SBP > 90 mm Hg, HR > 50 bpm
Jugdutt et al. [11]	< 8 h (mean 5.4)	11 controls 11 NG	24–63 h (mean 39)	Begin 5 μg/min; increased to reduce MBP by 10%, not < 80 mm Hg.
Nelson et al. [12]	< 14 h	14 controls 14 ISDN	90 min	Begin 50 μg/kg/h; up to 200 μg/kg/h to reduce MBP by 10 mm Hg
Jaffe et al. [13]	< 12 h	42 controls 43 NG	24 h (tapered over 4 h)	Begin 10 μg/min, up to 200 μg/min to reduce SBP by 10% or to 95 mm Hg, not < 90 mm Hg; HR not < 50 bpm; HR increase not > 20
Lis et al. [14]	< 12 h	76 controls 64 NG	48 h	Begin 10 μg/min; increased to reduce SBP ≤ 30 in those with SBP > 135 mm Hg, ≤ 15 in those with SBP 120–134 mm Hg, and ≤ 5 in those with SBP 95–119 mm Hg. (average dose 35 ± 26 μg/min)
Jugdutt and Warnica [15]	< 12 h (mean 5)	156 controls 154 NG	48 h	Begin 5 μg/min; increase to decrease MBP by 10% in normotensive and 30% in hypertensive patients but not < 80 mm Hg

HR, heart rate; ISDN, isosorbide dinitrate; MBP, mean blood pressure; NG, nitroglycerin; PCWP, pulmonary capillary wedge pressure; SBP, systolic blood pressure

5. Variable absorption by delivery systems.
6. Wide variation in infusion rates to produce a given hemodynamic effect in all reported studies where dose titration was done.
7. Absence of correlation between rate of infusion (37.5–175 µg/min) and plasma NG concentrations (less than 0.5–2.7 ng/ml) to achieve a 10% reduction in blood pressure [16].
8. Variation in degree of tolerance during infusions.

Thus, the need for individual dose titration is emphasized.

In our last study [15], NG was given by an infusion pump via a standard intravenous set. The infusion was started at 5 µg/min and increased by 5–10 µg/min until the target level or a dose of 200 µg/min was reached [15]. The infusion was continued for at least 48 h, and for up to 154 h [15]. It was slowed down or temporarily stopped whenever the MBP fell below the target level [15]. At the end, the infusion was tapered in decremental steps of 5–10 µg/min every 5 min [15].

Results of the Largest Randomized Study on Intravenous Nitroglycerin in Acute Myocardial Infarction

This recently published study [15] was designed to answer three questions about low-dose intravenous NG therapy over 48 h in AMI:

1. Does the effect of NG therapy on infarct size depend on the location of the AMI, timing of therapy, and the dose response?
2. Does NG therapy reduce the frequency of infarct expansion?
3. Is therapy safe in the conventional clinical setting without compulsory invasive hemodynamic monitoring?

The methodology is described in detail in our paper [15]. Briefly, 310 patients admitted within 12 h of onset of pain (mean 5 ± 4 h; range 0–12 h) and fulfilling the entry criteria were randomly allocated to therapy with intravenous NG or placebo (5% dextrose in water for 48 h). The patients were stratified on entry into early or late subgroups (≤ 4 h or > 4 h of onset of pain) and anterior or inferior AMI subgroups. Inclusion criteria were: definite AMI; clinical Killip class I, II, or III; systolic blood pressure > 100 mm Hg; heart rate > 50 beats/min and < 120 beats/min; age < 75 years. Exclusion criteria were: cardiogenic shock, heart block, severe hypertension and right ventricular infarction. All patients had standard coronary care therapy. Intravenous NG was given as described above.

The infusion was maintained at the final target level for 48 h unless MBP fell < 80 mm Hg, when it was slowed or temporarily stopped. After 48 h, the infusion was gradually tapered.

All patients were evaluated before, during, and after therapy for clinical parameters, hemodynamics (blood pressure by sphygmomanometer), ischemic injury by electrocardiograms, infarct size from creatine kinase levels,

Table 3. Results of the randomized study with prolonged low-dose intravenous nitroglycerin versus placebo

	Placebo ($n = 156$)	Nitroglycerin ($n = 154$)	p
Creatine kinase infarct size (gEq)	55.4 ± 42.3	40.6 ± 34.8	0.001
Left ventricular ejection fraction at 10 days (%)	42 ± 9	51 ± 9	0.001
Left ventricular asynergy at 10 days (%)	24 ± 9	16 ± 8	0.001
Change in infarct segment length (Δ%)	26 ± 24	− 4 ± 9	0.001
Change in infarct segment thickness (Δ%)	− 17 ± 11	+ 4 ± 9	0.001
Infarct expansion syndrome	15%	2%	0.0005
Left ventricular thrombus	22%	5%	0.0005
Cardiogenic shock	15%	5%	0.0005
Infarct extension	22%	11%	0.025
In-hospital deaths	26%	14%	0.01

Summarized from Jugdutt and Warnica [15]

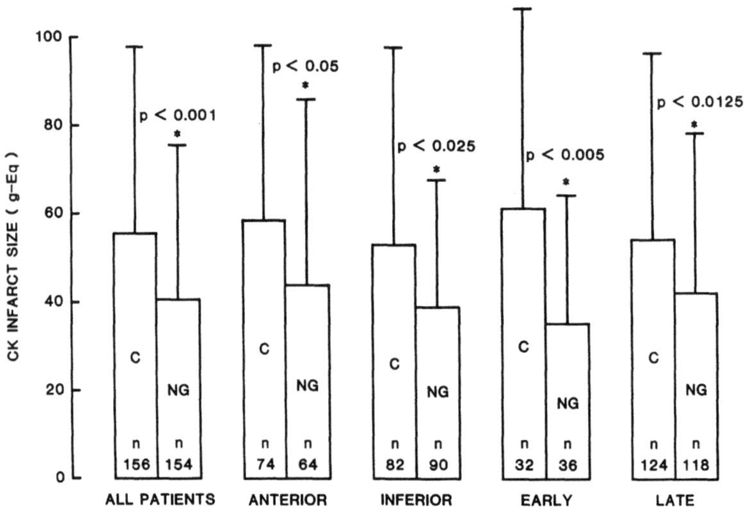

Fig. 1. Bar charts of creatine kinase *(CK)* infarct size in control *(C)* and nitroglycerin *(NG)* groups, anterior and inferior AMI subgroups, and early and late subgroups based on a 4-h cutoff for therapy. * Significant difference comparing NG with C. From [15]

left ventricular dimension and volume as well as indexes of left ventricular function and topography on two-dimensional echocardiograms, and complications including infarct expansion and extension. In addition, all patients were followed for at least 1 year for clinical parameters and complications.

The syndrome of infarct expansion was defined as a second acute clinical event associated with hypotension (systolic blood pressure < 90 mm Hg and peripheral hypoperfusion), left ventricular failure, evidence of left ventricular dilation, no evidence of new ischemic injury on electrocardiograms, no evidence of new necrosis (plasma creatine kinase elevation < 100 IU/l), and evidence of regional diastolic stretching, thinning, and dilation of the infarct segment on echocardiograms.

The main results are summarized in Table 3. Pretreatment parameters were similar in the two groups and all patients showed evidence of significant ischemic injury and impairment of global and regional left ventricular function. The MBP was 8% less and heart rate × MBP product was 6% less in the NG group for up to 10 h compared to placebo. After the infusions were stopped, these parameters increased within 1 h in the NG group, indicating that tolerance was not significant. In 48 patients (31%) who required thermodilution catheter data, NG produced prompt improvement in left ventricular filling pressure and late improvement in cardiac index. More important, NG reduced creatine kinase infarct size by 27% compared to placebo (Fig. 1). This effect was found with both anterior AMI (44 vs. 58 gEq, $p < 0.05$) and inferior AMI (39 vs. 53 gEq, $p < 0.025$). This beneficial effect was greater in patients treated early, especially within 4 h (Fig. 1). The beneficial effect on infarct size was abolished in the 14 patients whose MBP fell below 80 mm Hg in the first 12 h.

Adequate repeated echocardiograms were obtained in 128 patients (61 NG; 67 placebo). In these patients, NG promptly decreased the extent of left ventricular asynergy (akinesis plus dyskinesis) and improved ejection fraction compared to controls (Fig. 2). These differences persisted and were still

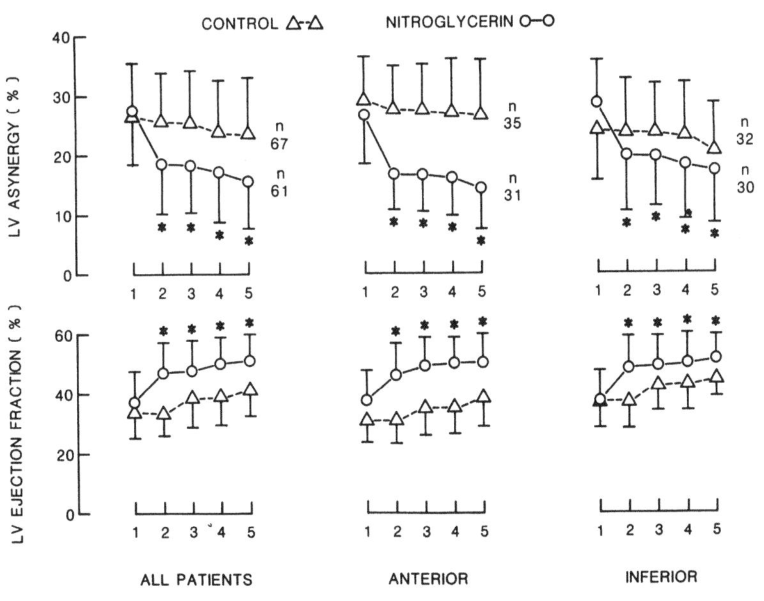

Fig. 2. Plots of changes in left ventricular *(LV)* asynergy and LV ejection fraction. *$p < 0.001$, significant difference of values at 6 *(2)*, 24 *(3)*, and 48 h *(4)* and 10 days *(5)* from baseline value (1) by multiple measures ANOVA. From [15]

present at 10 days and 6 months. Thus, left ventricular asynergy with NG was 29% less than with placebo at 6 h, 29% less at 48 h, and 40% less at 10 days. These beneficial effects were similar for anterior and inferior AMI subgroups, greater for early than late therapy subgroups, and less in the NG subgroup with MBP < 80 mmHg. Left ventricular internal dimension and volume on echocardiograms also decreased with NG but increased with placebo. In fact, after the NG infusions were stopped, the dimension and volume increased slightly within 1 h, again indicating that tolerance was not significant. More important, NG decreased infarct expansion and thinning (Fig. 3). Expansion index increased by 14% at 2–3 days and 31% at 7–10 days with placebo but did not change with NG. Also, the thinning ratio decreased by 6% at 2–3 days and 17% at 7–10 days with placebo but did not change with NG.

The frequency of infarct-related complications was also less with NG. Comparing NG to placebo, clinical infarct expansion was 87% less frequent, left ventricular thrombus 77% less frequent, cardiogenic shock 67% less frequent, infarct extension 50% less frequent, and all in-hospital deaths up to 38 days 46% less frequent. Although the frequency of deaths with NG was less for up to 12 months (Fig. 4), this beneficial effect was only seen with anterior Q-wave infarction.

In summary, low-dose intravenous NG therapy over the first 48 h of AMI limited indexes of infarct size, infarct expansion, infarct-related complications, and mortality up to 1 year. The results underscore the need for individualized titration to a discrete hemodynamic end point. Despite the use of standard intravenous sets in this study, target blood pressure reduction was achieved in all cases with modest doses of NG. Blood pressure was adequately monitored by the standard sphygmomanometer. Although tolerance is a concern with continuous and prolonged therapy [17], the dose of NG had to be increased to

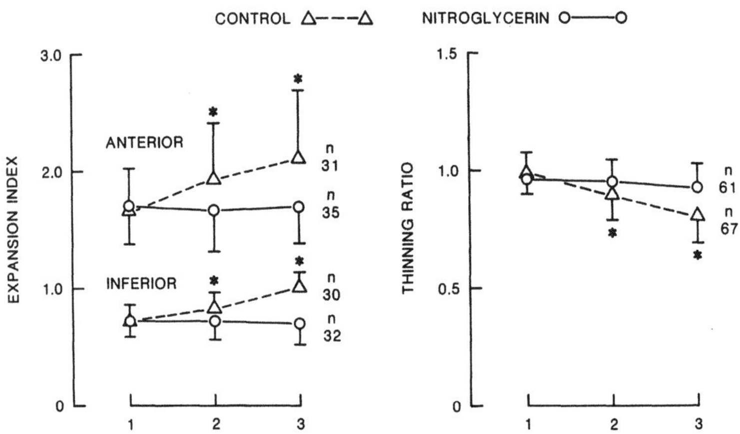

Fig. 3. Plots of changes in expansion index *(left panel)* and thinning ratio *(right panel)* for control and nitroglycerin groups. Expansion index is shown separately for anterior and inferior AMI subgroups. * $p \leq 0.05$, significance of difference comparing values at 2–3 days *(2)* and 10 days *(3)* with baseline. From [15]

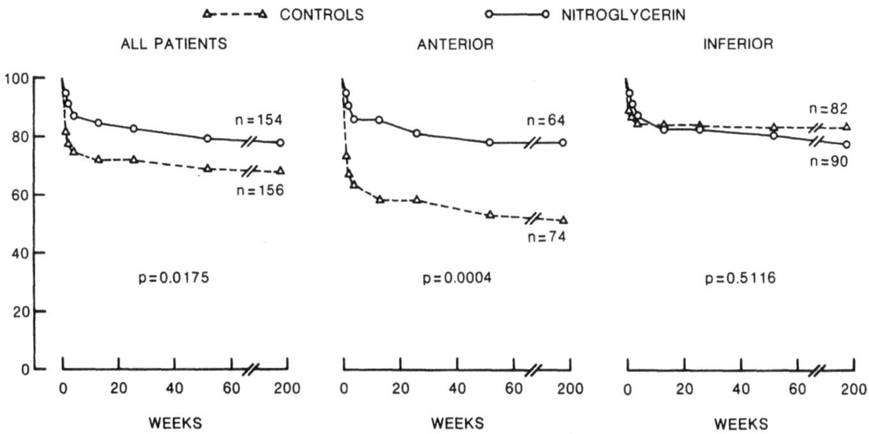

Fig. 4. Actuarial survival curves for nitroglycerin and control groups. Data over the 43 months or 186 weeks of the long-term follow-up are shown. *Left panel,* all patients; *middle panel,* anterior AMI subgroups; *right panel,* inferior AMI subgroups. From [15]

maintain the target MBP in only 12% of patients in the absence of chest pain and 12% in the presence of chest pain [18]. Greater sensitivity was found with inferior AMI but severe hypotension was relatively infrequent. The combined results suggest that low-dose intravenous NG therapy can be given safely in AMI and is an effective therapy for limiting infarct size and infarct expansion.

Mortality

Neither our study [15] nor other individual trials of NG in AMI [8–15] used mortality as an end point because the sample size was too small to detect a reduction. However, the high mortality and morbidity in our patients with anterior Q-wave AMI support the findings of the recent MILIS study [19]. The greater benefit in this high-risk subgroup might be related to limitation of topographic deterioration after anterior Q-wave AMI. Recently, meta analysis of pooled data from all published NG trials [20] indicated a 35% reduction in the odds of death ratio ($2p < 0.001$), suggesting a significant decrease in mortality. Nevertheless, a large multicenter mortality trial is needed to resolve the issue.

Prolonged NG Therapy and Cardiac Remodeling

The above study [15] clearly showed that low-dose intravenous NG therapy over 48 h during AMI is associated with a beneficial effect on early remodeling of the infarcted left ventricle. Thus, limitation of infarct expansion was already present at 2–3 days (15). In a dog model of AMI, low-dose NG therapy for the first 6 h of AMI was also found to reduce infarct expansion [21]. These findings led to the hypothesis that prolonged reduction of preload and afterload during the healing phase after AMI might preserve left ventricular geometry and function, prevent aneurysm formation, and prevent post infarction congestive heart failure.

This hypothesis was tested in a recent preliminary pathophysiologic study also reported at this meeting [22]. Briefly, 43 patients with a first anterior transmural AMI (or ATAMI), and who fulfilled similar inclusion and exclusion criteria as in our previous study [15], were treated with intravenous NG for 48 h and then randomly allocated to therapy with buccal NG (Nitrogard-SR, Syntex) or buccal placebo (Syntex) for 6 weeks. Buccal NG was given in a dose of 1–3 mg three times a day but on a 5-h schedule at 08:00, 13:00, and 18:00 h, with an 8-h washout phase to avoid vascular tolerance. Measurements included clinical data, electrocardiograms, and echocardiograms serially at 2, 4, 6, 8, and 10 days "and again" at 6, 12, and 24 weeks. The results indicated further beneficial remodeling between 48 h and 24 weeks with buccal NG compared placebo, resulting in smaller left ventricular volume, less asynergy, less infarct expansion, less infarct thinning, and higher ejection fraction.

Intravenous Nitroglycerin in the Reperfusion Era

It is known that prolonged ischemia prior to reperfusion, with thrombolytic agents or nonthrombolytic approaches, sets the stage for reperfusion injury, hibernation, myocardial stunning, and delayed recovery of function. Intravenous NG therapy prior to, during, and after successful reperfusion between 2 and 6 h from the onset of pain led to more prompt improvement of left ventricular function than reperfusion alone [23]. Thus, early NG therapy can be used as an adjunct with reperfusion therapy after AMI in order to accelerate recruitment of left ventricular function [23].

Increased Sensitivity Versus Nitrate Tolerance

Dose titration in our last study [15] was associated with initial decreases in MBP by 10% ± 8% among the 94 normotensive patients and 19% ± 19% among the 60 hypertensive patients. Increased sensitivity was a greater problem than tolerance [18]. It was more frequent in inferior AMI [18]. Thus, 15 patients (10%) developed an MBP below the 80 mm Hg cutoff within 30 min and 73% of them had inferior AMI. In another 11 patients (7%), the target MBP was maintained with a dose less than 6 μg/min, and 91% of them had inferior AMI. Finally, of the 14 other patients (9%) whose MBP fell below the blood pressure cutoff between 3 and 12 h, 64% had inferior AMI. Although the MBP averaged 75 ± 4 mm Hg and the dose of NG averaged only 25 ± 4 μg/min (range 10–35μg/min) in these 14 latter patients, they did have slightly larger infarct size than the placebo group and markedly larger infarct size than those in the NG group whose MBP was persistently above 80 mm Hg [15].

Tolerance developed in only 24% of the patients given low-dose NG therapy [15, 18]. "True" tolerance, when the dose was increased to maintain the hemodynamic effect in the absence of chest pain, occurred in 12%, and "apparent" tolerance, when the dose was increased in the presence of chest pain, occurred in 12% [18]. Tolerance was equally frequent in anterior and inferior AMI [18].

Detailed analysis of the hemodynamic, dose, and pain score profiles for the NG group in that study [15, 18] revealed a mild but progressive increase in blood pressure (mean, systolic and diastolic) beginning in early stages when pain was also frequent, and becoming more marked after 80 h, when episodes of pain were infrequent but more severe [18]. This analysis provided evidence for mild, early, and progressive attenuation of the hypotensive response, but more marked late attenuation of the hypotensive as well as, possibly, the anti-ischemic response during prolonged low-dose intravenous NG infusion [18]. Despite this pattern in grouped data, individual data showed varied patterns [18].

It is important to note that, despite evidence of tolerance on individual analysis, the group data in these patients indicated lasting beneficial effects on ischemic injury, regional left ventricular dysfunction, global systolic function, chamber size and volume, left ventricular topography, infarct-size, infarct-related complications as well as mortality [15]. In addition, upon discontinuation of the infusions, blood pressure and left ventricular volumes increased for the entire group [15], but not in the tolerant subgroup [18].

The low incidence of tolerance in that study [15] might be related to the use of a low dose to produce a mild decrease in blood pressure. The fluctuating dose during the infusion might have introduced an element of intermittency that prevented tolerance. Since the beneficial effects were mainly related to venodilation [15], it is possible that venodilation was less attenuated.

Thus, tolerance to oral nitrates is common but is promptly reversed by a nitrate-free interval [17]. This approach was used in the study with prolonged NG in AMI [22]. Tolerance is also seen with intravenous NG therapy in AMI especially beyond 24 h, and with high-dose therapy. Low-dose NG therapy in the 154 patients with AMI was associated with a low incidence of tolerance and without loss of major beneficial effects [15]. Intermittent dosing does not therefore seem necessary with low-dose intravenous NG therapy for 24–48 h. This appears to differ from the situation with continuous high-dose NG therapy for congestive heart failure where significant hemodynamic tolerance is seen regularly.

Summary

The scope of NG therapy in AMI has expanded on the basis of recent results. There is now a great deal of evidence that low-dose intravenous NG can be given safely in AMI. Several clinical trials of low-dose NG in AMI have shown benefit for multiple biologic and clinical end points. There is still a need to monitor blood pressure carefully with NG therapy in the early stages of AMI, but this can be done noninvasively. Prolonged NG therapy during healing after AMI might result in greater benefit on remodeling after AMI and prevent aneurysm formation. Although NG tolerance occurs, it did not appear to be a serious problem with intravenous NG in the first 48 h after AMI and can be avoided during prolonged administration by providing a nitrate-free interval. During reperfusion after AMI, intravenous NG appears to be an effective means for more rapid recruitment of ventricular function.

References

1. Prodger SH, Ayman D (1932) Harmful effects of nitroglycerin: with special reference to coronary thrombosis. Am J Med Sci 184: 480–491
2. Friedberg CK (1966) Acute coronary occlusion and myocardial infarction. In: Diseases of the heart, 3rd edn Saunders, Philadelphia, pp 913–914
3. Flaherty JT, Reid PR, Kelly DT, Taylor DR, Weisfeldt ML, Pitt B (1975) Intravenous nitroglycerin in acute myocardial infarction. Circulation 51: 132–139
4. Epstein SE, Kent KM, Goldstein RE, Borer JS, Redwood DR (1975) Reduction of ischemic injury by nitroglycerin during acute myocardial infarction. N Engl J Med 292: 29–35
5. Armstrong PW, Walker DC, Burton JR, Parker JO (1975) Vasodilator therapy in acute myocardial infarction: a comparison of sodium nitroprusside and nitroglycerin. Circulation 52: 1118–1122
6. Jugdutt BI, Becker LC, Hutchins GM, Bulkley BH, Reid PR, Kallman CH (1981) Effect of intravenous nitroglycerin on collateral blood flow and infarct size in the conscious dog. Circulation 63: 17–28
7. Jugdutt BI (1983) Myocardial salvage by intravenous nitroglycerin in conscious dogs: loss of beneficial effect with marked nitroglycerin-induced hypotension. Circulation 68: 673–684
8. Chiche P, Baligadoo SJ, Derrida JP (1979) A randomized trial of prolonged nitroglycerin infusion in acute myocardial infarction (abstract). Circulation 59 and 60 [Suppl II]: II–165
9. Bussmann WD, Passek D, Seidel W, Kaltenbach M (1981) Reduction of CK and CK-MB indexes of infarct size by intravenous nitroglycerin. Circulation 63: 615–622
10. Flaherty JT, Becker LC, Bulkley BH, Weiss JL, Gerstenblith G, Kallman CH, Silverman KJ, Wei JY, Pitt B, Weisfeldt ML (1983) A randomized prospective trial of intravenous nitroglycerin in patients with acute myocardial infarction. Circulation 68: 576–588
11. Jugdutt BI, Sussex BA, Warnica JW, Rossall RE (1983) Persistent reduction in left ventricular asynergy in patients with acute myocardial infarction by intravenous infusion of nitroglycerin. Circulation 68: 1264–1273
12. Nelson GIC, Silke B, Ahuja RC, Hussain M, Taylor SH (1983) Haemodynamic advantages of isosorbide dinitrate over frusemide in acute heart-failure following myocardial infarction. Lancet I: 730–733
13. Jaffe AS, Geltman EM, Tiefenbrunn AJ, Ambos HD, Strauss HD, Sobel BE, Roberts R (1983) Reduction of infarct size in patients with inferior infarction with intravenous glyceryl trinitrate: a randomised study. Br Heart J 49: 452–460
14. Lis Y, Bennett D, Lambert G, Robson D (1984) A preliminary double-blind study of IV nitroglycerin in acute myocardial infarction. Intensive Care Med 10: 179–184
15. Jugdutt BI, Warnica JW (1988) Intravenous nitroglycerin therapy to limit myocardial infarct size, expansion, and complications. Effect of timing, dosage, and infarct location. Circulation 78: 906–919
16. Wei JY, Reid PR (1979) Quantitative determination of trinitroglycerin in human plasma. Circulation 59: 588–592

17. Parker JO (1987) Nitrate tolerance. Am J Cardiol 16: 44 H–48 H
18. Jugdutt BI, Warnica JW (1988) Tolerance with intravenous nitroglycerin therapy in acute myocardial infarction (abstract). J Am Coll Cardiol 11 (Suppl A): 43 A
19. Stone PH, Raabe DS, Jaffe AS, Gustafson N, Muller JE, Turi ZG, Rutherford JD, Poole WK, Passamani E, Willerson JT, Sobel BE, Robertson T, Braunwald E (MILIS Group) (1988) Prognostic significance of location and type of myocardial infarction: independent adverse outcome associated with anterior location. J Am Coll Cardiol 11: 453–463
20. Yusuf S, Collins R, MacMahon S, Peto S (1988) Effect of intravenous nitrates on mortality in acute myocardial infarction: an overview of the randomised trials. Lancet I: 1088–1092
21. Jugdutt BI (1985) Delayed effects of early infarct-limiting therapies on healing after myocardial infarction. Circulation 72: 907–914
22. Michorowski BL, Tymchak WJ, Jugdutt BI (1987) Improved left ventricular function and topography by prolonged nitroglycerin therapy after acute myocardial infarction (abstract). Circulation 76 [Suppl IV]: IV–128
23. Tymchak WJ, Michorowski BL, Burton JR, Jugdutt BI (1988) Preservation of left ventricular function and topography with combined reperfusion and intravenous nitroglycerin in acute myocardial infarction (abstract). J Am Coll Cardiol 11 (Suppl A): 90 A

Effects of Intravenous Nitrates in Acute Myocardial Infarction

Dž. E. Rezaković Jr., G. Popović, M. Popadić, J. Štalec, and L. Pavićić

Introduction

Intravenous nitrate therapy is widely used in the treatment of acute heart failure as well as for the improvement of myocardial perfusion and ventricular function in acute myocardial infarction [1–3]. Although much research has been done to determine the hemodynamic action of intravenous nitrates, the indications for nitrate therapy in acute myocardial infarction with and without acute heart failure are still the subject of dicussion.

The goal of this study was to investigate the relationship between initial hemodynamic patient condition and the quality and quantity of nitrate response. Further, a hypothesis about the possibility and probability of predicting the nitrate response, based on initial patient's hemodynamic condition, was assessed. The hemodynamic mechanisms of intravenous nitrates, nitroglycerin (NTG) and isosorbide dinitrate (ISDN), were studied in 80 patients with different hemodynamic subsets during acute myocardial infarction.

Materials and Methods

Patients

The study was performed in 80 patients with an average age of 57.8 years (range, 39–69 years). An anterior and inferior location of myocardial infarct were found in 49 and 31 patients, respectively. Initially 84 patients were included, but 4 of them dropped out for different reasons (two, ventricular fibrillation; one, total atrioventricular block; one, rapid hemodynamic deterioration with pulmonary edema with the need to introduce supplementary appropriate therapy).

The aim of the study was to include a similar number of patients in all hemodynamic subsets (I, II, III, and IV), i.e., with and without acute heart failure. Patients were randomly chosen upon admission to the coronary care unit according to clinical signs of pulmonary congestion and peripheral hypoperfusion as well as to hemodynamic values of pulmonary wedge pressure (PWP) and cardiac index (CI). Hemodynamic subsets were determined according to Forrester's classification [4]: I subset: PWP ≤ 18 mmHg, CI

> 2.2 l/min/m^2; II subset: PWP > 18 mmHg, CI > 2.2 l/min/m^2; III subset: PWP ≤ 18 mmHg, CI ≤ 2.2 l/min/m^2; IV subset: PWP > 18 mmHg, CI ≤ 2.2 l/min/m^2. In hemodynamic subset I were included patients with persisting angina pectoris after sublingual NTG administration.

Exclusion criteria were: systolic arterial pressure < 100 mmHg; heart rate > 120 and < 55 beats/min; significant arrhythmia requiring intervention (supraventricular or ventricular tachycardia, ventricular fibrillation, heart block requiring pacemaker); hypertension over 180/110 mmHg; pre-existing valvular heart disease; mechanical complications (pericardial effusion or constriction); pulmonary obstructive disease; cardiomyopathy; age > 70 years; β-adrenergic blocking therapy administered within 48 h; severe life-threatening disease of other organ systems; acute intercurrent disease.

Measurements

The investigation was performed in the coronary care unit. A flow-directed, thermodilution catheter was inserted in the pulmonary artery under waveform monitoring control via a percutaneous approach to the right subclavian vein. All pressures were measured with the zero reference point at the axillary line. Systemic arterial pressures were measured via insertion of a cannula to the radialis artery or, if not possible, by sphygmomanometer. The following hemodynamic values were measured: heart rate (HR) (beats/min); systolic/diastolic/mean arterial pressures (SAP/DAP/MAP); central venous pressure (CVP); pulmonary artery systolic/diastolic/mean pressures (PASP/PADP/PAMP); pulmonary wedge pressure (PWP, mmHg); and cardiac output (CO, l/min). The values of cardiac index (CI, l/min/m^2) and systemic vascular resistance index (SVRI, dynes s/cm^5/m^2) were calculated.

Protocol

This was an open trial. The patients were randomized to be included in the NTG or ISDN investigation. The study was approved by the hospital committee on drug research and the patients gave informed consent.

Control hemodynamic parameters were measured three times over 45 min to confirm a stable hemodynamic condition and to determine hemodynamic subset. Patients with unstable hemodynamic condition, indicating a deterioration of acute myocardial infarction, were excluded from the study.

A titration period was started with 0.5–1.0 mg/h NTG (8–16 µg/min), with dosage increases for 1.0 mg/h every 10–20 min. ISDN administration was started with a dosage of 2.0 mg/h (33 µg/min), increasing the dosage up to 1.0 mg/h. When hemodynamic values approached the prestated criteria of nitrate efficacy, a continuous infusion was started at a fixed dose. The infusion period was continued over 2.0 h with NTG dosage in the range of 2.0–10.0 mg/h (mean, 6.8 mg/h) and with ISDN dosage in the range of 3–12 mg/h (mean, 9.4 mg/h). Nitrate infusion was not interrupted during titration and infusion periods, i.e., for approximately 4.0 h. Hemodynamic measurements were performed every 15 min.

Optimal dosage of nitrate infusion was determined during the titration period according to the following prestated criteria of clinical and hemodynamic patient improvement: a disappearance or significant attenuation of angina pectoris in subset I; decrease in elevated pulmonary wedge pressure at 16–18 mmHg and/or an increase in low cardiac index by at least 15% or over 2.2 l/min/m^2. The following additional end points were also taken into account: a 10% decrease in mean arterial pressure but not below 80 mmHg; a 10% decrease in normal and 30–40 mmHg in elevated systolic arterial pressure but not below 90 mmHg; and an increase in heart rate of not more than 20 beats/min and a decrease of not less than 50 beats/min.

If hemodynamic effects during nitrate infusion were considered clinically deleterious, the study was discontinued and other vasodilator and inotropic therapy was introduced. Unfavorable effects include increasing severe angina pectoris; an increase in pulmonary wedge pressure by 20%; a decrease in cardiac index by $>15\%$; a decrease in systolic arterial pressure below 90 mmHg; and either marked tachycardia or bradycardia. Whenever MAP fell below 80 mmHg, the nitrate infusion rate was decreased or completely stopped.

All hemodynamic parameters were measured twice in a standardized manner by a physician, to improve the accuracy and reproducibility of the results that were finally calculated as a mean of at least two values in each time period. No other drugs were given during nitrate hemodynamic investigation, except previous or newly started antiarrhythmic or anticoagulant therapy. Long-acting nitrates and calcium antagonist therapy were withdrawn within 6–12 h and diuretics 4–6 h prior to the beginning of trial. When sublingual NTG was administered for persistent and severe angina at the beginning of the investigation, the first hemodynamic measurement was performed after 30 min at the earliest. All patients were continuously ECG monitored, and CK activity measurements were repeated serially every 3 h.

Statistics

Statistical analysis of the quantitative and qualitative changes during nitrate administration was performed by multivariate analysis of differences, the model of differences being based on the analysis of the matrix of differences between nitrate and control hemodynamic results. In this analysis each subject served as its own control.

The quantitative and qualitative relationships between baseline control and NTG or ISDN hemodynamic condition were performed by a canonical correlation analysis. This multivariate analysis determines canonical factors, i.e., significant combinations of two systems of variables, as well as the probability of their occurrence. The correlation coefficient of each variable determined its order of importance in forming a factor [5–7]. The significance of all analyses was determined at the $p < 0.01$ level.

Results

The analysis of differences between control and nitrate values demonstrated significant changes in hemodynamic subsets; however, they were different in quantity and quality of NTG and ISDN response.

Nitroglycerin demonstrated significant changes in all hemodynamic subsets with major effects on pulmonary artery mean pressure and pulmonary wedge pressure in subsets II and IV. It also induced a moderate decrease in systemic vascular resistance in subsets III and IV, but without a parallel increase in cardiac index; on the contrary it effected a significant decrease in mean arterial pressure in subset III. Furthermore, NTG showed a major decrease in preload values, cardiac index, and mean arterial pressure in subset I (Table 1).

Isosorbide dinitrate also induced a decrease in pulmonary artery mean pressure and pulmonary wedge pressure in subsets II and IV, as well as systemic vascular resistance in subsets III and IV, but with a parallel increase in cardiac index (Table 2).

Testing the hypothesis that nitrate response is dependent on initial-control hemodynamic values of treated patients, all NTG and ISDN results were analyzed by canonical correlation analysis. Analysis of NTG results defined three factors. Canonical factor 1 demonstrated that the higher the initial values of central venous pressure, pulmonary artery mean pressure, and pulmonary wedge pressure the greater their decrease during NTG therapy. These two systems of variables have a very high correlation coefficient (0.92) and 85% of

Table 1. Nitroglycerin results in hemodynamic subsets I–IV

Variable	I	II	III	IV
PAMP	19.2 ± 5.1	29.4 ± 2.7	14.2 ± 3.7	32.3 ± 4.3*
	15.4 ± 2.8	20.7 ± 2.1	12.5 ± 2.6	20.8 ± 2.9**
	− 20%	− 30%		− 36%***
PWP	14.1 ± 2.2	24.6 ± 2.3	11.4 ± 1.7	27.3 ± 2.5
	10.5 ± 1.6	16.5 ± 1.3	9.8 ± 2.6	17.2 ± 1.4
	− 26%	− 33%		− 37%
CI	2.99 ± 0.24	2.98 ± 0.24	2.12 ± 0.15	1.88 ± 0.16
	2.64 ± 0.15	2.83 ± 0.29	2.01 ± 0.28	1.98 ± 0.18
	− 12%			
MAP	97.7 ± 4.8	98.2 ± 7.6	88.5 ± 4.6	84.7 ± 2.8
	87.7 ± 2.1	83.5 ± 3.9	82.3 ± 2.8	82.0 ± 2.5
	− 10%	− 15%	− 7%	
SVRI	2197.8 ± 236.5	2307.5 ± 218.4	3574.4 ± 306.5	3543.9 ± 227.3
	2241.0 ± 232.5	2242.8 ± 106.9	3176.8 ± 99.9	3143.3 ± 114.7
			− 11%	− 15%

Arithmetic mean ± standard deviation. PAMP, pulmonary artery mean pressure; PWP, pulmonary wedge pressure; CI, cardiac index; MAP, mean arterial pressure; SVRI, systemic vascular resistance index; I–IV, hemodynamic subsets; *, control values; **, NTG values; ***, percentage change between control and nitrate values. Underlined values, $p < 0.01$ in multivariate analysis of differences

Table 2. Isosorbide dinitrate results in hemodynamic subsets I–IV

Variable	I	II	III	IV
PAMP	18.7 ± 5.8	29.0 ± 1.7	13.6 ± 4.1	31.9 ± 3.9*
	16.9 ± 3.9	21.9 ± 1.9	11.5 ± 2.2	21.6 ± 3.4**
		− 24%		− 32%***
PWP	13.6 ± 2.9	24.3 ± 2.1	10.8 ± 2.2	26.7 ± 3.1
	10.8 ± 2.4	17.1 ± 1.9	9.1 ± 2.5	17.1 ± 2.1
		− 30%		− 36%
CI	2.91 ± 0.27	2.94 ± 0.31	2.04 ± 0.12	1.83 ± 0.13
	2.95 ± 0.22	3.01 ± 0.23	2.40 ± 0.19	2.47 ± 0.21
			+ 18%	+ 35%
MAP	95.6 ± 6.1	97.6 ± 8.2	91.3 ± 4.9	86.4 ± 3.4
	89.7 ± 4.9	94.6 ± 5.8	95.7 ± 7.4	88.7 ± 4.1
SVRI	2487.9 ± 185.6	2433.1 ± 221.2	3564.3 ± 347.6	3489.4 ± 339.8
	2432.1 ± 198.3	2489.4 ± 145.6	2786.5 ± 218.8	2698.5 ± 205.4
			− 22%	− 23%

Abbreviations as in Table 1

Table 3. Analysis of relationship between control and NTG hemodynamic values in 40 patients

Variables	Factor I		Factor II		Factor III	
	Control	NTG	Control	NTG	Control	NTG
CVP	0.54	− 0.48	0.14	0.16	0.20	− 0.18
PAMP	0.65	− 0.54	0.24	− 0.28	0.26	− 0.29
PWP	0.70	− 0.69	0.30	− 0.60	0.38	− 0.41
HR	0.14	0.20	0.12	0.44	0.12	0.12
MAP	0.18	− 0.22	− 0.55	− 0.53	0.14	− 0.10
CI	0.11	− 0.13	− 0.40	− 0.38	− 0.36	0.50
SVRI	0.06	− 0.09	0.13	− 0.09	0.46	− 0.30
Correlation coefficient	0.92		0.70		0.61	
Percentage of common variability	85%		58%		32%	
p	$p < 0.01$		$p < 0.01$		$p < 0.01$	

Matrix of factor correlations

	Factor 1	Factor 2	Factor 3
Factor 1	1.00		
Factor 2	0.69	1.00	
Factor 3	0.38	− 0.06	1.00

Percentage of common variability, p = probability; underlined factor correlation coefficient = $p < 0.01$; not underlined factor correlation coefficient = NS. Abbreviations as in Table 1

the common variability. Canonical factor 2 shows a relationship of initial low mean arterial pressure, low cardiac index, and high pulmonary wedge pressure with a decrease in mean arterial pressure, cardiac index and pulmonary index paralleled by augmented mean arterial pressure, as well as a decrease in systemic vascular resistance and pulmonary wedge pressure. This relationship

Table 4. Analysis of relationship between control and ISDN hemodynamic values in 40 patients

Variables	Factor I		Factor II		Factor III	
	Control	ISDN	Control	ISDN	Control	ISDN
CVP	0.50	-0.27	0.11	-0.15	0.27	-0.20
PAMP	0.69	-0.38	-0.21	-0.24	0.31	-0.30
PWP	0.74	-0.44	-0.29	-0.30	0.44	-0.44
HR	0.17	0.36	-0.21	-0.17	0.10	0.36
MAP	-0.44	0.32	-0.39	0.19	-0.49	0.40
CI	-0.04	0.63	-0.60	0.39	-0.47	0.58
SVRI	0.05	-0.07	0.33	-0.77	0.39	-0.53
Correlation coefficient	0.88		0.76		0.69	
Percentage of common variability	80%		63%		54%	
p	$p<0.01$		$p<0.01$		$p<0.01$	

Matrix of factor correlations

	Factor 1	Factor 2	Factor 3
Factor 1	1.00		
Factor 2	0.29	1.00	
Factor 3	0.76	0.31	1.00

Abbreviations as in Tables 1 and 3

has a moderately high correlation coefficient (0.69) and a moderate proportion of the common variability (54%) (Table 3).

Canonical factors 1 and 3 have a high correlation coefficient, indicating that an initial combination of high values of pulmonary wedge pressure, low mean arterial pressure, low cardiac index, and elevated systemic vascular resistance are related to ISDN decrease in preload values as well as to an increase in cardiac index with an increase in mean arterial pressure and a decrease in systemic vascular resistance (Table 4).

Discussion

This investigation has demonstrated that NTG and ISDN administered to patients with and without acute heart failure during acute myocardial infarction achieve their action via complex and different hemodynamic mechanisms, which are related in quantity and quality to initial hemodynamic conditions.

Nitroglycerin response, being predominantly at the preload level, could be predicted with 85% probability when pulmonary wedge pressure and pulmonary artery mean pressure were elevated at baseline. Initially low mean arterial pressure and reduced cardiac index predicted a further decrease in mean arterial pressure as well as cardiac index in 58% of these patients, due to a decrease in pulmonary wedge pressure. Moreover, the decrease in initially elevated preload values of NTG was closely related to a reduction in mean arterial pressure and cardiac index, regardless of whether these values were initially decreased or not. When high pulmonary wedge pressure was combined with elevated systemic vascular resistance and low cardiac index, then an increase in cardiac index could be predicted with 32% probability, being dependent on a decrease in systemic vascular resistance.

These results are supported by significant NTG responses found in hemodynamic subsets. Predominant changes were a decrease in pulmonary wedge pressure and pulmonary artery mean pressure; the higher the initial values were, the greater were these changes. These changes were combined with a decreased cardiac index and mean arterial pressure in subset I and a decrease in mean arterial pressure in subsets II and III, probably due to exaggerated decrease of preload values [8]. However, no significant increase in cardiac index was found, probably due to a modest decrease in systemic vascular resistance in subsets III and IV. As supposed by factor analysis, the changes in baseline pulmonary wedge pressure and pulmonary artery mean pressure in hemodynamic subsets were the most important variables for the prediction of NTG response at the preload and afterload levels. However, NTG response could be predicted with a higher probability at preload (85%) than at afterload level (32%–58%).

The effects of isosorbide dinitrate seem to be different in quantity and quality from NTG response. ISDN decrease in pulmonary wedge pressure and pulmonary artery mean pressure as well as an increase in cardiac index and mean arterial pressure could be predicted with 80% probability when preload values were initially elevated and combined with decreased mean arterial pressure. It is of interest that the increase in cardiac index is the most important change in this mechanism. However, isolated initial changes at afterload, low cardiac index, mean arterial pressure, and elevated systemic vascular resistance predicted ISDN decrease in systemic vascular resistance and an increase in cardiac index with 63% probability. Furthermore, a combination of initially low mean arterial pressure and cardiac index with high systemic vascular resistance and preload values predicted with 54% probability a major increase in cardiac index and a decrease in systemic vascular resistance with parallel lowering of preload values and moderate increase in mean arterial pressure. However, this response actually has a greater probability of prediction, since it is shown to be highly correlated with the pattern of ISDN action with high baseline preload and low mean arterial pressure, having an 80% predictive probability. It is obvious that ISDN responses are closely related to initial condition at preload and afterload, and also that the main patterns of its action induce hemodynamic changes at both levels.

The results from isosorbide dinitrate in hemodynamic subsets confirmed most of these assumptions. An increase in baseline preload values in subset II, high pulmonary wedge pressure and pulmonary artery mean pressure, were followed by an ISDN decrease in these values without significant changes in cardiac index or mean arterial pressure. ISDN induced a decrease in systemic vascular resistance and an increase in cardiac index with unchanged mean arterial pressure values in subset III, demonstrating an action at initially abnormal afterload values, without affecting preload. In subset IV, ISDN demonstrated a decrease in initially high pulmonary wedge pressure and pulmonary artery mean pressure, with a major increase in low cardiac index, due to a significant reduction in systemic vascular resistance. ISDN demonstrated in subsets III and IV major action at afterload level in decreasing high systemic vascular resistance, to such an extent not only to override preload effects in reducing cardiac index, but to achieve a significant increase in cardiac index (18%–22%). ISDN hemodynamic responses showed fairly balanced action at preload and afterload level, which could be predicted with 54%–80% probability.

These varieties in NTG and ISDN action could not be attributed to differences in dosages, since they were similar in all subsets. These results explained variability of nitrate response in a relatively high percentage of patients, 32–85% and 54%–80% of patients who received NTG and ISDN, respectively. They proved the hypothesis that NTG and ISDN actions are not unsystematic and incidental events, but are mechanisms with a predictable magnitude and a defined pattern on response, mainly dependent on baseline hemodynamic condition.

Studies of intravenous nitrates in heart failure and acute myocardial infarction have frequently shown contradictory results. Many authors have shown that NTG had a major influence on pulmonary wedge pressure but with different responses of cardiac output. Some authors have confirmed that cardiac output remained unchanged or increased in patients with pulmonary wedge pressure >18 mm Hg, but tended to decrease with lower preload values [9–11]. Similar results were shown by Flaherty [12], indicating that changes in cardiac output were related to the presence and severity of acute heart failure in acute myocardial infarction, either with a decrease or no changes in the patients without heart failure, and an increase in values in those with severe heart failure. Baligadoo [13] demonstrated that NTG increased with cardiac output only when initial pulmonary wedge pressure and systemic vascular resistance were elevated, while in patient subsets with low values it tended to decrease. However, other authors have shown that arterial hypotension occurred in 20% of patients during NTG therapy [14, 15] and, moreover, that pulmonary wedge pressure, mean arterial pressure, and cardiac output decreased in all patients with acute myocardial infarction, regardless of initial values of pulmonary wedge pressure [16, 17].

Intravenous ISDN trials also produced different hemodynamic results in heart failure. Gwilt et al. [18] found only a decrease in pulmonary wedge pressure without other changes at the dosage of 12 mg/h, while other authors showed a reduction in preload values and an increase in cardiac index, while a change in mean arterial pressure occurred only at initially high values [19]. Some authors demonstrated, besides a significant decrease in preload values, an increase in cardiac index and a reduction in systemic vascular resistance, with no changes in mean arterial pressure. The changes in cardiac index and systemic vascular resistance were greater the lower their values had been at baseline [20, 21].

In comparative studies, some authors found similar hemodynamic effects between ISDN and NTG [22], except that mean arterial pressure decreased less with ISDN than with NTG [23]. Other authors also showed that NTG successfully reduced perioperative hypertension in more patients, in 83% versus 63% with ISDN therapy [24]. Moreover, several trials showed that ISDN induced an increase in cardiac index at lower pulmonary wedge pressure and at higher systemic vascular resistance levels [25]. A recent study in patients with pulmonary congestion during acute myocardial infarction showed that ISDN and NTG lowered pulmonary wedge pressure and mean arterial pressure with an increase in cardiac output to the same degree in acute measurements. However, during sustained therapy, cardiac output was significantly higher with ISDN at 24 h than with NTG at 18 h of infusion, while pulmonary wedge pressure was significantly higher at the 6- and 18-h intervals with NTG, indicating partial tolerance. Furthermore, NTG needed more upward titrations and dose adjustments to maintain the effects over 24 h [3]. These results showed that ISDN effects may offer the benefit of a more stable hemodynamic condition.

These trials mostly support the findings of other studies, showing that NTG and ISDN demonstrate different mechanisms of action and that the final effect depends significantly on initial hemodynamic disturbance, which also explains the variety of the experimental findings. [1, 10, 11, 12, 13, 20, 22, 23]. Further, numerous data indicate that NTG has a major effect at preload, while ISDN demonstrates a more balanced action at preload and afterload [17, 19, 25, 26].

These studies confirmed that intravenous nitrates are a successful treatment in the patients with and without acute heart failure during acute myocardial infarction, respecting the differences in the mechanisms of action between NTG and ISDN and its significant relationship with baseline hemodynamic disturbance. The results were of help in clinical decision-making in choosing the appropriate intravenous nitrate for individual patients and improving the probability of preventing potential deleterious effects, such as a decrease in mean arterial pressure and cardiac index, expecially in patient subsets III and IV [2, 27].

Summary

The relationship of baseline hemodynamic disturbance with mechanisms of intravenous nitroglycerin (NTG) and isosorbide dinitrate (ISDN) response was investigated in 80 patients with and without acute heart failure (hemodynamic subsets I–IV) during acute myocardial infarction. NTG affected mainly preload and to a lesser extent afterload, while ISDN exerted more balanced action at preload and afterload; however, these changes were dependent on baseline hemodynamic status. NTG changes in mean arterial pressure and cardiac index were mainly related to initial pulmonary wedge pressure and to a smaller extent to systemic vascular resistance. NTG action could be predicted with 85% probability as a decrease of pulmonary wedge pressure; with 58% probability as a decline in mean arterial pressure and cardiac index; and with 32% probability as an increase in cardiac index and a reduction in systemic vascular resistance. ISDN effects on cardiac index and mean arterial pressure were predominantly dependent on systemic vascular resistance. ISDN decline in pulmonary wedge pressure, as well as an increase in cardiac index and mean arterial pressure, could be predicted with 80% probability; a reduction in systemic vascular resistance and a rise in cardiac index with 63% probability; an increase in cardiac index and mean arterial pressure, with a decrease in systemic vascular resistance and preload values with 54% probability. The data prove NTG being the optimal treatment in patients with high pulmonary wedge pressure and normal cardiac index, while ISDN is more reliable in patients with increased preload values, as well as in those with a low cardiac index and high systemic vascular resistance.

References

1. Flaherty JT (1988) Role of nitrates in acute myocardial infarction. Eur Heart J 9 [Suppl A]: 165–170
2. Jugdutt BI, Warnica JW (1988) Intravenous nitroglycerin therapy to limit myocardial infarct size, expansion and complications. Circulation 78: 906–919
3. Cintron GB, Glasser SP, Weston BA, Linares E, Conti R (1988) Effect of intravenous isosorbide dinitrate versus nitroglycerin on elevated pulmonary arterial wedge pressure during acute myocardial infarction. Am J Cardiol 61: 21–25
4. Forrester JS, Diamond GA, Chatterjee K, Swan HJC (1976) Medical therapy of acute myocardial infarction by application of hemodynamic subsets. N Engl J Med 295: 1356–1413
5. Cooley WW, Lohnes PR (1971) Multivariate data analysis. Wiley, New York
6. Mulaik SA (1972) The foundation of factor analysis. McGraw Hill, New York
7. Zakrajšek E, Štalec J, Momirović K (1973) SS-program system for multivariate analysis of data. Proceedings of international symposium computer on university, Zagreb, p C8: 1–16
8. Crexells C, Chatterjee K, Forrester JS et al (1973) Optimal level of filling pressure in the left side of the heart in acute myocardial infarction. N Engl J Med 289: 1263–1270
9. Bussmann WD, Schöfer H, Kaltenbach M (1976) Wirkung von Nitrglycerin beim akuten Myokardinfarkt. II. Intravenöse Dauerinfusion von Nitroglycerin bei Patienten mit und ohne Linksinsuffizienz und ihre Auswirkungen auf die Infarktgrösse. Dtsch Med Wochenschr 101: 642–648
10. Gold HK, Leinbach RC, Sanders CA (1972) Use of sublingual nitroglycerin in congestive heart failure following acute myocardial infarction. Circulation 46: 839–845
11. Korewicki J, Kraska T, Opolski G, Ostrzycki A, Palynyczko Z (1984) Beneficial effects of intravenous nitroglycerin on hemodynamics and enzymatically estimated infarct size. Eur Heart J 5: 697–704
12. Flaherty JT, Come PC, Baird MG et al. (1976) Effects of intravenous nitroglycerin on left ventricular function and ST-segment changes in acute myocardial infarction. Br Heart J 38: 612–621
13. Baligadoo S, Ingrand JC, Maiti D, Derrida JP, Savier CH, Chiche P (1979) La trinitrine intraveineuse chez les malades avec et sans insuffisance cardiaque mise en évidence par une analyse multifactorielle (analyse des correspondances) de 3 types de réponse hémodynamique dépendant de l'état initial. Nouv Presse Méd 8: 283–292
14. Jaffe AS, Geltman EM, Tiefenbrunn AJ, Ambos HD, Strauss HD (1983) Reduction of infarct size in patients with inferior infarction with intravenous glyceryl trinitrate. A randomized study. Br Heart J 49: 452–460

15. Lis Y, Bennet D, Lambert G, Robson D (1984) A preliminary double blind study of intravenous nitroglycerin in acute myocardial infarction. Intensive Care Med 10: 179–184
16. Williams DO, Amsterdam EA, Mason DT (1975) Hemodynamic effects of nitroglycerin in acute myocardial infarction. Decrease in ventricular preload at the expense of cardiac output. Circulation 51: 421–427
17. Kötter V, von Leitner ER, Schröder R (1978) Comparison of effects on hemodynamics and myocardial metabolism of phentolamine, sodium nitroprusside and glyceryl trinitrate in acute myocardial infarction. In: Kaltenbach M, Lichtlen P, Balcon R, Bussmann WD (eds) Coronary heart disease. Thieme, Stuttgart, pp 273–278
18. Gwilt DJ, Petri M, Reid DS (1983) Intravenous isosorbide dinitrate in acute left ventricular failure, a dose response study. Eur Heart J 4: 712–717
19. Behrenbeck DW, Niehus B, Pöhler E, Hötzel J, Lechler E (1978) The therapy of acute left ventricular failure by infusion of isosorbide dinitrate. Herz 3: 15–18
20. Wermuth G, Dettmar F, Schnitz J, Schlauch D, Klaus D (1980) Effect of isosorbide dinitrate as continuous infusion on the hemodynamics in severe chronic congestive heart failure. In: Rudolph W, Schrey A (eds) Nitrate 2. Urban and Schwarzenberg, Munich, pp 246–249
21. Rabinowitz B, Tamari I, Elazar E (1982) Intravenous isosorbide dinitrate in patients with refractory pump failure and acute myocardial infarction. Circulation 65: 771–778
22. Luther M, Röken U (1976) Activity of isosorbide dinitrate given intravenously to patients with angina pectoris and recent myocardial infarction. Herz 8: 654–649
23. Bussmann WD (1976) Wirkung von Nitroglyzerin i. v. und ISDN oral beim akuten Myokardinfarkt. In: Rudolph W, Siegenthaler W (eds) Nitrate I. Urban and Schwarzenberg, Munich
24. Durkin MA, Thys D, Morris RB, Kaplan J, Cahalan M, Barash PG (1988) Control of perioperative hypertension during coronary artery surgery. A randomized double-blind study comparing isosorbide dinitrate and nitroglycerin. Eur Heart J 9 [Suppl A]: 181–185
25. Baligadoo S, Chiche P (1978) The influence of initial hemodynamic parameters on the hemodynamic response to isosorbide dinitrate and intravenous nitroglycerin. Herz 3: 206–216
26. Rezaković Dž, Rutishauser W, Pavicic L, Bloch A, Popadić M, Imhoff EW (1983) Different hemodynamic actions of trinitroglycerin and isosorbide dinitrate in patients with acute myocardial infarction. Eur Heart J 4: 718–723
27. Forrester JS, Diamond GA, Swan HJ (1977) Correlative classification of clinical and hemodynamic function after acute myocardial infarction. Am J Cardiol 39: 137–145

Hemodynamic Effects of a Subchronic Therapy with 120 mg Isosorbide Dinitrate Slow-Release in Coronary Artery Disease and Left Heart Failure

R. Wolf, A. Nötges, U. Traber, and R. Sinn

The hemodynamic effects of nitrates in patients with left ventricular dysfunction and both acute and chronic heart failure are well documented [1, 12]. However, the development of hemodynamic tolerance during long-term treatment with oral nitrates remains a subject of debate [6, 7]. This has two clinically important implications [2–4]:

1. Cardiac unloading by long-acting nitrates would increase exercise tolerance.
2. The progression of the degree of heart failure would be prevented or delayed.

Thus, we investigated the hemodynamic effect of a subchronic (3-week) treatment with 120 mg isosorbide dinitrate (ISDN) in slow-release (SR) form in patients with mild or moderate chronic left heart failure after myocardial infarction. In a subgroup of patients hemodynamic measurements were performed after a 1-year treatment with 120 mg ISDN SR.

Patients and Methods

Eighteen patients (13 male; age: range 48–73, mean 62.2 years) with proven myocardial infarction in the chronic stage were investigated. All patients were in a stable clinical condition with symptoms of chronic left heart failure in NYHA functional classes II–III. No patient showed signs of persistent postinfarction ischemia, and there was no indication for bypass grafting or percutaneous transluminal coronary angioplasty by angiographic and nuclear studies in any case.

After Swan-Ganz catheterization, the hemodynamics were measured at rest and during symptom-limited exercise by supine bicycle ergometry before, 2 and 10 h after intake of 1 × 120 mg ISDN SR orally. In 16 patients hemodynamic measurements were performed after a 3-week therapy with 120 mg ISDN per day using an identical exercise protocol.

No patient received long-term nitrates or vasodilators within 10 days before both hemodynamic investigations; basic drug therapy was unchanged.

Diuretics were given at the end of each investigation when indicated. In six of 16 patients hemodynamic measurements could be performed after a 1-year treatment with 120 mg ISDN SR.

Results

Left Ventricular Filling Pressure at Rest

Before the subchronic treatment was initiated, ISDN significantly reduced left ventricular filling pressure (LVFP = mean capillary wedge pressure) from 18.9 ± 8.1 to 12.1 ± 5.9 and 12.2 ± 4.6 mm Hg 2 and 10 h after intake, respectively ($p < 0.0005$). After a 3-week therapy a decrease of LVFP from 20.6 ± 8.9 before ISDN to 11.7 ± 5.8 and 13.0 ± 3.8 mm Hg after application was observed ($p < 0.025$) (Fig. 1).

LVFP During Exercise

Mean exercise capacity was 52 ± 25 watt. LVFP increased to 31.2 ± 10.5 mm Hg and was significantly reduced to 19.0 ± 10.1 and 23.5 ± 11.0 mm Hg 2 and 10 h after 120 mg ISDN SR, respectively ($p < 0.25$). LVFP 2 h after intake was significantly decreased compared to LVFP 10 h after ISDN ($p < 0.05$). After 3 weeks LVFP significantly decreased from 33.1 ± 14.4 mm Hg to 20.5 ± 10.4 and 25.8 ± 11.7 mm Hg 2 and 10 h after ISDN,

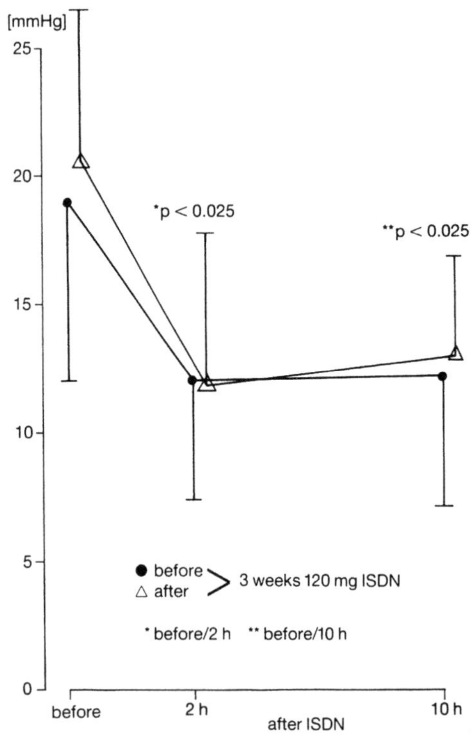

Fig. 1. LVFP at rest before *(circles)* and after *(triangles)* a subchronic (3-week) therapy with 120 mg ISDN SR

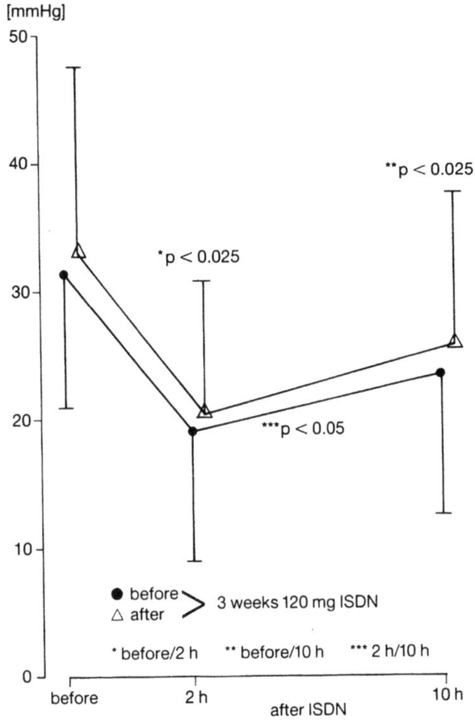

Fig. 2. LVFP during exercise before *(circles)* and after *(triangles)* a subchronic (3-week) therapy with 120 mg ISDN SR

respectively ($p < 0.0025$). LVFP 10 h after ISDN SR was significantly increased compared to 2 h after intake ($p < 0.05$) (Fig. 2).

Mean Pulmonary Artery Pressure at Rest

Before subchronic therapy was initiated, ISDN significantly decreased mean pulmonary artery pressure (MPAP) from 29.4 ± 11.1 to 20.6 ± 6.5 and 21.2 ± 5.5 mm Hg 2 and 10 h after application, respectively ($p < 0.0005$). After 3 weeks a significant reduction of MPAP from 31.4 ± 11.4 mm Hg before ISDN SR to 21.7 ± 6.5 and 23.3 ± 4.8 mm Hg after intake was registered ($p < 0.0025$) (Fig. 3).

MPAP During Exercise

MPAP increased to 46.8 ± 12.9 mm Hg and was significantly reduced to 32.7 ± 12.7 and 39.4 ± 11.6 mm Hg 2 and 10 h after 120 mg ISDN SR, respectively ($p < 0.0005$). MPAP 10 h after intake was significantly increased compared to 2 h after ISDN ($p < 0.05$). After a 3-week treatment MPAP was significantly reduced by ISDN from 53.3 ± 19.9 mm Hg to 39.2 ± 15.5 and

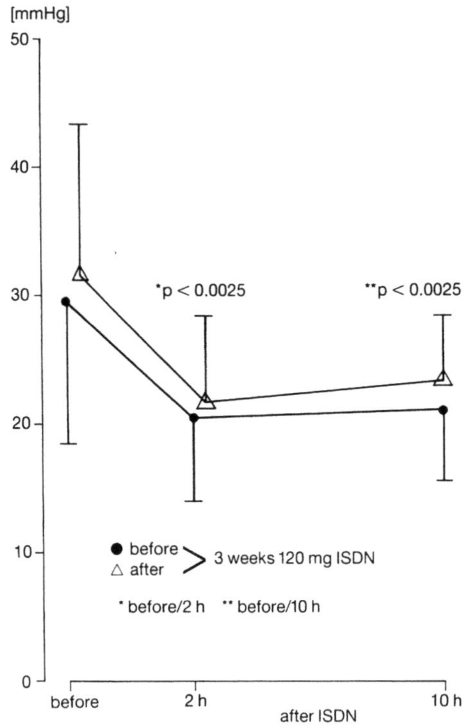

Fig. 3. MPAP at rest before *(circles)* and after *(triangles)* a subchronic (3-week) therapy with 120 mg ISDN SR

46.0 ± 18.6 mm Hg 2 and 10 h after intake, respectively ($p < 0.0005$). Ten hours after ISDN application MPAP significantly increased compared to 2 h after intake ($p < 0.05$) (Fig. 4).

Cardiac Output and Heart Rate at Rest and During Exercise

Figure 5 demonstrates that the reduction of LVFP by ISDN was not associated with significant changes of cardiac index both at rest and during exercise. However, the relationship between LVFP and cardiac index can be expressed by two log functional curves. This indicates "optimal" LV filling pressures at rest and during exercise. No significant changes of heart rate at rest and during exercise were observed both before and after a 3-week treatment (Table 1).

LVFP During Exercise after 1-Year Treatment with 120 mg ISDN SR

In six patients ISDN markedly reduced LVFP from 38.7 ± 14.6 mm Hg to 27.0 ± 7.0 and 32.1 ± 7.0 mm Hg after 2 and 10 h, respectively, after 1 year. This decrease was comparable to LVFP reduction before long-term treatment was initiated (Table 2).

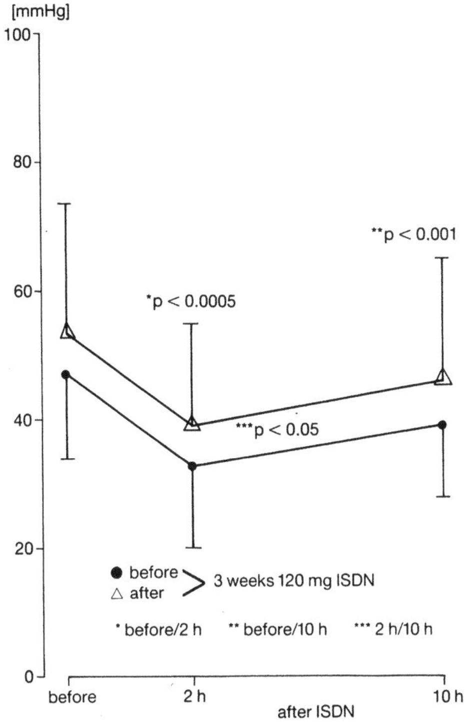

Fig. 4. MPAP during exercise before *(circles)* and after *(triangles)* a subchronic (3-week) therapy with ISDN SR

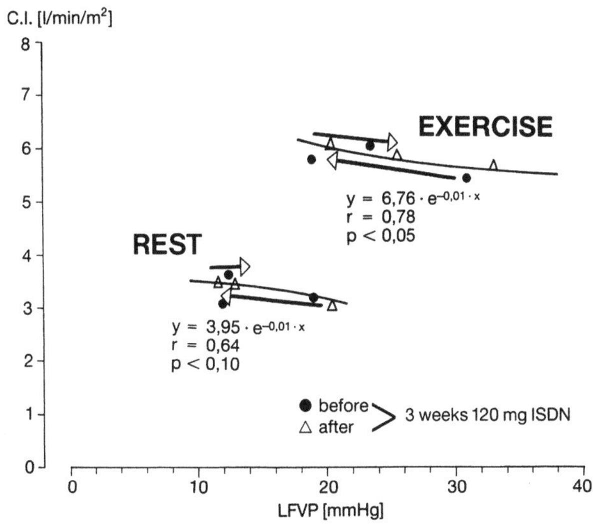

Fig. 5. Cardiac index and LVFP at rest and during exercise before (*circles*) and after (*triangles*) a subchronic (3-week) therapy with ISDN SR. Arrows indicate changes of the control values before ISDN intake 2 and 10 hours after ISDN. For all variables log functional curves can be calculated

Table 1. Heart rate before and after a subchronic (3-week) therapy with 120 mg ISDN SR

		Before	After 3 weeks
Rest	Before	70.7 ± 14.7	68.0 ± 12.8
	2 h after ISDN	73.1 ± 12.5	71.4 ± 12.8
	10 h after ISDN	73.3 ± 12.3	73.1 ± 11.8
Exercise	Before	115.6 ± 16.0	112.1 ± 14.3
	2 h after ISDN	111.1 ± 15.1	110.1 ± 12.5
	10 h after ISDN	115.4 ± 17.0	115.9 ± 15.5

Table 2. LVFP during exercise before and after 1-year treatment with 120 mg ISDN SR ($n = 6$)

	Before	After 1 year
Before ISDN	39.3 ± 11.0	38.7 ± 14.6
2 h after ISDN	29.7 ± 13.5	27.0 ± 7.0
10 h after ISDN	33.0 ± 11.8	32.1 ± 7.0

Discussion

Progression of LV dilatation and of the clinical syndrome of congestive heart failure provides important prognostic relevance [10, 11]. Both the combination of vasodilators and ISDN and the treatment with angiotensin-converting enzyme inhibitors in patients with congestive heart failure could improve LV function, exercise tolerance, and prognosis [2–5, 11]. The results presented indicate that a chronic preload reduction at rest and during exercise by ISDN may represent the hemodynamic mechanism for improving LV dysfunction. Cardiac unloading by 120 mg ISDN in SR form could be documented over 10 h after intake and was reproducible after a 1-year treatment in a smaller subgroup. The vasodilating effect of nitrates with increasing arterial compliance and venous capacitance in mild heart failure might be preferable to a long-term treatment with diuretics [1, 4, 8, 9]. In conclusion, our results indicate that there is no significant evidence for the development of nitrate tolerance or reflex sympathetic activity using this treatment regimen in chronic left heart failure.

References

1. Abrams J (1985) Hemodynamic effects of nitroglycerin and long-acting nitrates. Am Heart J 110: 216–224
2. Cohn JN, Archibald DG, Ziesche S, Franciosa JA, Harston WE, Tristani FE, Dunkman WB, Jacobs W, Francis GS, Flohr KH, Goldman S, Cobb FR, Shah PM, Saunders R, Fletcher RD, Loeb HS, Hughes VC, Baker B (1986) Effect of vasodilator therapy on mortality in chronic congestive heart failure. Results of a Veterans Administration Cooperative Study. N Engl J Med 314: 1547–1552
3. Cohn JN, Archibald DG, Francis GS, Ziesche S, Franciosa JA, Harston WE, Tristani FE, Dunkmann WB, Jacobs W, Flohr KH et al (1987) Veterans administration cooperative study on vasodilator therapy of heart failure: influence of prerandomization

variables on the reduction of mortality by treatment with hydralazine and isosorbide dinitrate. Circulation 75 IV: 49–54
4. Cohn JN (1987) Role of nitrates in congestive heart failure. Am J Cardiol 60: 39–43
5. The Consensus Trial Study Group (1987) Effects of enalapril on mortality in severe congestive heart failure. N Engl J Med 316: 1429–1435
6. Franciosa JA, Cohn JN (1980) Sustained hemodynamic effects without tolerance during long-term isosorbide dinitrate treatment of chronic left ventricular failure. Am J Cariol 45: 648–654
7. Kulick D, Roth A, McIntosh N, Rahimtoola SH, Elkayam U (1988) Resistance to isosorbide dinitrate in patients with severe chronic heart failure: incidence and attempt at hemodynamic prediction. J Am Coll Cardiol 12: 1023–1028
8. Nelson GI, Silke B, Ahuja RC, Hussain M, Taylor SH (1983) Hemodynamic advantages of isosorbide dinitrate over frusemide in acute heart failure following myocardial infarction. Lancet 1: 730–733
9. Nelson GI, Silke B, Forsyth DR, Verma SP, Hussain M, Taylor SH (1983) Hemodynamic comparison of primary venous or arteriolar dilatation and the subsequent effect of furosemide in left ventricular failure after acute myocardial infarction. Am J Cardiol 52: 1036–1040
10. Pfeffer MA, Lamas GA, Vaughan DE, Parisi AF, Braunwald E (1988) Effect of captopril on progressive ventricular dilatation after anterior myocardial infarction. N Engl J Med 319: 80–86
11. Pfeffer MA, Pfeffer JM (1987) Ventricular enlargement and reduced survival after myocardial infarction. Circulation 75 (Suppl IV): 93–97
12. Williams DO, Bommer WJ, Miller RR, Amsterdam EA, Mason DT (1977) Hemodynamic assessment of oral peripheral vasodilator therapy in chronic congestive heart failure: prolonged effectiveness of isosorbide dinitrate. Am J Cardiol 39: 84–90

Isosorbide Dinitrate and Nitroglycerin Oral Spray in Heart Failure

A. Schneeweiss, A. Marmor, and L. Reisin

Introduction – Nitrate Oral Spray

Nitrates are widely used for the treatment and prevention of heart failure and angina pectoris. Even the traditionally most rapid method of administration, the sublingual tablet, takes at least 2 min until the onset of effect. Usually this takes more time because 2 min is the time required from the moment of dissolution for the nitrates to reach peak plasma levels [1]. An additional period is, however, required for dissolution of the tablet which must take place, at least partially, before absorption begins. This period shows interpatient variability and further delays the onset of action.

Several forms of oral spray, in which the active compound (isosorbide dinitrate, ISDN, or nitroglycerin) is kept as a solution in a spray container, are available. These were planned to avoid the delay required for dissolution of sublingual tablets. ISDN spray was found effective in a variety of cardiovascular diseases: it relieves pain in patients with chronic stable or unstable angina pectoris and produces hemodynamic and symptomatic improvement in patients with acute exacerbation of congestive heart failure and pulmonary edema.

Nitrate Oral Spray in Congestive Heart Failure

The superiority of nitrate oral spray is in the immediate treatment of acute episodes of the diseases for which nitrates are used. In congestive heart failure, an acute episode usually presents as acute pulmonary edema. Nitrates have substantially improved the management of acute pulmonary edema. The nitrate-induced hemodynamic improvement in this condition results mainly from venodilation and is accompanied by clinical improvement [1–6].

As the rapidity of drug administration plays a major role in the management of acute pulmonary edema, the usual routes of administration of nitrates are the intravenous and sublingual ones. Sublingual administration of a nitrate tablet is often not practical in very dyspneic patients. Moreover, it takes about 2 min before any significant absorption begins [7].

ISDN spray appears to be the optimal method for very rapid treatment. We studied two patients with acute pulmonary edema in whom hemodynamic improvement occurred within 1 min after administration of the ISDN spray. The patients, both with chronic congestive heart failure, were studied by Swan-Ganz catheter for hemodynamic evaluation. Both developed acute pulmonary edema while the catheter was in place and during chronic treatment with diuretics only. In both, ISDN spray produced hemodynamic improvement within 60 s, evident by decreases in wedge pressure and increases in cardiac output. Further hemodynamic improvement, associated with symptomatic improvement, was observed at 5 min.

In a larger and more recent study [8], the hemodynamic effects of ISDN spray were evaluated in eight patients with acute deterioration of chronic congestive heart failure. All cardiovascular drugs were discontinued at least 3 days prior to the study, and a Swan-Ganz catheter was introduced. During hospitalization, the patients developed symptoms and hemodynamic signs of acute deterioration of the chronic condition. All were tachypneic, dyspneic, and orthopneic. One squirt of ISDN spray (equal to 1.25 mg ISDN) produced hemodynamic improvement which became evident in all patients within 1 min and reached a peak within 5.2 min after administration. The drug decreased the pulmonary arterial pressure from $57 \pm 13/29 \pm 10$ mmHg to $40 \pm 7/22 \pm 5$ mmHg and pulmonary capillary wedge pressure from 25 ± 10 mmHg to 17 ± 6 mmHg and increased cardiac index from 2.1 to 3.0 l/min/m^2. Systemic and pulmonary vascular resistances were reduced. All the changes were statistically significant. They were associated with symptomatic relief. The fact that hemodynamic effect was achieved more rapidly than the time required to start an intravenous line indicates that the ISDN spray may be beneficial for emergency treatment of acute deterioration in chronic congestive heart failure.

In a recent trial, we compared the magnitude and time course of the hemodynamic effect of a new formulation of an aqueous solution of the ISDN spray to that of sublingual nitroglycerin tablets in 12 patients with chronic congestive heart failure. The patients received, in random order, ISDN spray, 2.5 mg, or sublingual NTG, 0.8 mg. Hemodynamic measurements were performed before and at 1, 3, 5, 10, 20, 30, and 60 min after each drug. The second drug was given only after the hemodynamic parameters had returned to baseline, plus a washout period of two hours. The hemodynamic variables measured were comparable at baseline. Both drugs produced hemodynamic improvement, including a decrease in pulmonary capillary wedge pressure (PCWP), right atrial pressure (RAP), and systemic and pulmonary vascular resistances. Only the ISDN spray significantly increased cardiac output. The onset of action of the ISDN spray was significantly more rapid than that of nitroglycerin. This was primarily evident in the decrease in PCWP. With the ISDN spray, the decrease started at 1 min after administration, and at 3 min a decrease of 8.6 mmHg was already found. The corresponding value for sublingual nitroglycerin was 1.6 mmHg. The difference between ISDN and nitroglycerin was highly significant ($p < 0.02$). The peak effect of ISDN spray on PCWP and RAP was greater than that of nitroglycerin.

Nitrates can also produce arterial dilation but this requires moderate to high doses [8–11]. Therefore, it is obvious that nitrates would be especially effective in patients with heart failure whose main hemodynamic impairment is pulmonary congestion, that is, elevated left ventricular filling pressure in the presence of only a minimal decrease in cardiac output. Nevertheless, most studies of nitrates in heart failure have involved mainly patients with both elevated left ventricular filling pressure and reduced cardiac output. In a previous study [10], we demonstrated the superiority of ISDN oral spray over sublingual tablets in patients with this combined hemodynamic impairment. However, no report on the comparative effect of these two formulations in patients with predominant pulmonary congestion (where venodilation is the main expected beneficial effect) has been reported.

We compared the time course of the hemodynamic effect of ISDN, 5 mg, in the form of a sublingual tablet and an oral spray, in 15 patients with isolated chronic pulmonary congestion (pulmonary arterial end-diastolic pressure of 15 mm Hg or more in the presence of normal or only slightly reduced cardiac index). Both formulations produced significant reductions in the pulmonary arterial end-diastolic pressure. The effect of ISDN tablets (sublingually) became evident at 10 min after administration and was maximal at 30 min. The effect of ISDN oral spray became evident at 3 min and reached a peak at 10 min. The magnitude of the hemodynamic responses was similar. These findings indicate that ISDN oral spray is superior to ISDN sublingual tablets for rapid relief of pulmonary congestion.

Nitrate Oral Spray in Angina Pectoris

The same principles that apply to the use of nitrate oral spray in heart failure are valid also for angina pectoris. The antianginal effect of ISDN spray was assessed in patients with exercise-induced angina pectoris in order to determine and quantify its time course. Ten patients (mean age 58.9 ± 6.2 years) were included in the study. All had chronic stable angina pectoris. The length of time to onset of angina and to appearance of a 1.0-mm ST-segment depression was determined; the time to disappearance of pain and of ST-segment depression after discontinuation of exercise was also measured. ISDN spray delayed the onset of angina and electrocardiographic ischemic changes in all patients. Exercise time to pain was 5.1 ± 1.4 min with placebo and 7.2 ± 1.3 min with ISDN ($p < 0.001$). Time to appearance of ST-segment depression was 7.1 ± 1.5 min with placebo and 10.2 ± 1.2 min with ISDN. Time to disappearance of pain after discontinuation of exercise was shortened from 3.2 ± 0.8 min on placebo to 2.1 ± 0.8 min with ISDN ($p < 0.001$). Time to disappearance of ST-segment depression was shortened from 4.2 ± 0.6 to 2.5 ± 0.8 min, respectively ($p < 0.005$). Two patients experienced transient headache, and one patient had transient dizziness with slight decrease in blood pressure after ISDN. All patients completed the study without experiencing major adverse effects.

In a recent trial, Bachmann and Gansser (unpublished data) found that the new formulation of the hydrophilic ISDN spray has been shown to act more rapidly on the coronary arteries than the older lipophilic spray, as evaluated by quantitative coronary angiography.

Conclusion

Nitrate oral spray, and particularly the new formulation of the hydrophilic ISDN spray, are the most rapid methods for achieving the therapeutic effect of nitrates in heart failure.

References

1. Armstrong PW, Armstrong JA, Marks GS (1979) Blood level after sublingual nitroglycerin. Circulation 59: 585–588
2. Morrison RA, Wiegand UW, Jähnchen DE et al (1983) Isosorbide dinitrate kinetics and dynamics after intravenous, sublingual, and percutaneous dosing in angina. Clin Pharmacol Ther 33: 747–756
3. Doyle E, Chasseaud LF, Taylor T (1980) Measurement of plasma concentrations of isosorbide dinitrate. Biopharm Drug Dispos 1: 141–147
4. Robinson BF (1968) Mode of action of nitroglycerin in angina pectoris. Correlation between hemodynamic effects during exercise and prevention of pain. Br Heart J 30: 295–302
5. Sandler G, Ilani MA, Lawson CW (1963) Glyceryl trinitrate in angina pectoris. Lancet 1: 1130–1136
6. Kattus AA, Alvaro AB, Coulson A (1975) Effectiveness of isosorbide dinitrate and nitroglycerin in relieving angina pectoris during uninterrupted exercise. Chest 67: 640–646
7. Detry JR, Bruce RA (1971) Effect of nitroglycerin on "maximal" oxygen intake and exercise electrocardiogram in coronary heart disease. Circulation 43: 155–163
8. Marmor A, Schneeweiss A, Plich M, Blondheim D, Sharir T, Albert JS (1987) Isosorbide Dinitrate Oral Spray in Acute Exacerbation of Chronic Congestive Heart Failure Secondary to Coronary Artery Disease. Am J Cardiol 59: 1199–1200
9. Chiche P, Baligadoo S (1979) Nitrates, "pharmacological phlebotomy" and pulmonary edema. Am Heart J 97 (3): 408
10. Strauer BE, Scherpe A (1978) Ventricular function and coronary hemodynamics after intravenous nitroglycerin in coronary artery disease. Am Heart J 95: 210
11. Franciosa JA, Mikulic E, Cohn JN, Jose E, Fabie A (1974) Hemodynamic effects of orally administered isosorbide dinitrate in patients with congestive heart failure. Circulation 50: 1020–1024

Nitrates in Heart Failure

B. Rabinowitz

Nitrates in Use: Tolerance and Pseudotolerance

The three nitrates in clinical use are nitroglycerin (NTG), isosorbide dinitrate (ISDN), and isosorbide mononitrate (ISMN); the first two are in wide use all over the world and the mononitrate preparations have gained momentum in Europe and, more recently, in Israel. All three nitrates are available intravenously by intravenous bolus or continuous infusion; NTG and ISDN are available for traditional sublingual use or in the most recent method of immediate use by spray; short- and long-term, various dosages of oral NTG, ISDN, and ISMN are of widespread application and, in spite of the known tolerance problems, many transdermal preparations are still in general use. There are methods of circumventing hemodynamic and clinical tolerance. A good example of applying the recently accumulated knowledge on basic mechanisms to immediate clinical application is the one described by Packer et al. [1], as follows: "the tolerance to the effects of nitrates partly relates to neurohumoral activation (pRA) and partly to depletion of vascular smooth muscle sulfhydryl groups, which impair the activation of guanylate cyclase through which the nitrates achieve vasodilation." The development of tolerance was reduced by intermittent infusion of NTG and was reversed by the concurrent administration of oral *N*-acetylcysteine.

Although true tolerance has been clearly demonstrated and nitrate-free intervals help in avoiding it, less is known about the phenomenon of pseudotolerance.

What is pseudotolerance? Pseudotolerance is a state of apparent tolerance caused by extrinsic factors which change actual drug inflow with reference to the expected dose. This effect may be caused by: (a) gradual chemical degradation in containers, (b) adsorption to plastic bags and sets, and (c) change in drop size while the drug is delivered by drop-counting infusion pumps.

In order to assess this last possibility, we studied two brands of NTG, presently commercially available, one containing 18% ethanol in the vehicle, and another, which is an aqueous NTG solution and does not contain ethanol [2]. As the concentration of alcohol increased (Fig. 1), the number of drops required to form a milliliter of fluid rose, from 59.5 ± 8.5 with saline to 78.7 ± 8.8 with alcoholic NTG in saline 1:1. There was no change with aqueous NTG. The above effect appears to be an additional cause of apparent tolerance, when using alcohol containing NTG infusions. A way of cir-

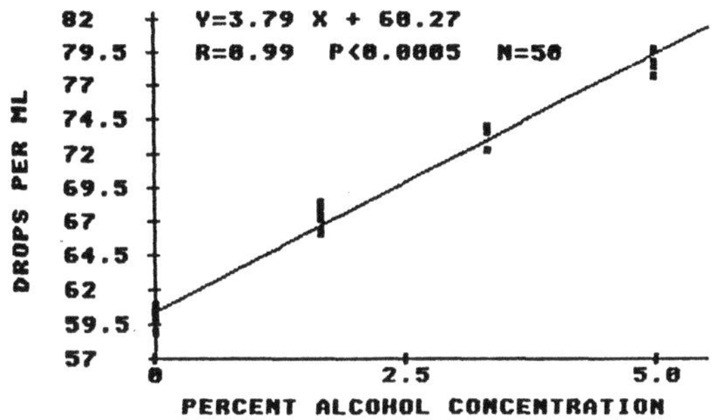

Fig. 1. Relationship between the number of drops/ml and alcohol concentration of ALNTG-saline solutions at flow of 15 drops/min. The vertical axis is in drops/ml and the horizontal axis in percentage alcohol concentration. *ALNTG*, alcoholic nitroglycerin

cumventing this problem is by using a volume pump instead of a drop-counting device.

Potential Effects of Nitrates

The effect of nitrates may be judged by their effects on:

1. Symptomatology, both short and longterm.
2. Hemodynamics, currently a controversial issue.
3. Exercise capacitance, quantitated by exercise tests.
4. The heart itself, or its parameters of dysfunction, as the ejection fraction, which is not likely to change with nitrates, or the ventricular volumes which might be influenced by therapy with nitrates. The modern concept of preventing postmyocardial infarction expansion, or modifying the topography, was a recently demonstrated favorable effect not only of the ACE inhibitor, captopril [3], but also of the nitrates [4].
5. Survival, the ultimate criterion on which we still only have the data provided by the Veterans Administration Cooperative Study [5] on the combined use of nitrates and hydralazine.

Although there is a tendency today to judge the use of drugs in general and vasodilators, in particular, solely by their potential to prolong life, and no one doubts this major goal, we should not forget that improvement in symptomatology correlates better with the hemodynamic effects and peripheral changes than with any change in the function of the heart itself; this has to do with the quality of life.

Left Ventricular Function and Peripheral Mechanisms

It is currently well accepted that the outcome of patients with heart failure depends on two factors: (a) the degree of left ventricular (LV) dysfunction and (b) the peripheral mechanisms, altered by neurohormonal factors.

Although LV dysfunction may be static or only slowly deteriorating for months and sometimes years, particularly in ischemic chronic heart failure, the peripheral mechanisms change continuously and vary both individually from patient to patient and with the course of the disease.

There is much controversy over the use of initial hemodynamics as predictors of clinical short- and long-term response and the relation of these hemodynamic variables to the neurohumoral parameters. Franciosa et al. [6, 7] remarked long ago on the uncertain relationship between the hemodynamic state and the functional capacity of patients with heart failure. On the other hand, the same author [7] concluded that long-term therapy with nitrates improves exercise capacity in patients with congestive heart failure as well as in patients with angina pectoris. Packer [8] remarked: "It is uncertain why we should be titrating vasodilating drugs to achieve specific hemodynamic goals, if such hemodynamic end-points have not been shown to correlate with clinical efficacy." However, should we not have at least an idea if any of the "desirable" hemodynamic effects are achieved at all, when using a certain vasodilator? Are we certain that by discarding any hemodynamic evaluation, we are not possibly risking a deleterious or untoward effect, particularly in the patient with ischemic failure?

Nitrates in Acute Myocardial Infarction

In order to assess the potential value of initial hemodynamic parameters and of circulating catecholamines in predicting the response to nitrate in acute myocardial infarction [9–12], we have analyzed the data obtained from 65 patients. This population (Table 1) included the following subgroups of AMI patients: (a) patients with heart failure manifested by pulmonary capillary wedge pressure above 20 mm Hg and systolic arterial pressure of at least 90 mm Hg, treated either with an individually titrated continuous infusion of ISDN, NTG, and ISMN or with oral ISMN and ISDN, in a crossover study, (b) patients belonging to the hyperkinetic subset of acute myocardial infarction with high arterial pressures, high heart rates, and normal filling pressures who had recurrent episodes of ischemia, treated with intravenous ISDN.

The data demonstrated the following: (a) A high initial systemic vascular resistance in patients with acute myocardial infarction complicated with heart

Table 1. Nitrates in acute myocardial infarction: pretreatment variables for a population of 65 patients

Patients with heart failure	
(initial PCWP > 20 mm Hg; systolic AP > 90 mm Hg)	
Treated with:	i. v. ISDN (*n*, 18)
	i. v. NTG (*n*, 10)
	i. v. ISMN (*n*, 12)
	Oral ISMN and ISDN
	(crossover, *n*, 10)
Patients from hyperkinetic subgroup treated with i. v. ISDN (*n*, 15)	

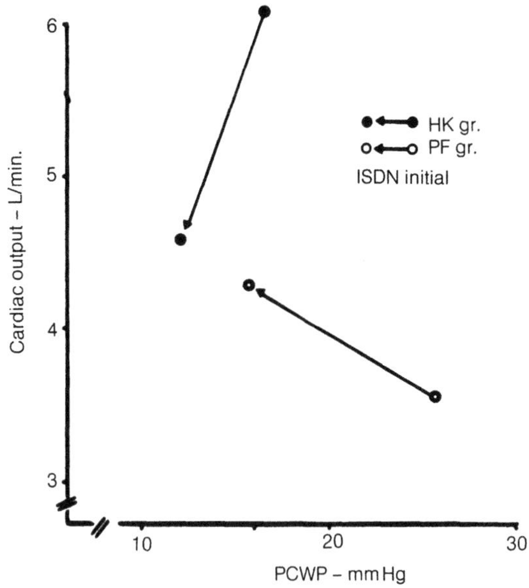

Fig. 2. Left ventricular function curves in patients with acute myocardial infarction *(AMI)*, belonging to the hyperkinetic subgroup *(HK)* and in AMI patients, complicated by pump failure *(PF)*: effects of isosorbide dinitrate *(ISDN)* infusion. *HK* (*n*, 15), mean levels; *PF* (*n*, 15), mean levels

failure has predictive value for a moderate increase in cardiac output by intravenous nitrates (dinitrate, mononitrate, and NTG), and less by oral ISDN and ISMN. In addition to the expected marked decrease in the filling pressures, cardiac output increased by more than 25% in patients with a calculated systemic vascular resistance above 2000 dynes s/cm^{-5}. (b) A positive association was found between high initial systemic vascular resistance and high circulating catecholamines (r, 0.63; $p < 0.05$) and both parameters decreased significantly during nitrate infusions. (c) A good correlation between initial levels of plasma catecholamines and simultaneous heart rates was demonstrated particularly in the hyperkinetic subgroup (r, 0.8, $p < 0.001$). (d) The specific hemodynamic profile of the hyperkinetic group predicted a characteristic hemodynamic response to i.v. ISDN, in direct contrast to that observed in acute myocardial infarction complicated by failure (Fig. 2). (e) Randomized comparisons between the various nitrates showed no significant differences in the hemodynamic response obtained, provided that the increase in the dose of the intravenous continuous infusion was limited by the decrease in the transmyocardial gradient.

Thus, nitrates are efficient and the hazards of using them in patients with acute myocardial infarction might be diminished by assessing initial parameters, knowing their relative immediate predictive value, and by adjusting the dose of the drug according to the hemodynamic and clinical aim in the various subgroups of patients.

Nitrates in Chronic Heart Failure

The issue of pretreatment or initial treatment variables in chronic failure is certainly more complicated and still unsettled. The hemodynamic end points do not correlate well with clinical efficiency and we still need to know which are desirable end points in the chronic use of nitrates in failure, as compared with other vasodilators.

In the earlier studies on the use of nitrates in chronic failure, we have used them in severely ill patients (NYHA classification IV or at least III). Recently, some investigators have attempted to use vasodilators for long-term purposes in patients in earlier functional classifications [13], with the aim of treating LV dysfunction and not only overt severe congestive heart failure. For this purpose, it is imperative to know whether we can achieve a beneficial effect and that we are not inducing unpleasant untoward effects. The drug could induce too much of a decrease in the filling pressures and, in conjunction, an untoward decrease in cardiac output; in some cases this hemodynamic effect might trigger ischemia. Data from one of our earlier studies showed that a high initial systemic vascular resistance in patients with chronic pump failure (NYHA class III–IV) had predictive value for an increase in cardiac output induced either by ISDN or by NTG in addition to the expected decrease in primary capillary wedge pressures (PCWP) ($p < 0.005$). A randomized comparison between NTG and ISDN [14] revealed no difference in response to these drugs if the dose of the intravenous continuous infusion was limited by a decrease in the transmyocardial gradient (TMG).

On the basis of present knowledge, we should look at patients with chronic heart failure according to different subgroups and adjust therapy rationally. A good practical example is a special subgroup of patients in class IV, the patients who are in renal failure, in addition to severe "intractable" congestive heart failure; these patients have a tendency to suffer from hyperkalemia and, according to the data presented by Packer et al. [15], they do not respond well to acetonitrile (ACE) inhibitors. Even if potassium is kept in check, Packer et al. [15] have shown that patients with serum creatinine levels above 1.5 mg and right atrial pressure above 12 mmHg are nonresponders to ACE inhibitors. These patients constitute a specific subgroup of chronic failure patients, in whom nitrates could and should be used as the first-choice alternative. Long-term studies, relating to the effect on survival of the various preparations of nitrates and their potentially various modes of administration in subsets of patients with heart failure, are, therefore, much needed.

Summary

The rational use of nitrates in heart failure is based on the understanding of the complex mechanisms by which they act and the etiology of heart failure. The existing methods of clinical evaluation and the difference between acute and long-term effects should be taken into consideration.

The factors influencing hemodynamic response and the long-term effect on symptomatology, hemodynamics, exercise capacity, and survival are related to mode and dose of administration, the neurohumoral setting, and the state of the LV function. In acute ischemic heart failure, nitrates appear to be the vasodilator of choice. In chronic advanced ischemic failure, in the subgroup of patients with renal failure, nitrates might also be the vasodilator of choice. Long-term survival studies, using various preparations of nitrates are currently needed.

The rational use of nitrates in heart failure should be based on the recent knowledge of the basic mechanisms by which they act. It should take into consideration the etiology of heart failure, the proper methods of clinical evaluation, and the difference between acute and chronic failure.

References

1. Packer M, Lee WH, Kessler PD, Gottlieb SS, Medina N, Yushak M (1987) Prevention and reversal of nitrate tolerance in patients with congestive heart failure. N Engl J Med 317: 799–804
2. Leor R, Feinstein M, Luria MH, Hod C, Rabinowitz B (1989) Influence of ethanol on the delivery of intravenous nitroglycerin solutions. Eur J Clin Pharmacol 36: 307–310
3. Pfeffer MA, Lamas GA, Vaughan DE, Parisi AF, Braunwald E (1988) Effect of captopril on progressive ventricular dilatation after anterior myocardial infarction. N Engl J Med 319: 80–86
4. Jugdutt BI, Warnica JW (1988) Intravenous nitroglycerin therapy to limit myocardial infarct size, expansion, and complications: effect of timing, dosage and infarct location. Circulation 78: 906–919
5. Cohn JN, Archibald DG, Ziesche S, Franciosa JA, Harston E, Tristani EE, Donkman WB, Jacobs WI, Francis GS, Flohr KH, Goldman S, Cobb FR, Shah PM, Saunders R, Fletcher RD, Loeb HS, Highes VC, Baker B (1986) Effect of vasodilator therapy on mortality in chronic congestive heart failure. Results of a Veterans Administration Cooperative Study. N Engl J Med 314: 1547–1552
6. Franciosa JA, Dunkman B, Leddy CH (1984) Hemodynamic effects of vasodilators and long-term response in heart failure. J Am Coll Cardiol 3: 1521–1530
7. Franciosa JA (1985) Isosorbide dinitrate and exercise performance in patients with congestive heart failure. Am Heart J 110: 245–250
8. Packer M (1986) The role of vasodilator therapy in the treatment of severe chronic heart failure. Drugs 32 [Suppl 5]: 13–26
9. Rabinowitz B, Tamari I, Elazar E, Neufeld HN (1982) Intravenous isosorbide dinitrate in patients with refractory pump failure and acute myocardial infarction. Circulation 65: 771–778
10. Rabinowitz B, Hod H, Chouraqui P, Rath S, Agranat O, Neufeld HN (1987) Hemodynamic effects of oral isosorbide-5-mononitrate and dinitrate in ischemic heart failure. Clin Cardiol 10: 603–608
11. Rabinowitz B, Katz A, Shotan A, Chouraqui P, Neufeld HN (1988) Hemodynamic effects of intravenous isosorbide-5-mononitrate in acute and chronic left heart failure of ischemic etiology. Eur Heart J 9 [Suppl A]: 175–180
12. Rabinowitz B, Hod H, Elazar E, Klein R, Neufeld HN (1988) Hemodynamics, circulating catecholamines and response to intravenous nitrate therapy in a specific subset of acute myocardial infarction: the hypertensive-hyperkinetic-coronary active group. Cardiology 75: 260–268

13. Parmley WW (1989) Pathophysiology and current therapy of congestive heart failure. J Am Coll Cardiol 13: 771–785
14. Rabinowitz B, Elazar E, Klein R, Chouraqui P, Neufeld HN (1986) Hemodynamic profiles and changes in circulatory catecholamines in patients with ischemic heart disease on intravenous nitrates. J Mol Cell Cardiol [Suppl A] 18: 87
15. Packer M, Kessler PD, Gottlieb SS (1986) Adverse effects of converting-enzyme inhibition in patients with severe congestive heart failure: pathophysiology and management. Postgrad Med J 62 [Suppl I]: 179–182

MIX
Papier aus verantwortungsvollen Quellen
Paper from responsible sources
FSC® C105338

If you have any concerns about our products,
you can contact us on
ProductSafety@springernature.com

In case Publisher is established outside the EU,
the EU authorized representative is:
Springer Nature Customer Service Center GmbH
Europaplatz 3, 69115 Heidelberg, Germany

Printed by Libri Plureos GmbH
in Hamburg, Germany